Applied Aerodynamics
for Private and Commercial Pilots

Applied Aerodynamics

for Private and Commercial Pilots

by Steve Pomroy

SkyWriters Inc.
Portage la Prairie, MB, Canada

Printed in Canada

Cover Art by Kevin Pomroy

Canadian Cataloging in Publication Data

Pomroy, Steve, 1974-

Applied aerodynamics

Includes bibliographical references and index.
ISBN 0-9687214-0-0

1. Aeronautics.   2. Aerodynamics.   I. Title

TL546.P65 2000   629.132'3   C00-950122-3

This book is dedicated to the memory of
John Hugh Robinson

He will be dearly missed by a loving sister, Anne, and Parents, Edward and Janet.

www.SkyWritersPublishing.com

Table of Contents

Page #

Page #

Page #

Page #

Part 2 - Application

Page #

Appendices

Introduction to <u>*Applied Aerodynamics*</u>

In flight training and in regular flight operations, an understanding of aerodynamics, flight mechanics, and their applications is important to a pilot. This understanding allows flight crew members to operate their aircraft to the limits of it's flight envelope - without the risk of stepping beyond these limits. To pilots who fly in commercial operations, this ability is critical. Even to recreational flyers who are likely to lean towards less demanding flying, this knowledge can be invaluable.

The purpose of this book is to introduce the fundamental aspects of aerodynamics to flight crew - students and veterans alike - so they can understand the operation of their aircraft more thoroughly. The book starts from scratch and is intended to bring the student from the pre-solo training stage right up through the commercial licence and the instructor course. As such, some of the information presented is more in-depth than may be necessary for a private or recreational pilot. This shouldn't discourage beginning students, but should allow them - with the assistance of an instructor - to pick and choose what they *must* know and then to come back at a later date to learn more on their own.

As a student and instructor, I noticed a recurring problem among training manuals - they tended to be practical in the extreme or theoretical in the extreme. There was little or no opportunity presented to bridge the gap between theory and practice. As a result, it has become a common practice for a student to learn one thing in the classroom and something else entirely in the airplane. This is unfortunate since all "theory" is developed to allow some practical application or use, and all "practice" is based on some theory. There clearly needs to be a way to link the two. This book is intended to do just that.

Part I (Chapter #'s 1 through 9) of *Applied Aerodynamics for Private and Commercial Pilots* is developed to present the theoretical background

that is needed to move on to Part II. Part II (Chapter #'s 10 through 23) of the book is built on the foundation laid in Part I and is intended to establish practical methods for handling the aircraft. The two parts of the book are semi-independent of one another. They can be used separately but are intended to work best together. As well, this book can be used independently of others, but I recommend referencing your studies to other related texts.

Since this book is intended to introduce the fundamental elements of flight theory, I have intentionally left some things out. These ideas may be of interest to some readers - particularly those who wish to pursue a career in aviation. The items that I have ignored are jet engine propulsion and performance, multi-engine aerodynamics, and high speed aerodynamics. For those interested, these topics are covered very well in several other texts - including some listed in the bibliography.

Throughout this book, abbreviations are used regularly. Reading the book in order should allow the reader to become familiar with these abbreviations as they are introduced. However, if you are using this text as a reference manual, some abbreviations may be unfamiliar to you. They are presented with their meanings in the Glossary. As well, for readers using this text as a reference, a comprehensive index is included at the back of the book.

At the end of each chapter, there are some questions and problems for readers to work with. Most of these questions can be answered directly out of the text. However, some of the questions require some interpretation on the part of the reader. The answers are provided in the back of the book.

Many of the concepts discussed in this book (particularly in Part I) have been simplified for our purposes. These simplifications reflect the needs of student pilots, as opposed to engineers. The discrepancies are acceptable to us as pilots, and they allow us to avoid the more complicated mathematics while still grasping the important concepts which we need to fly our aircraft.

I have made every effort to make this book as comprehensive, accurate, and understandable as possible. However, with a project of this size, there will always be imperfections. Readers wishing to provide me with feedback can write to me via e-mail at the following addresses:

steve@skywriters.aero

In writing this book, there have been many people who have earned my thanks and gratitude. I could not possibly name them all here, but I will name a few that stand out from the crowd.

First and foremost, my family has always supported my many endeavors. Thank-you to my parents, Kevin and Elizabeth, and my brother and sister, Kevin and Tina.

A few of my school teachers stand out from the rest: Phyllis Singleton, Kelvin Hollihan, and Andy Byrd.

Air Cadet instructors started me on the road of aviation. In particular Dianne Batten, Sharon Clarke, Jason Bridger, Jeff Andrews, and Claire Vavasour.

My friends who have supported and encouraged me along the way. In particular, Tony Gallagher, Kevin Cadigan, Craig Richard, Allison Walsh, Anne Robinson, and Jennifer Armstrong.

Thank-you to all of my flight instructors who originally trained me. In particular, Steve Blackwood, Rod Krieger, Kirby Short, Bjorn Sunde, and Glen Loder.

My students all deserve a mention. Each of them has had something to teach me. In particular, Todd MacLeod, Kathie Sanderson, Lynwood Webber, Phil Safire, Herb Bottomley, Noel Machat, Jim Fraser, Andrew Sorenson, Lee Cameron, John Aucoin, Jackie Andrews, Wilfred Wright, Paul MacInnis, Dave Hiscock, Diane Dumas, Greg McNeil, Dale Langthorne, and Suzanne Maclean.

Last, but not least, a special thanks goes to my friend and colleague, Shane Wilson, whose proofreading and editing has added immeasurably to this text.

That's it for now. I hope you enjoy the book.

Steve Pomroy
April 20th, 2000

Chapter 1
Basic Physics

During the course of flying training, flight theory forms a cornerstone in the knowledge and skill required to pilot an aircraft. Without this knowledge, it would be difficult or impossible to fly an aircraft safely to the limits of it's flight envelope.

To begin any discussion of aerodynamics or flight mechanics, we first have to discuss some basic concepts of physics. These concepts are the background needed to understand exactly what is happening under various conditions of flight. The topics covered in this chapter can be covered more thoroughly from most any high school or first year university physics text. Namely, the ideas we'll cover shall be: Newton's Laws, linear motion, vectors, couples, gravity, work and power, friction, the laws of conservation, simple machines, and circular motion. At the end of the chapter, we will have a look at the forces that are acting on an aircraft in equilibrium.

It is important for the reader to note that this book is not intended as a physics text. This 1st chapter is written only as a review of concepts which are applicable to aviation, and not as an initial primer. If you do not have any background in basic physics, it may be a good idea to refer to a basic physics text.

Newton's Laws

Newtons three laws of motion govern the motion of objects and their reaction to various forces that act upon them. With Newton's laws, an object's behavior in relation to it's surrounding environment can be predicted and even controlled.

Newton's First Law – A body at rest will tend to remain at rest. A body in motion will tend to remain in motion at a constant velocity.

Newton's First Law relates to us a property of matter called inertia. Inertia is the tendency for an object to maintain a constant state of motion. The amount of inertia that an object possesses is it's mass. One of the units that we use to measure mass is the slug.

Due to inertia, a car that is traveling at 60 MPH will continue to travel at 60 MPH when you put the clutch in and remove the driving force provided by the engine. This is, of course, ignoring the effects of friction, which we will take into consideration later. Briefly stated, friction is one of the many possible outside forces. The effects of these outside forces are detailed by Newton's second law.

Newton's Second Law – A body at rest, or in motion, when acted upon by an outside force, will accelerate in the direction of the force by an amount proportional to the magnitude of the force and inversely proportional to the mass of the body.

Newton's second law details the relationship between the mass (the amount of inertia) of a body, the size of the force acting upon it, and the amount of acceleration that results. In the form of an equation, this law indicates that the force acting on an object is equal to the mass of the object multiplied by the acceleration experienced by the object:

$$F = ma \qquad (1.1),$$

and slightly rearranged:

$$a = F/m \qquad (1.1).$$

What Newton's second law tells us is that the larger the force acting on a body, the larger the acceleration. At the same time, the more massive the body, the smaller the acceleration will be.

An acceleration is a change of velocity. Velocity can be measured in feet per second (fps). A change in velocity (acceleration) can be measured in units of feet per second per second - or feet per second squared (ft/s^2). The unit of force we will be dealing with in this book is the pound. One pound is the force required to accelerate one slug of mass at 1 ft/s^2. (Note - for the reader interested in metric units, the unit for mass is the kilogram, velocity is measured in meters per second, and acceleration in meters per second squared. A force that accelerates one kilogram of mass at one meter per second squared is called one Newton.)

Newton's Third Law – For every force that acts on a body, a force that is equal in magnitude, but opposite in direction, will act on another body.

In other words, if object 1 exerts a force, F1, on object 2, object 2 will in turn exert a force, F2, on object 1. F1 and F2 will be equal and opposite:

$$F1 = -F2 \qquad \text{(Fig 1-1)}$$

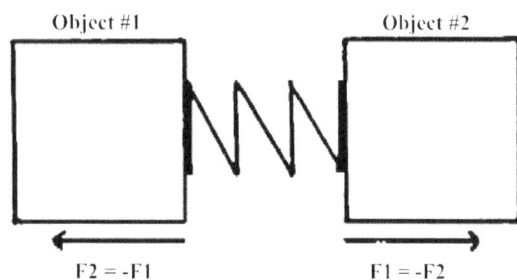

Fig 1-1. - *Newton's Third Law*

According to Newton's third law, forces always occur in pairs that are equal in magnitude and opposite in direction. Note here that these forces don't cancel each other out because they act on different objects, and each object will experience an acceleration in opposite directions. For example, when you fire a rifle, the bullet is subject to a force that accelerates it down the barrel and towards the target. At the same time, the rifle is subjected to an equal force in the opposite direction, and, as a result, you experience the recoil of the rifle.

Linear Motion

As forces are applied to an object and accelerations occur, it would be nice to know the distance that has been covered, or the time that is needed to cover a given distance. The equations of motion are intended for this purpose. These equations provide us with the necessary tools to relate, distance, velocity, acceleration, and time.

The most fundamental equation of motion is the equation that relates distance, speed, and time:

$$d = vt \qquad (1.2),$$

or, rearranged:

$$v = d/t \qquad (1.2)$$

where (v) is the average velocity, (d) is the distance covered, and (t) is the time lapse.

As well, we have the equation which gives us acceleration:

$$a = (v_f - v_i)/t \qquad (1.3)$$

where (v_f) is the final velocity, and (v_i) is the initial velocity. This gives us the change in velocity per unit of time.

Given the acceleration of an object, we want to determine the distance covered in a certain period of time. We will assume here that we are dealing with a constant acceleration. With a constant acceleration, the average velocity can be determined by the following:

$$v_{av} = (v_f + v_i)/2 \qquad (1.4),$$

or:

$$v_{av} = v_i + \tfrac{1}{2}at \qquad (1.5).$$

So if we substitute equation (1.5) into (1.2), we will get:

$$d = v_i t + \tfrac{1}{2}at^2 \qquad (1.6).$$

In the event that the object is starting from rest ($v_i = 0$), equation (1.6) becomes:

$$d = \tfrac{1}{2}at^2 \qquad (1.7).$$

Vectors

If a quantity has both magnitude and direction, it is considered to be a vector quantity (Fig 1-2). Vectors can be added and subtracted, but not in the conventional sense. Simply adding two numbers won't do the trick. To add two vectors, their directions must be taken into consideration, as well as their magnitudes.

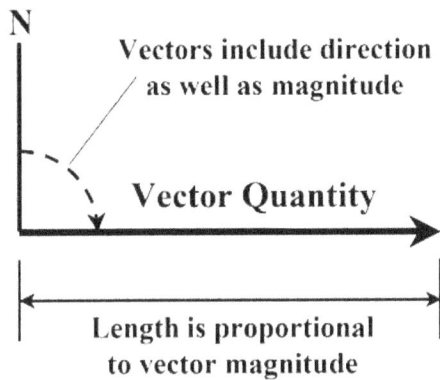

Fig 1-2. - *Vector Representation* - Vector Quantities are represented by arrows. The direction of the arrow represents the direction of the quantity (i.e. - force, velocity, etc.) and the length of the arrow represents the magnitude of the quantity.

Any vector quantity can be represented graphically by an arrow. The arrow's direction represents the direction of the vector, and its length represents the magnitude of the vector. To add vectors, we simply line them up head to tail and draw a new vector from the first tail to the last head (Fig 1-3).

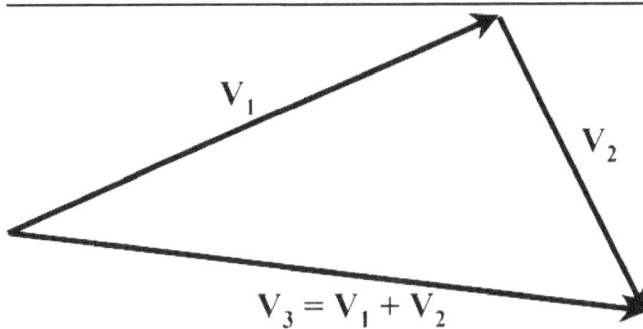

Fig 1-3. - *Vector Addition* - To determine vector sums, the vectors can be lined up head to tail.

A similar method can be used to divide vectors into components. The components will usually be perpendicular to each other. The components are arranged so that there sum is equal to the original vector. See Fig 1-4 for an

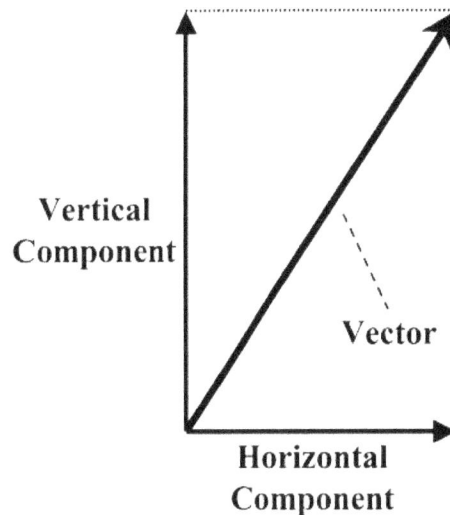

Fig 1-4. - *Vector Components* - Vector components are normally perpendicular to one another. The sum of the components is equal to the original vector.

example.

Some examples of vector quantities are force, velocity, and acceleration.

At some points in this book, basic trigonometry will be used to manipulate vectors. If you are not familiar with the concepts of trigonometry, you may want to refer to some of the books in the bibliography or to almost any high school or first year university math or physics text.

Systems of Forces

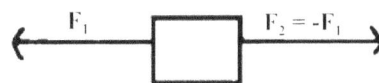

Net Force = F_2 + F_1 = 0

Net Acceleration = 0

(a) Equal and opposite forces create a net force of zero.

Net Force = F_2 + F_1 > 0

Acceleration > 0

(b) With imbalanced forces, the net force will create an acceleration.

Fig 1-5. - *Multiple Forces Acting on an Object*

As discussed in Newton's Second Law, when the net force acting on a body is anything other than zero, an acceleration will result. This net force is the resultant of all of the forces on the body and can be determined by using vector addition.

If two opposing forces of equal magnitude are acting on an object, they will cancel each other out, and the net force is zero (Fig 1-5a). In this case, the object will be in a state of equilibrium. If one of these forces is increased, the forces are no longer balanced, and an acceleration will result. This acceleration will be in the direction of the net force (Fig 1-5b).

If two forces acting on an object are equal and opposite, but they don't pass through the same point, they form a couple (Fig 1-6). This couple causes a torque, or a turning moment, which in turn will cause a rotational acceleration, upsetting the object's equilibrium.

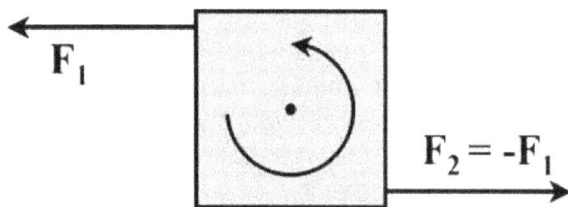

Fig 1-6. - *A Couple* - The two forces cancel, but because they are offset, they create a moment—which will rotate the object.

Gravity

Gravity, as we all know, is the unfortunate entity that creates the need for aerodynamic lift. Gravity is a force that acts on all object to accelerate them towards the center of the earth. As a result of this force, everything has a tendency to fall to the surface of the earth and remain there.

Gravity will accelerate an object in freefall at approximately 32 ft/s². Weight is the force that occurs as a result of this acceleration. Equation (1.1) becomes:

$$w = mg \qquad (1.8),$$

where w is the weight and g is the acceleration due to gravity. For an object with a mass of 1.5 slugs, it's weight is found with equation (1.8):

$$w = (1.5)(32) \text{ therefore,}$$

$$w = 48 \text{ lbs.}$$

Force, Work, and Power

When a force is acting on an object and causing an acceleration, the object is gaining kinetic energy. Kinetic energy is energy due to motion, and it is proportional to half the object's mass and the square of the object's velocity:

$$KE = \tfrac{1}{2}mv^2 \qquad (1.9).$$

This energy has got to come from somewhere. It is being transferred from the object that is applying the force. This transfer of energy is called work. The amount of work done on an object (the energy transferred) is determined by multiplying the net force by the distance over which it is applied:

$$W = Fd \qquad (1.10).$$

The units of work or energy that we will be dealing with in this book will be foot-pounds (ft-lbs). 1 ft-lb is the work done (energy transferred) when a force of 1 lb is exerted over a distance of 1 ft.

If you apply 20 lbs of force to a crate that is sitting on a frictionless surface over a distance of 10 feet, the work done can be determined using (1.10):

$$W = (20)(10) \text{ therefore,}$$

$$W = 200 \text{ ft-lbs.}$$

Since work is a measure of energy transferred, we know that this crate will now have 200 ft-lbs of kinetic energy. Using (1.9), we can now determine the velocity of the crate. If the mass of the crate is 25 slugs:

$$200 = \tfrac{1}{2}(25)v^2 \qquad \text{therefore,}$$

$$v^2 = 16 \qquad \text{and therefore,}$$

$$v = 4 \text{ fps.}$$

Energy gained by an object doesn't have to be kinetic energy. An object can also possess potential energy, or energy due to it's state or position. For our purposes, we will restrict our discussion of potential energy to gravitational potential energy, which is a result of an object's height above the earth's surface (other types of potential energy include chemical, electrical, and elastic). Gravitational potential energy is a function of an object's mass, gravitational acceleration, and the object's height:

$$PE = mgh \qquad (1.11),$$

and using equation (1.8):

$$PE = wh \qquad (1.12).$$

If a force acts on an object and causes that object to rise, a gain in potential energy is occurring. For example if a crane raises an 1100 lb steel beam to a height of 30 ft., the crane has done 33,000 ft-lbs of work on the beam, and the beam has gained 33,000 ft-lbs of potential energy. It is important to note here that this energy has to come from somewhere - in this case, the energy came from the crane, which gets its energy from its engine, which in turn, gets its energy from the burning of fuel.

If the beam was then released and allowed to fall to the ground, gravity would cause an acceleration, and the beam would then gain 33,000 ft-lbs of kinetic energy. Using (1.9), we can determine the velocity of the steel beam in the same way as the crate.

Notice that when discussing work, time is not an issue. It could take 10 seconds or 10 years to raise the beam to 30 ft., and the work done will still be 33,000 ft-lbs. If we want to know that a certain amount of work can be accomplished in a certain amount of time, we need to discuss power. Power is the rate of doing work:

$$Pwr = W/t \qquad (1.13).$$

Using equation (1.10):

$$Pwr = Fd/t \qquad (1.14),$$

or, using equation (1.2):

$$Pwr = Fv \qquad (1.15).$$

The unit of power that we will be dealing with is the horsepower. One horsepower is defined as the ability to do 550 ft-lbs of work in one second, or 33,000 ft-lbs of work in one minute. This gives us the following equations for horsepower:

$$1HP = 550 \text{ ft-lbs/s} \qquad (1.16).$$

Therefore:

$$Pwr(HP) = w/550t \qquad (1.17).$$

So if the 33,000 ft-lbs of work done on the steel beam were done in 20 seconds, we could simply use equation (1.17) to find the horsepower used:

$$Pwr = 33,000/(550)(20)$$

therefore:

$$Pwr = 3 \text{ HP}.$$

We could also determine the time required to do the work if the power available to use is known, if we have 2 HP to use, the time needed to lift the beam to 30 ft can be found as follows:

$$2 = 33,000/550t \qquad \text{therefore:}$$

$$t = 30 \text{ s}.$$

Friction

According to Newton's second law, if you apply a force to an object, and cause an acceleration, the object should continue to move at it's final velocity once you have removed the force. However, we know from daily experience that this is just not the case. When you slide a book across the table, it soon comes to a stop. When you push a chair across the floor, it quickly comes to rest. In this we see a contradiction, all apparent forces have been removed, yet an acceleration is still occurring. Is Newton's second law incorrect, or have we missed something altogether?

The answer lies in friction. Friction is the force that occurs when two objects are in contact with one another. It is caused by the interactions of the molecules at or near the surface of each object. It is also a result of the fact that, regardless of how smooth we think an object is, on a microscopic level, the objects surface is extremely rough.

Friction is a resistive force that always acts exactly opposite the direction of motion. Because of this, friction will always slow an object down and eventually bring it to a stop. This is the reason that the book stops when sliding across the table, or the chair stops when sliding across the floor.

The amount of friction present depends on the coefficient of friction - which depends on the materials that are in contact - and the normal force. The normal force is the force that is holding the two objects together, it is perpendicular to the contact surface (Fig 1-7).

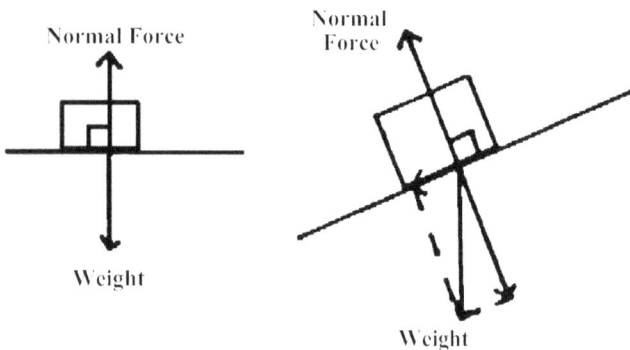

Fig 1-7. - *The Normal Force* - The normal force acts perpendicular to the contact surface.

There are two types of friction coefficients, static and kinetic. The static coefficient of friction is associated with two objects that are in contact and stationary relative to one another. The kinetic coefficient of friction, on the other hand, refers to two objects that are moving against one another. The static coefficient of friction between two materials will invariably be higher than the kinetic coefficient of friction.

For static friction:

$$F_{fr} = \mu_s N \qquad (1.18),$$

or, for kinetic friction:

$$F_{fr} = \mu_k N \qquad (1.19).$$

μ (mu) is the Greek letter that is used to represent the coefficient of friction. The coefficient of friction would be high between two surfaces that are rough, or that are made of materials that tend to stick. The coefficient of friction would be low between two surfaces that are smooth. Lubricating oil is used in many machines to reduce the coefficient of friction between moving parts.

Friction is the reason that we need to continue supplying a force - after an object is moving - just to maintain a constant velocity. The force that you supply to keep a box moving across the floor is needed to cancel out the force of friction

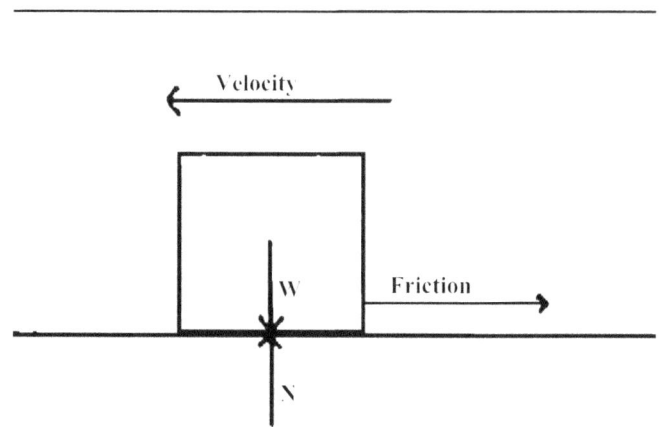

Fig 1-8. - *Friction* - Friction always acts opposite to the direction of motion.

(Fig 1-8). If it weren't for friction, a cars engine would be unnecessary - once the car is moving. On the other hand, if it weren't for friction, the brakes on that same car wouldn't work.

Conservation of Mass

The defining property of all matter is the fact that it has mass. Mass determines how much inertia an object has, as detailed by Newton's first and second laws. One of the fundamental laws of nature is called the Conservation of Mass. According to the conservation of mass, matter (mass) can be neither created nor destroyed. Matter can to a certain extent change forms in chemical reactions, but the amount of matter in the universe is a constant and cannot be changed. As well, in a closed system, matter cannot be created or destroyed, the total amount remains a constant at all times.

Conservation of Energy

Another fundamental law of nature, equivalent to the Conservation of Mass, is called the conservation of Energy. According to the Conservation of Energy, the total energy of a closed system cannot change. Energy can change forms (kinetic, potential, thermal, etc.) through a variety of processes, but the change in the amount of energy will always be zero.

For total energy in a system:

$$E_{tot} = KE + PE + TE + other \qquad (1.20).$$

And for the change in the energy of a system, according to the Conservation of Energy:

$$\Delta E_{tot} = \Delta KE + \Delta PE + \Delta TE + \Delta other = 0 \quad (1.21).$$

Δ (delta) is the Greek letter that is used to represent an amount of change. Therefore, ΔKE means the change in kinetic energy.

For the sake of simplicity in this book, most discussions of energy will be restricted to kinetic and potential energy (usually expressed by pilots as airspeed and altitude). So the other types of energy can be ignored for our purposes, and equation (1.21) becomes:

$$\Delta KE + \Delta PE = 0 \qquad (1.21),$$

therefore:

$$\Delta KE = -\Delta PE \qquad (1.21).$$

In other words, using (1.9) and (1.11):

$$mgh + \tfrac{1}{2}mv^2 = Constant \qquad (1.22).$$

Bear in mind the effect that friction has on our lives. Friction will always slow an object down, hence we often view friction as an energy robber. The kinetic energy which we invariably lose to friction is converted into thermal energy. So even though energy often seems to disappear, the Conservation of Energy still holds true.

For example, if you were to calculate the kinetic energy of a moving car, and then bring the car to a stop using the brakes, the energy seems to have disappeared. However, the brake pads and disks of the car have now been heated up by friction. If you were to add up the increase in the individual kinetic energies of the molecules in the brakes, you would find that it would be equal to the initial energy of the car.

As an object falls, it is gaining kinetic energy, where is this energy coming from? The object began with gravitational potential energy due to it's height, this potential energy is converted to kinetic energy as the object accelerates.

In the atmosphere, freefall never really occurs. Aerodynamic drag builds up as speed increases. This upward force slows down the object's acceleration until the object reaches equilibrium (Fig 1-9). This equilibrium occurs at the

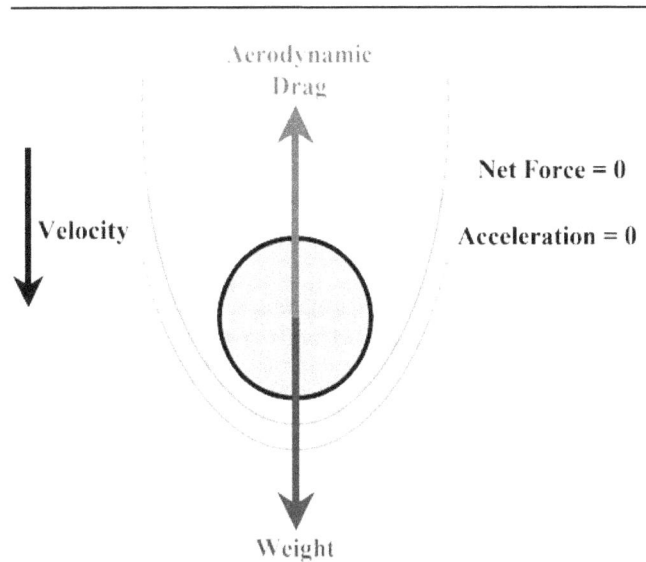

Fig 1-9. - *Atmospheric "Freefall"* - As an object accelerates in freefall, aerodynamic drag increases until equilibrium is reached. The speed at which this occurs is the objects terminal velocity.

the object is now being transferred into the objects terminal velocity. The potential energy of atmosphere. Initially the atmosphere gains kinetic energy as the air is put into motion. As this motion dampens out due to internal friction, the atmosphere loses kinetic energy and gains heat energy.

Simple Machines

A machine is a device used to convert one form of energy or work into another, more useful, form of energy or work. We use many different types of machines from day to day, a common one - in cars and in airplanes - being the reciprocating engine. Any engine that burns fuel is converting the chemical potential energy of the fuel into some other form of energy. This can be potential or kinetic mechanical energy (lifting or accelerating an object), or some other type of energy such as electrical energy from an alternator.

A simple machine is a single device used to convert or transfer energy. Some examples of simple machines are levers, pulleys, inclined planes, and gears. These devices will convert mechanical energy of one form into mechanical energy of another form or, depending on their configuration, they may simply convey the energy to another location. The main method for a conversion is to vary the distance over which a force is applied - thus varying the amount of force.

We know from the conservation of energy that energy cannot be created from nothing - nor can it disappear. This means that if we input energy into a system, that energy must be either stored or passed out of the system. We also know from (1.10) that W=Fd. Since work is the transfer of energy, if we do work on a system the energy must go somewhere.

Let's take for example a lever (Fig 1-10). A lever is essentially a rod with a pivot, or fulcrum, somewhere along it's length. If the fulcrum is at the center of the rod, both ends will move the same distance. Since work going into the lever is equal to the work coming out, the force applied to each end will be the same as well.

$$W_{in} = W_{out}$$

Therefore, from (1.10):

$$F_1d_1 = F_2d_2 \qquad (1.23).$$

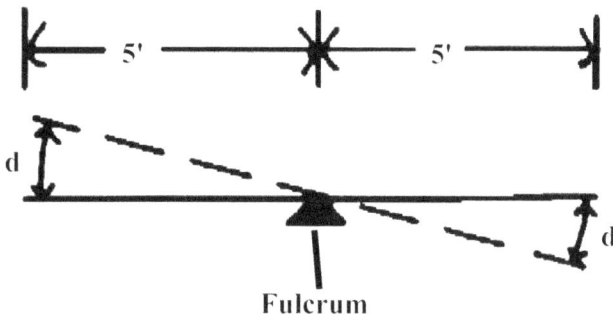

Fulcrum

Fig 1-10. - *A Lever*

Keep in mind that d is not the length of the arm, which is measured from the fulcrum, but the distance that the end of the arm travels, which will in turn be proportional to the length of the arm. If the arm to the right of the fulcrum is five feet long, and the arm to the left is five feet long, then d_1 and d_2 will be equal and cancel each other out. If this is the case,

$$F_1 = F_2.$$

If, however, we were to shift the fulcrum to the left so that the arm to the right was eight feet long and the arm to the left was two feet long (Fig 1-11), things would change somewhat. Equation (1.23) will still hold true, but the input and output forces will no longer be equal—since the arm lengths are no longer equal. If the arm lengths are not equal, the distances that the forces are applied over will not be equal. The proportion between the arm lengths will be the same as the proportion between the distances.

$$A_{right}/A_{left} = d_1/d_2 \qquad (1.24).$$

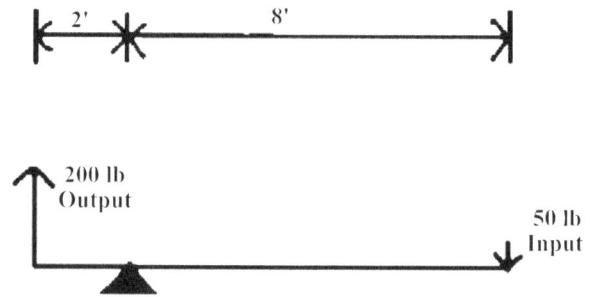

Fig 1-11. - *A Lever With a Mechanical Advantage*

If we apply a force of 50 lbs to the right side of the lever in Fig 1-11 over a distance of 4 feet, what happens on the left side? Using equation (1.24), we have,

$$8/-2 = 4/d_2 \qquad \text{therefore,}$$

$$d_2 = -1 \text{ foot}$$

The negative sign is a result of the fact that the arm is measured from the fulcrum. The fulcrum is the zero point, an arm to the right is positive, and an arm to the left will be negative. As you can see from the diagram, the negative makes sense, since a downward force on the right of the lever will result in an upwards force on the left.

Continuing with the problem, we use (1.23),

$$(50)(4) = F_2(-1) \qquad \text{therefore,}$$

$$F_2 = -200 \text{ lbs.}$$

Once again, we see a negative sign - indicating that the force is applied in the opposite direction. So, even though only 50 lbs of force was applied to the lever on the right, 200 lbs of force was delivered on the left side. This gain in force was offset however by the loss in distance traveled. The gain in force is called mechanical advantage. In this particular example, we have a mechanical advantage of 4 since the force applied was multiplied by 4.

The problem we just solved could also be solved by using moments. The moment is equal to the force multiplied by the arm:

$$M = FA \qquad (1.25).$$

The moment on both sides of the fulcrum must be equal, therefore:

$$F_1 A_1 = F_2 A_2 \qquad (1.26).$$

So, returning to the previous problem using (1.26):

$$(50)(8) = F_2(-2) \qquad \text{therefore,}$$

$$F_2 = -200 \text{ lbs.}$$

As you can see, we get the same answer with a simpler calculation. The negative sign once again, indicates that the output force is being applied in the direction opposite the input force.

A third scenario that we should discuss in relation to the lever is an arrangement in which the input and output forces are both on the same side of the fulcrum (Fig 1-12). In this case, equation (1.23) still holds true, the work done on the lever will be equal to the work done by the lever. As well moments can still be used to calculate the output force, equation (1.26) can still be applied.

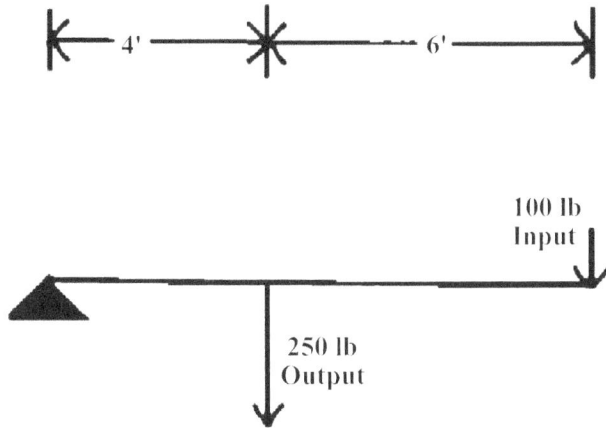

Fig 1-12. - *Input an Output Forces on the Same Side of the Lever*

If the distance between the fulcrum and an input force of 100 lbs is 10 ft, and the output force is taken 4 ft from the fulcrum, what will the magnitude of the output force be? Using equation (1.26):

$$(100)(10) = F_2(4) \qquad \text{therefore,}$$

$$F_2 = 250 \text{ lbs.}$$

In this example, we have a mechanical advantage of 2.5. Remember that mechanical advantage means that we gain in the force applied, but we lose in the distance which the force is applied over. So if the 100 lb force was applied over a distance of 20 ft, the 250 lb force would only be applied over a distance of 8 ft (20 divided by 2.5).

This brings us to the idea of placing the output of the lever further from the fulcrum than the input (Fig 1-13). With this arrangement, the force supplied by the lever is less than the force applied to the lever, but this force is applied over a greater distance. The moment calculations and the concept of mechanical advantage are applied in the same way. In this case, however, the mechanical advantage will be less than 1.

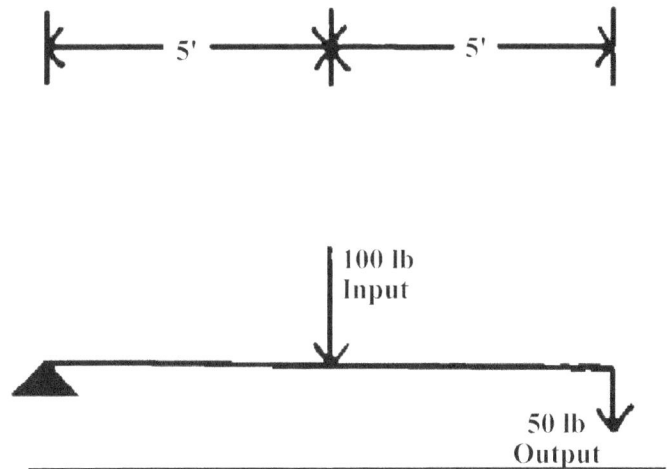

Fig 1-13. - *Mechanical Advantage Less than One.*

Up to this point we have neglected friction. Obviously, some energy is going to be lost at the fulcrum due to friction converting the mechanical energy to heat energy. This is the unfortunate cost of using machines. However, the benefit far outweighs the cost in most cases. Imagine trying to change a flat tire without a jack, or trying to lift a boulder without a crowbar.

Other types of machines will not be detailed here, but the same basic concept applies to them all. Energy (work) applied to the machine is the energy supplied by the machine. I chose to detail the lever in this section because the concept of leverage and moments will be recurring throughout this book.

Uniform Circular Motion

As we discussed with Newton's first law, an object in motion will tend to remain in motion at a constant velocity. Velocity is a vector quantity, and as such, it involves direction as well as speed. As well, we discussed Newton's second law, which relates to us the effect that outside forces will have on an object's equilibrium.

Until now, we have worked with the assumption that the forces we are dealing with are parallel to the motion of our object. As a result of this, the object will either speed up or slow down, according to the direction of the force.

Now we need to consider what will happen when the accelerating force is not parallel to the direction of motion. To keep this discussion simple,

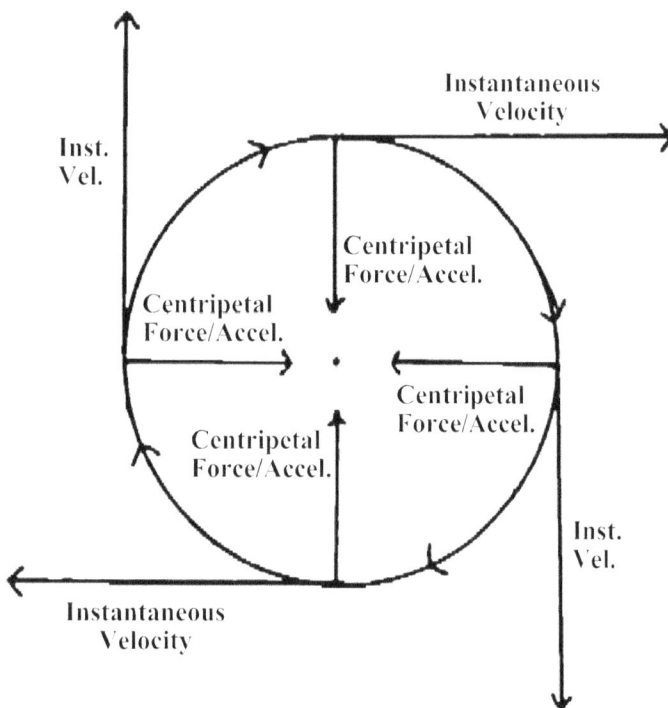

Fig 1-14. - *Centripetal Force and it's Result* - As an object travels around a circle, it's acceleration is always directed toward the center of the circle.

we will only look at one scenario of a non-parallel force. We will consider a force that is perpendicular to the direction of motion (Fig 1-14). This force is called the centripetal force, and it acts toward the center of a circle. Centripetal acceleration will occur in the same direction as centripetal force, toward the center of a circle. Even though the speed of an object doesn't change, an acceleration is occurring because the direction of motion is changing, therefore velocity is not constant.

We now need some way of relating the size of the force to the amount of acceleration. Equation (1.1) still applies:

$$a = F/m \qquad (1.1).$$

As well, we now have another equation that will tell us the radius of the circle:

$$a = V^2/r \qquad (1.27).$$

If we combine equations (1.1) and (1.27), we get:

$$F/m = V^2/r \qquad (1.28)$$

or

$$F = mV^2/r \qquad (1.28).$$

Using the above formulae, we can calculate the radius of a circle, given the size of the centripetal force, the mass of the object, and the speed of the object. If an object with a mass of 5 slugs is travelling at 10 fps, and a force of 50 lbs act towards the center of a circle, what is the radius of that circle?

Using equation (1.28):

$$50 = (5)(10)^2/r \qquad \text{therefore,}$$

$$r = 10 \text{ ft.}$$

The Myth of Centrifugal "Force"

In relation to circular motion, a term that is often used is "centrifugal force". Centrifugal force is often explained as "the force that pulls you to the outside of a turn" and is considered to be equal and opposite to centripetal force. Considering this for a moment, it clearly cannot be true. If the centrifugal force was exactly equal and opposite to the centripetal force, the net force would be zero and the acceleration would therefore be zero. Without an acceleration, the object that is supposed to be traveling around in a circle will simply continue to go straight.

Centrifugal force is what physicists refer to

as a pseudo force - meaning that it only exists from certain reference points. Consider for example a car that is traveling around a corner. An observer outside the car sitting on the curb will observe the centripetal force provided by the friction between the tires and the road. However, a passenger *inside* the car (not in equilibrium) will observe a "force" pulling them to the outside of the turn. This "force" is in fact an illusion - it is our interpretation of the physical sensation of the car seat pushing us *into* the turn.

The concept of centrifugal force does have some usefulness however. It is primarily a mathematical tool based on Newton's Third Law which allows us to determine some of the internal forces experienced by an object during an acceleration. As such, the centrifugal force will be referred to at various points throughout this book.

Aircraft in Equilibrium

Newton's first law details for us a situation known as equilibrium. In equilibrium, all of the forces acting on an object are balanced so that the net force is zero. As well, all of the torques or moments acting on the object must be zero.

For an aircraft in flight, we have several forces and moments to account for (Fig 1-15). First

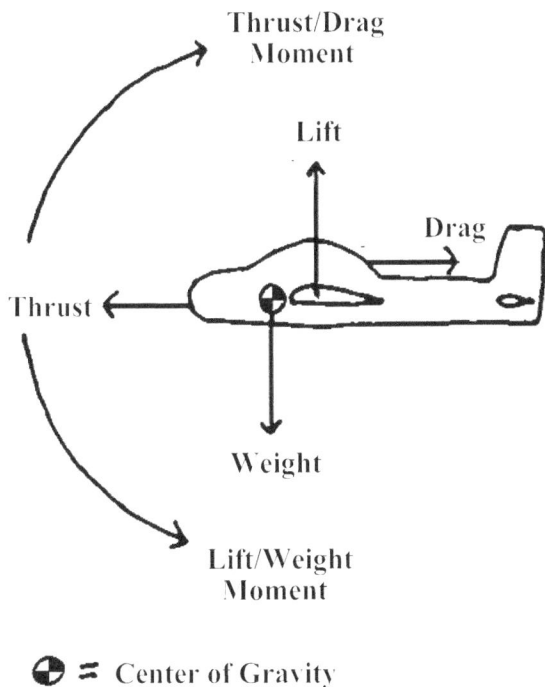

Fig 1-15. - *An Aircraft in Equilibrium, Straight and Level*

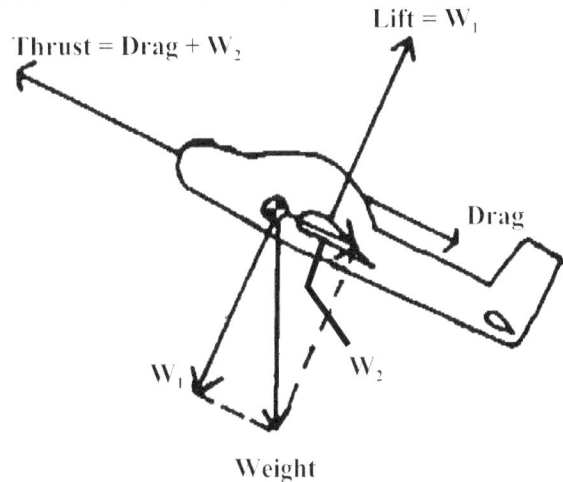

(a) Equilibrium in a climb - Weight is divided into components that are perpendicular and parallel to the flight path.

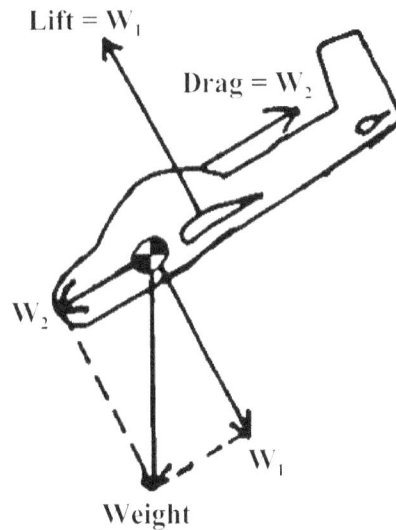

(b) Equilibrium in a descent - As with the climb, weight is divided into components. This time however, the component of weight that is parallel to the flight path will contribute to "thrust" instead of "drag".

Fig 1-16. - *Aircraft in Equilibrium, Climbs and Descents* - During climbs and descents, the four forces (lift, weight, thrust, and drag) can be resolved into components that are parallel and perpendicular to the flight path.

we have weight, which is the force that holds us on the ground. To counteract weight, we need to produce aerodynamic lift. In equilibrium, weight and lift are equal, and therefore they cancel each other out. To produce aerodynamic lift, an aircraft has to move through the air.   This motion unfortunately, will cause aerodynamic drag which will tend to slow us down. To counteract drag and maintain motion through the air, we need thrust. In equilibrium, thrust and drag will be equal and will therefore cancel each other out.

As you can see from Fig 1-15, there are also some torques, or turning moments. These turning moments are caused by the couples formed between the four forces. The thrust/drag couple causes a nose up moment, while the lift/weight couple causes a nose down moment. These two moments will cancel each other - when the aircraft is in equilibrium.

This particular example of equilibrium is an aircraft in straight and level flight. Steady climbs and descents are also equilibrium conditions once they are established. Once again, for equilibrium to exist, no net forces or moments must exist (Fig 1-16).

This picture of the forces and moments acting on an aircraft is greatly simplified. Throughout this book we will have a more detailed look at each force and how it is produced, as well as it's effects.

Chapter Summary

Newtons laws govern the motion of objects and the way they react to various forces that act on them. Newton's first law defines for us a property of matter known as inertia, or mass. Newton's second law allows us to determine how mass will react to forces acting upon it. And Newton's third law details action-reaction force pairings.

Force, velocity, and acceleration are all vector quantities, and as such they can be represented by arrows. These vectors can be added together graphically or mathematically - provided their directions are taken into consideration.

When multiple forces are acting on an object, the net force can be determined by using vector addition. The objects reaction according to Newton's second law is based on the net force. If two forces are opposite in direction but do not pass through the same point, they form a couple. This couple will result in a rotational acceleration due to torque.

Gravity is the force that attracts objects toward the center of the earth. The rate of acceleration experienced as a result of gravity is 32 ft/s^2. Weight is the magnitude of the force of gravity. Weight is determined by multiplying the mass of an object by the acceleration due to gravity.

Work is the transfer of energy and is done by applying a force over a distance. There are several forms of energy including potential energy from an objects state or position, and kinetic energy from an objects motion. Power is the rate of doing work. One unit of power is the horsepower. One horsepower is the ability to do 550 ft-lbs of work in one second.

Friction is a force that resists motion. It is caused by the interaction of molecules at or near the surface of materials. The force of friction will increase with a higher coefficient of friction which is dependant on the materials in contact. As well, the force of friction will increase when the force holding the two objects together - called the normal force - is increased. The two types of friction are static and kinetic friction. The coefficient of static friction will always be higher than the coefficient of kinetic friction.

The conservation of mass dictates that the total mass in a closed system cannot be changed. The conservation of energy dictates that the total amount of energy in a closed system cannot be changed. Energy can change forms, but a decrease in one form will always result in an increase in another form, and the total will remain the same.

Machines are used to transfer energy from one place to another or to change the form which the energy takes. A method which is often used to vary the form of the energy is to change the distance over which a force is applied. This variation in distance will also cause a variation in the amount of force. Some energy is lost to friction in a machine, but the advantage gained by the machine will usually be worth the loss.

If a force is applied perpendicular to an objects direction of motion, the resulting acceleration will be a change in direction instead of a change in speed. As long as the force remains perpendicular to the new direction of motion, a circular path will result.

Centrifugal force is in fact not a force, but a reaction to centripetal acceleration. This reaction manifests itself as an apparent force acting to the outside of a circular path.

For an aircraft in equilibrium, the forces of weight, lift, drag and thrust combine to create a net force of zero. As well, the lift/weight couple and the thrust/drag couple combine to produce a net torque of zero. With no net force and no net torque, there will be no acceleration, thus the aircraft is in equilibrium.

Summary of Formulae

<u>Newton's Laws</u>

$$F = ma \qquad (1.1)$$

<u>Linear Motion</u>

$$d = vt \qquad (1.2)$$

$$a = (v_f - v_i)/t \qquad (1.3)$$

$$v_{av} = (v_f + v_i)/2 \qquad (1.4)$$

$$v_{av} = v_i + {}^1\!/_2 at \qquad (1.5)$$

$$d = v_i t + {}^1\!/_2 at^2 \qquad (1.6)$$

$$d = {}^1\!/_2 at^2 \qquad (1.7)$$

<u>Gravity</u>

$$w = mg \qquad (1.8)$$

<u>Force, Work, and Power</u>

$$KE = {}^1\!/_2 mv^2 \qquad (1.9)$$

$$W = Fd \qquad (1.10)$$

$$PE = mgh \qquad (1.11)$$

$$PE = wh \qquad (1.12)$$

$$Pwr = fd/t \qquad (1.13)$$

$$Pwr = W/t \qquad (1.14)$$

$$Pwr = Fv \qquad (1.15)$$

$$1HP = 550 \text{ ft-lbs/s} \qquad (1.16)$$

$$Pwr(HP) = w/550t \qquad (1.17)$$

<u>Friction</u>

$$F_{fr} = \mu_s N \qquad (1.18)$$

$$F_{fr} = \mu_k N \qquad (1.19)$$

<u>Conservation of Energy</u>

$$E_{tot} = KE + PE + TE + other \qquad (1.20)$$

$$\Delta KE + \Delta PE = 0 \qquad (1.21)$$

$$\Delta KE = -\Delta PE \qquad (1.21)$$

$$mgh + {}^1\!/_2 mv^2 = constant \qquad (1.22)$$

<u>Simple Machines</u>

$$F_1 d_1 = F_2 d_2 \qquad (1.23)$$

$$A_{right}/A_{left} = d_1/d_2 \qquad (1.24)$$

$$M = FA \qquad (1.25)$$

$$F_1 A_1 = F_2 A_2 \qquad (1.26)$$

<u>Uniform Circular Motion</u>

$$a = v^2/r \qquad (1.27)$$

$$F = mv^2/r \qquad (1.28)$$

Questions and Problems

1) What is inertia?

2) How does the magnitude of a force relate to an objects acceleration?

3) If two people on skates are facing each other and one pushes the other, what will happen? Why?

4) What is a vector? How would you go about adding vectors?

5) What is a couple? What is the effect of a couple?

6) What is work, and how does it relate to power?

7) Why does friction occur? What determines the amount of friction present?

8) What is the law of conservation of energy?

9) What is potential energy? Kinetic energy?

10) How does the conservation of energy apply to machines?

11) When a box sliding across the floor comes to rest, where did it's kinetic energy go?

12) What is mechanical advantage, and how is it determined?

13) What is the force that acts towards the center of a circle called?

14) What are the four forces acting on an aircraft in equilibrium? How do these forces create couples?

15) A crate with a mass of 2 slugs is resting on a floor with a friction coefficient of 0.15. If you were to push on this crate with a force of 20 lbs, what would the crate's acceleration be?

16) If you pushed the crate in Q#15 for a distance of 30 feet, how much work would you do? How much kinetic energy would the crate have? Where did the other energy go?

17) How much time would it take the crate in Q#16 to come to rest once you stop pushing? How much distance would it cover from the point at which you released it?

18) At the point just prior to releasing the crate in Q#16, how much power were you producing?

19) For this question, ignore the effects of aerodynamic lift and drag. If a 2000 lb aircraft lands at a speed of 60 knots (1 knot = 6080 ft per hour) compare the stopping distances required with maximum braking and with the brakes "locked" (i.e. - the tires are skidding). Assume μ_s = 0.4 and μ_k = 0.08 for contact between rubber and pavement.

20) A 3000 lb. aircraft is flying at a speed of 150 knots. If this aircraft is turning with a radius of 2000 ft., what is the centripetal force acting on the aircraft?

Chapter 2
The Production of Lift

As we saw in the previous chapter, for an aircraft to maintain equilibrium, lift must be equal to weight. If lift does not equal weight, the aircraft won't be able to sustain flight. As pilots, we need to be able to control the lift that our aircraft is developing in order to effectively pilot the craft. In learning to do this, the first question to answer is, where does this lift come from?

The purpose of this chapter is to answer that question. We will investigate the production of lift and the variables that affect it. Although some would argue successfully that you can fly a plane with little or no knowledge of aerodynamics, a thorough understanding of the basic principles of flight will allow you to fly the plane safely to it's limits. As well, simply flying by rote will not necessarily prepare you for situations which you have not previously encountered.

Airflow and Streamlines

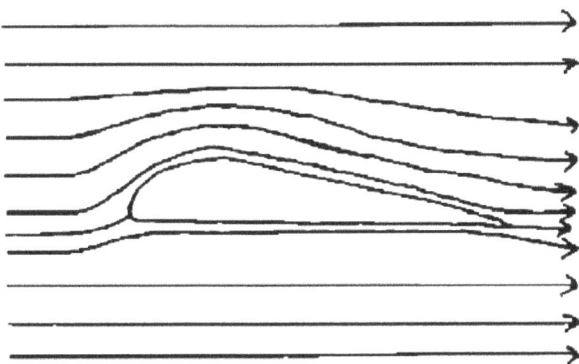

Fig 2-1. - *Streamlines* - A method of airflow visualization. No two streamlines can cross. Each line indicates the path of an air particle.

In order to discuss airflow about any object, it is important that we understand some basic concepts of air flow and the representation of this flow.

Airflow is represented graphically by streamlines (Fig 2-1). These streamlines represent the path followed by individual parcels of air as they flow around an object. Two streamlines can never cross, since this would indicate that two parcels of air are occupying the same space at the same time—clearly an impossibility.

When the streamlines are spaced widely, this indicates that the air velocity is relatively low and the air pressure is relatively high. When the streamlines are spaced closely, it indicates the opposite. The reason for this will become apparent after the next three sections of this chapter.

Steady State Flow

In a moment, we will consider the continuity equation. This equation is used by engineers and physicists to determine the mass of air (or any other fluid) that is flowing through a given point in space in a given period of time. For our consideration, we will be making one simplifying approximation—a condition known as steady state flow is in effect.

This assumption restricts the usefulness of the equation for engineers, but it suits our purposes just fine and it allows us to avoid some more complicated mathematics.

For someone who is accustomed to dealing with solid objects, a steady state is considered to exist when no net forces are acting on the object. In other words, the object is in equilibrium. However, when studying the flow of fluids, steady state does not apply to any individual particle of the fluid but to a point in space which the fluid is passing through.

Referring to Fig 2-2, two items are pointed out. The first is an air molecule that is traveling along a surface. This molecule may be traveling at a constant velocity or it may be accelerating—changing it's direction and/or speed. In the case of an acceleration, the molecule itself is not in a steady state. It is possible, however, for the flow in general to have what is considered to be a steady state flow.

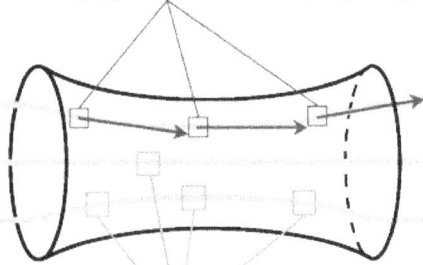

As an element of fluid moves along a streamline, it's velocity (speed and/or direction) may change.

If we look instead at a fixed volume of space, it's possible that each distinct element of fluid passing through that space is moving at the same velocity. If this is the case for every point in the flow, the flow is steady. Otherwise, the flow is unsteady.

Fig 2-2. - *Steady State Flow* - If each particle crossing a given point is at the same velocity, that point is steady state. If all of the points in a flow are steady state, it is a steady state flow.

The second item pointed out in Fig 2-2 is a stationary point within the flow. Many parcels of air are passing through this point in a given period of time. If each of these parcels of air possess the same velocity (speed *and* direction), a steady state flow exists at this point. If all of the points in a given flow are steady state, then the flow is said to be a steady state flow.

The Continuity Equation

In the previous chapter we discussed the Conservation of Mass, which states that matter can be neither created nor destroyed—in the universe as a whole, or in a closed system where nothing is added or taken away. The continuity equation relates the Conservation of Mass to fluid dynamics.

First, we need to define the term density. The density of a material is the amount of mass present per unit of volume:

$$\rho = m/V \qquad (2.1).$$

ρ (rho) is the Greek letter that is used to represent density. Density is measured in units of slugs per cubic foot (or, in metric, kilograms per cubic meter).

Now let's start with a pipe that has a steady state airflow through it (Fig 2-3). If we take the section of pipe from *a* to *b* (or any section for that matter) and measure the airflow, the Conservation of Matter must be observed. If any mass of air is introduced to section *a-b*, an equal mass of air must be passed out of the section.

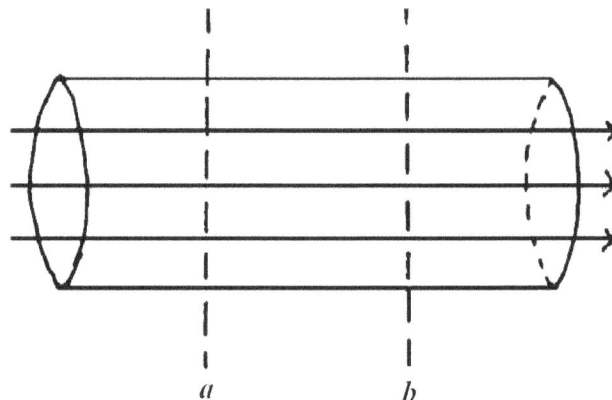

Fig 2-3. - *Pipe With Steady State Airflow*

The distance between *a* and *b* can be reduced and the Conservation of Matter must still hold true. If we reduce the distance between *a* and *b* to zero, we are left with just a cross-section (Fig 2-4). Along the length of the pipe, there are an infinite number of cross sections.

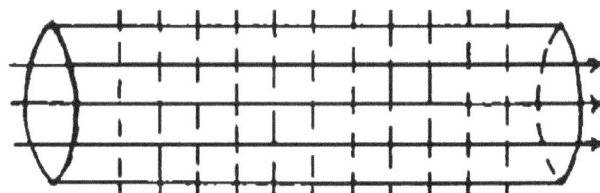

Fig 2-4. - *Cross-Sections Along the Pipe*

The mass of air that flows into a given cross section must also flow out of that cross section and into the next one. The mass flow through a cross section is determined by multiplying the density of the air by the volume of air passing through the section in a given period of time:

$$MF = \rho V/t$$

where V/t is the volume flow of the air. The volume

flow can be determined by multiplying the air velocity by the area of the cross section:

$$V/t = vA.$$

Therefore:

$$MF = \rho vA \qquad (2.2).$$

If the mass flow from each cross section must pass into the next, we know from the conservation of mass that the mass flow through any given section must be constant along the length of a pipe. Therefore:

$$\rho_1 v_1 A_1 = \rho_2 v_2 A_2 \qquad (2.3)$$

for any two given cross sections along the length of the pipe. This is the equation of continuity.

For the purpose of low speed aerodynamics (below ~250 kts), there is no appreciable change in air density. As a result of this, the continuity equation can be reduced to:

$$v_1 A_1 = v_2 A_2 \qquad (2.4).$$

For a straight pipe, as seen in Fig 2-3 and Fig 2-4, the continuity equation is nothing profound. However, we could also consider a pipe with a restriction in it. Such a pipe is known as a venturi tube (Fig 2-5).

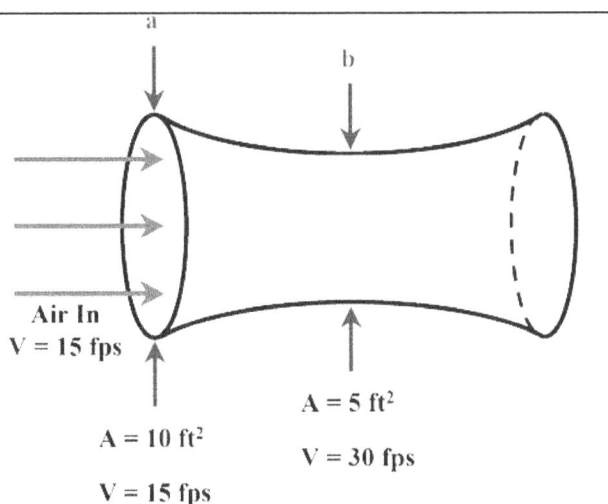

Fig 2-5. - *Venturi Tube* - Cross sectional areas vary along the length of the pipe. This leads to changes in velocity as the air flows through.

Because the cross sectional area changes along the length of the venturi, the velocity of the

airflow will also change along it's length. If cross section *a* has an area of 10 ft^2, and cross section *b* has an area of 5 ft^2, we can determine what the velocity at *b* would be given that the velocity at *a* is 15 fps. Using equation (2.4):

$$(15)(10) = v_b(5)$$

therefore:

$$v_b = 30 \text{ fps.}$$

This same calculation can be applied to any two cross sections along the length of the pipe. Thus we have a maximum velocity at the narrowest point of the pipe, and a minimum velocity at the widest point.

Bernoulli's Theorem

Daniel Bernoulli was a Swiss mathematician who lived during the 1700's. Bernoulli developed a theorem that relates the Conservation of Energy to fluid dynamics. As we have seen with the conservation of energy, discussed in Chapter 1, energy cannot be created from nothing. We have also seen from the continuity equation that air will speed up as it passes through a restriction. This increase in velocity is also an increase in kinetic energy. This kinetic energy must come from somewhere

The two concepts that are important to Bernoulli's Theorem are those of static pressure and dynamic pressure. Pressure is defined as force per unit area:

$$P = F/A \qquad (2.5).$$

Static pressure is the pressure exerted by the air when it is at rest, and it acts in all directions at once (Fig 2-6). Static pressure is a result of the random motion of the air molecules, as well as the density of the air. The motion of the air molecules is a result of their temperature, which in turn is a function of the kinetic energy (energy due to motion) of individual molecules. The more kinetic energy possessed by an individual molecule, the more of a force it will exert against an object that it strikes. As well, for a given temperature (kinetic energy per molecule), the higher the density, the more collisions will occur with that surface per second. If we were to multiply the energy per molecule (or, more correctly, the energy per unit of mass—which is in turn a function of the number of molecules) by the density, we will get the energy per unit of volume. This is static pressure.

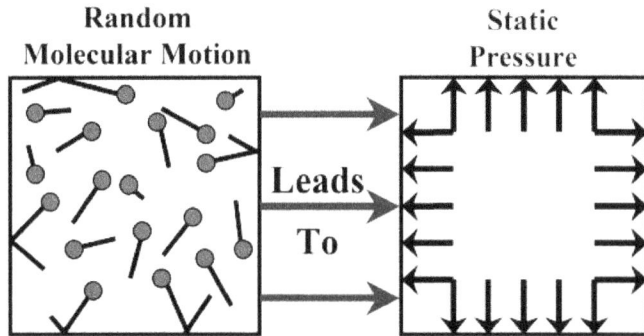

Fig 2-6. - *Static Pressure* - Pressure resulting from the motion of the air molecules. Static pressure acts in all directions.

Dynamic pressure is a result of the motion of the air and acts in the direction of motion (Fig 2-7). As we know, motion of an object means that the object possesses kinetic energy. With airflow, it is easier however to consider dynamic pressure. From Chapter 1, we know how to determine an objects kinetic energy:

$$KE = \tfrac{1}{2}mv^2 \qquad (1.9).$$

Dividing equation (1.9) by volume, we get:

$$KE/V = \tfrac{1}{2}mv^2/V.$$

We know from previous discussions that energy per unit volume is pressure and from equation (2.1) that mass per unit volume is density. Therefore:

$$P_d = \tfrac{1}{2}\rho v^2 \qquad (2.6).$$

The static pressure of the air is a form of potential energy, while the dynamic pressure is a form of kinetic energy. As a result of this, we know from the conservation of energy that the two must always sum up to be a constant in a closed system:

$$P_s + P_d = \text{Constant} \qquad (2.7),$$

and if we substitute in (2.6):

$$P_s + \tfrac{1}{2}\rho v^2 = \text{Constant} \qquad (2.8).$$

So to sum up Bernoulli's Theorem, if the velocity of a fluid changes, the static pressure of

that fluid will change as well. An increase in velocity will result in a decrease in pressure and vice versa.

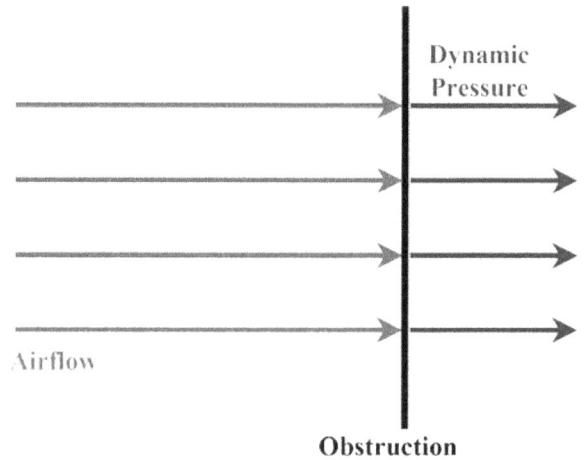

Fig 2-7. - *Dynamic Pressure* - Pressure resulting from the overall motion of the air. Dynamic pressure only acts in the direction of the motion.

Flow About an Airfoil

What on Earth has all of this got to do with flying??

Let's go back to the Continuity Equation and look again at a venturi tube (Fig 2-8). We know that the air traveling through the restriction in the venturi is accelerated to maintain a constant mass flow. We also know now that the speed of the air is increased at the expense of static pressure. This means that at narrower points along the length of the venturi, a drop in static pressure occurs. This can be seen graphically by looking at the streamlines—they are closer together in the restriction of the venturi.

Now let's erase the top half of the venturi

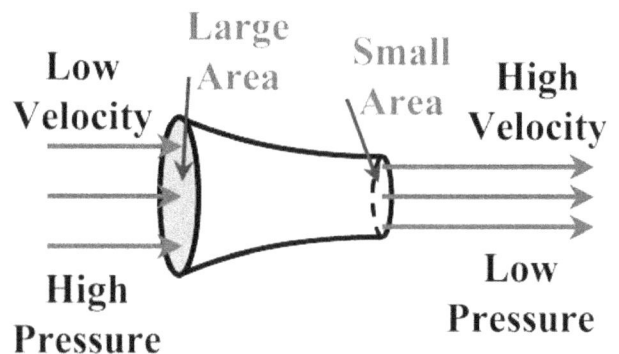

Fig 2-8. - *Venturi Tube with Airflow* - The increase in flow velocity at narrower points leads to a decrease in static pressure.

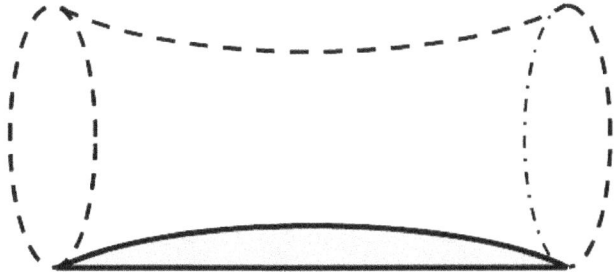

Fig 2-9. - Converting a Venturi Tube to a Wing Cross Section or "Airfoil".

tube and connect the lower points at the openings (Fig 2-9).

What we are left with here is a cross section of a wing—an airfoil. Looking at the streamlines around the airfoil (Fig 2-10), we can see that some distance above, the air is undisturbed by the presence of the airfoil—it travels in a straight line. Effectively, this creates a restriction due to the curvature of the wing, and the airflow near the wing must travel through this restriction.

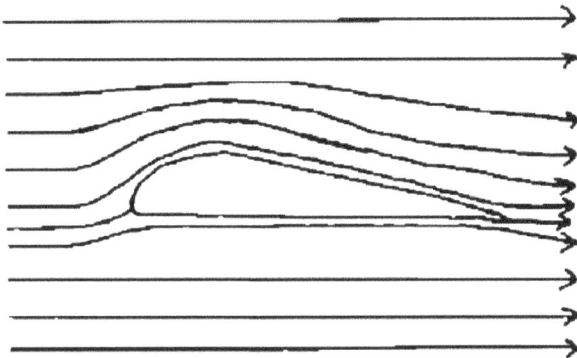

Fig 2-10. - Streamlines Around an Airfoil - The curvature of the upper surface of the airfoil creates a restriction for air to flow through. Due to the continuity equation and Bernoulli's Theorem, this creates a pressure drop on top of the wing which in turn produces lift.

As seen with the Continuity Equation, the restriction will cause an acceleration. Then, as seen with Bernoulli's Theorem, the increased velocity will cause a decrease in static pressure.

Underneath the airfoil, it can be seen that the restriction which is present is much less significant. Because of this, the acceleration of air below the wing is smaller and the drop in pressure will be less pronounced. It is this pressure difference (Fig 2-11) that provides us with

aerodynamic lift.

It is noteworthy to mention that airfoils can have varying curvatures above *and* below the wing. For more information on the shape and designations of airfoils, refer to Appendix I.

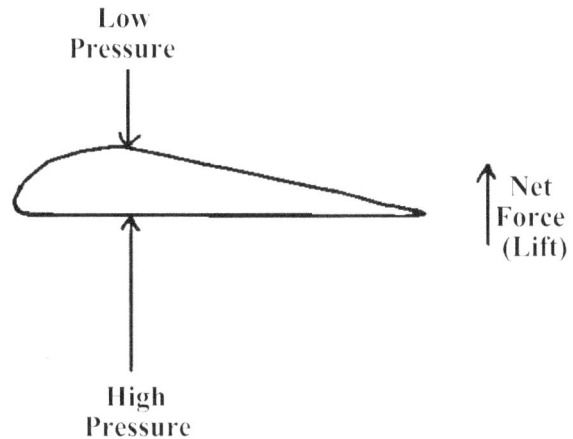

Fig 2-11. - Pressure Differences Above and Below an Airfoil.

Stagnation Point

As we can see in Fig 2-10, the air that is flowing around an airfoil must split up at some point near the leading edge. This point is called the stagnation point. At the stagnation point, the airflow divides and flows over and under the airfoil.

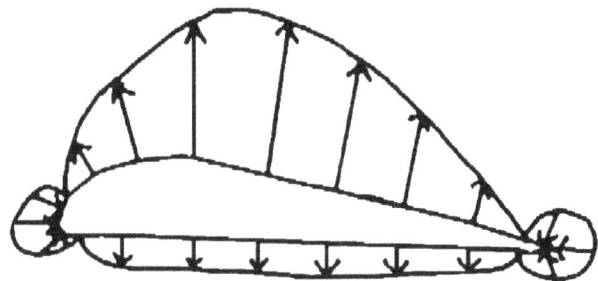

Fig 2-12. - Pressure Distribution Around an Airfoil - Note the pressure increases at the leading and trailing edges. These are regions where the air velocity is decreased and as a result the pressure increases. At the points of highest pressure, the airflow is completely stopped. These are the stagnation points.

Something else that happens at the stagnation point is a complete cessation of airflow. Since the flow is completely stopped, all of the dynamic pressure of the freestream air is converted to static pressure at the stagnation point. A typical pressure distribution about an airfoil is shown in Fig 2-12. Notice the increase in pressure at the leading edge, as well as at the trailing edge. At the trailing edge, when the airflow comes together again, there is another stagnation point where the flow is stopped altogether.

Airspeed

Since the difference in pressure above and below the wing is a result of the dynamic pressure, it stands to logic that if we can increase the dynamic pressure, we can increase the pressure difference—thus increasing lift.

The drop in static pressure that occurs on top of the airfoil is a result of an increase in dynamic pressure. So if we can increase the dynamic pressure over the airfoil, we should also decrease the static pressure. The two variables which determine the freestream dynamic pressure are air density and velocity, as seen in equation (2.6):

$$P_d = \tfrac{1}{2}\rho v^2 \qquad (2.6).$$

Air density is a constant for a given altitude, so according to (2.6) as we increase our velocity, the dynamic pressure will increase. Recall from (2.5) that pressure is a force per unit area. It is this force that we are converting to the lifting force. How well the airfoil converts this force to lift depends on a dimentionless coefficient called the lift coefficient. The lift coefficient will be discussed in detail in the next two sections.

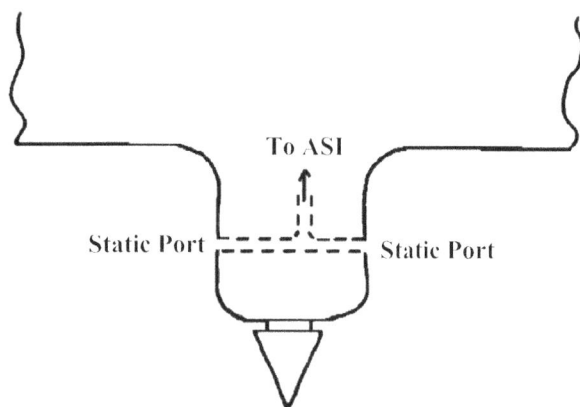

Fig 2-13. - *The Static Port* - This port vents the ambient atmospheric pressure into the airspeed indicator casing.

Now we need to think about methods of quantifying airspeed. There are several types of airspeed that are of concern to us. They are True Airspeed, Equivalent Airspeed, Calibrated Airspeed, and Indicated Airspeed.

True airspeed is the actual speed of the freestream airflow around the aircraft. The (v) in equations (2.2) and (2.6) represents true airspeed.

Equivalent airspeed is the airspeed that our airspeed indicator is meant to measure. It is a function of the dynamic pressure of the freestream airflow. The airspeed indicator does not in fact measure speed, it measures dynamic pressure.

The static port (Fig 2-13) is used to measure the static pressure of the ambient atmosphere. The pitot tube (Fig 2-14), on the other hand measures "ram" pressure, which is a combination of static and dynamic pressure. This combination of static and dynamic pressure occurs because of the stagnation point that forms at the tip of the pitot tube. Recall from the previous section that at the stagnation point, all of the dynamic pressure of the airflow is converted into static pressure. So the stagnation point contains the static pressure of the freestream airflow plus the dynamic pressure of the freestream converted into static pressure.

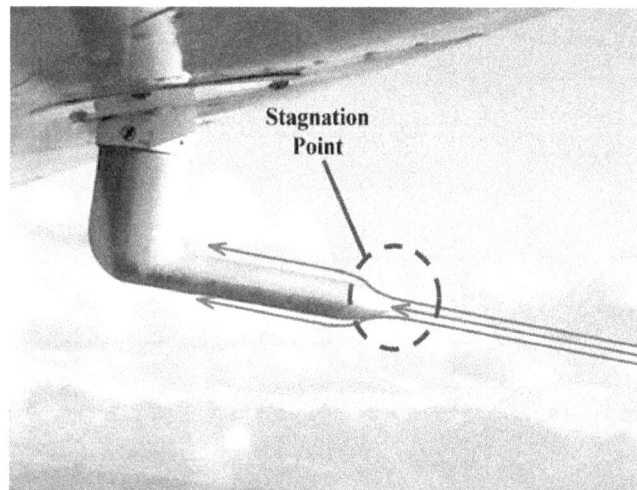

Fig 2-14. - *The Pitot Tube* - This tube vents "ram" pressure into the airspeed indicator's aneroid capsule. Note the stagnation point on the tip. At the stagnation point, all of the dynamic pressure is converted into static pressure.

As we can see from Fig 2-15, the static pressure is vented into the casing of the airspeed indicator, while the pitot pressure is vented into the aneroid capsule. Because of this, there will be an imbalance of pressure anytime the aircraft is moving. This imbalance will be equal to the dynamic pressure, since the static pressure taken in by the static port will cancel the static pressure taken in by the pitot tube.

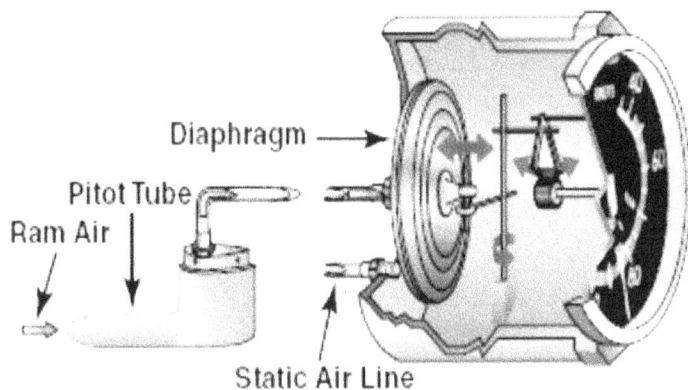

$$\tfrac{1}{2}(0.001756)v^2 = 33.9$$

therefore:

$$v = 196.5 \text{ fps.}$$

In knots:

$$v = \frac{(196.5)(3600)}{(6080)}$$

$$v = 116 \text{ kts.}$$

Fig 2-15. - *The Airspeed Indicator* - The ASI measures the difference between static and "ram" pressure. This is the dynamic pressure.

This imbalance in pressure will cause the aneroid capsule to expand. The expansion will increase as the aircraft speeds up, and through linkages the airspeed needle will indicate an increase in airspeed.

Let's look at an example. Suppose an aircraft is travelling at 100 kts TAS at sea level. The first thing we need to do is to convert 100 kts to fps:

$$\frac{(100)(6080)}{(3600)} = 168.9 \text{ fps}$$

Now referring to Appendix II, we can see that standard sea level air density is 0.002377 slugs per cubic foot. Using equation (2.6):

$$\tfrac{1}{2}(0.002377)(168.9)^2 = 33.9 \text{ psf.}$$

So when the freestream dynamic pressure is 33.9 pounds per square foot, our airspeed indicator should read 100 kts.

Our airspeed indicator is calibrated to read accurately at sea level. What happens though, at higher altitudes? Referring once again to Appendix II, we can see that the air density at 10,000 ft is only 0.001756 slugs per cubic foot. If our airspeed indicator is reading 100 kts again, how fast are we actually travelling? Using (2.6) again:

So even though our airspeed indicator reads 100 kts, our aircraft is actually travelling at 116 kts. The difference between TAS and EAS is a result of density changes at different altitudes, and is called density error.

With a TAS of 100 kts, our EAS would be lower, indicating to us that our dynamic pressure is lower. The exact value of our EAS could be determined using (2.6). We won't do the calculation here , but it turns out that our EAS is 86 kts.

One way to look at it is this, EAS is the speed that we would have to travel at to achieve the same dynamic pressure at standard sea level. In other words, with our previous example of a TAS at 10,000 ft of 100 kts, the dynamic pressure acting on the aircraft has the same value as that acting on the aircraft at sea level and 86 kts. Our airspeed indicators are calibrated for sea level, so at sea level there is no density error—that is to say, TAS and EAS are the same at standard sea level. However, as we get higher and the air density gets further from sea level density, the difference between TAS and EAS gets larger.

Calibrated Airspeed is the term that we usually use to indicate EAS. In training aircraft this is acceptable because at low speeds and low altitudes, the difference between EAS and CAS is negligible. However, as the airspeed is increased, compressibility effects gradually come into play.

With high airspeeds, the air is "packed" into the pitot tube. This results in a change in air density. Remember from (2.6) that air density is one of the variables contributing to dynamic pressure, so this variation in density will also cause a variation in dynamic pressure.

The error caused by this change in density is called compressibility error. Fig 2-16 is a chart that is used to determine the compressibility error for a given altitude and airspeed. This chart is applicable to all aircraft. As can be seen, at higher altitudes and higher speeds the compressibility error is larger. At low altitudes and low speeds, the compressibility error is negligible. Because compressibility error is so small at low altitudes and low airspeeds, we generally ignore it. Hence, the fact that we often use CAS in place of EAS is not unreasonable—in low and slow aircraft.

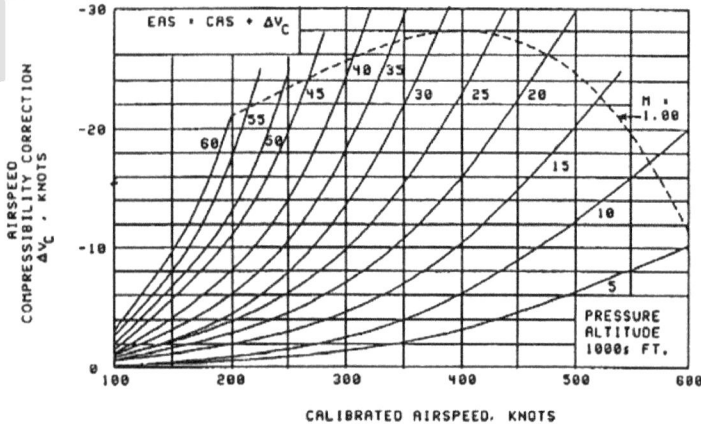

Fig 2-16. - *Compressibility Error Correction Chart.* - This graph can be used to convert between CAS and EAS at various altitudes and airspeeds. Reference to the graph will show that the difference is negligible at low altitudes and airspeeds. This is because of the low compressibility in this flight regime. The graph is adapted from *Dole, C. – Flight Theory for Pilots* (Ref. 3).

From CAS, we have two more errors to consider. Instrument and position error are both inherent in the design of the airspeed indicator. Applying these two errors to CAS, we end up with Indicated Airspeed, which is the speed that we actually read off the ASI dial.

Instrument error is inherent in the airspeed indicator, and will vary with the design of the indicator.

Position error is partially a result of airflow over the static port. It is almost impossible to position the static port to prevent a difference between the local velocity and the freestream velocity. We already know that a variation in velocity will cause a variation in pressure. The change of velocity around the static port means that the static pressure measured by the port will be lower than that of the ambient atmosphere.

Another factor contributing to position error is the angle at which the airflow meets the pitot tube. As we can see from (Fig 2-17), if the airflow is not flowing directly into the pitot tube, the stagnation point moves away from the tip of the

tube. When this is the case, the stagnation pressure is not applied fully to the pitot tube, and an airspeed error will result.

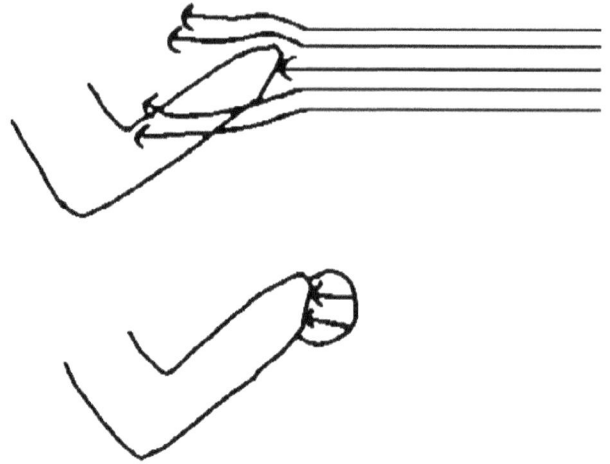

Fig 2-17. - *The Pitot Tubes Contribution to Position Error.*

Corrections for instrument and position errors vary from one aircraft to the next. The conversion from CAS to IAS can be found in the *POH* or Flight Manual.

Angle of Attack

Obviously, airspeed is not going to be the only variable that controls how much lift is being produced. If this was the case, we would not be able to maintain equilibrium at different airspeeds. So there must be another variable that allows us to maintain a constant amount of lift with variations in speed.

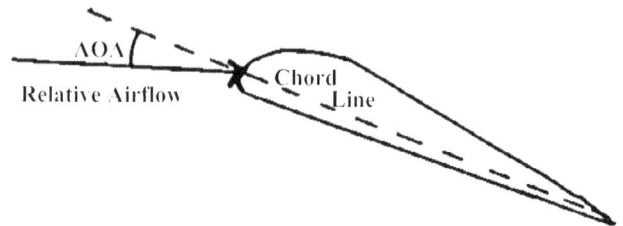

Fig 2-18. - *Angle of Attack.* - The angle between the chord line of the airfoil and the relative airflow.

The angle of attack allows us to maintain a constant amount of lift at different airspeeds—or to vary the amount of lift at a constant airspeed. The angle of attack (Fig 2-18) is the angle between the airfoil's chord line—which runs from the leading edge to the trailing edge—and the relative airflow.

At a low angle of attack (AOA), the stagnation point is on the nose of the airfoil, and the airflow is flowing over and under the airfoil as we previously discussed (Fig 2-19).

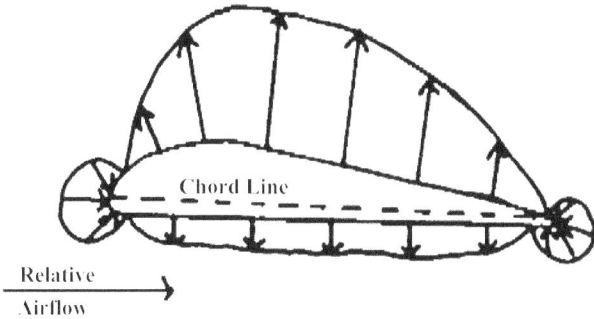

Fig 2-19. - An Airfoil at a Low AOA.

If we increase the angle of attack (Fig 2-20), the stagnation point begins to move down on the nose of the airfoil. When this happens, the airflow travelling over the top of the wing has to follow a more pronounced curvature. As you can see, this causes the airflow over the top of the wing to travel through a smaller "restriction". Thus, this airflow experiences a larger acceleration and gains more dynamic pressure. Once again returning to Bernoulli's Theorem, the larger increase in dynamic pressure causes a larger decrease in static pressure.

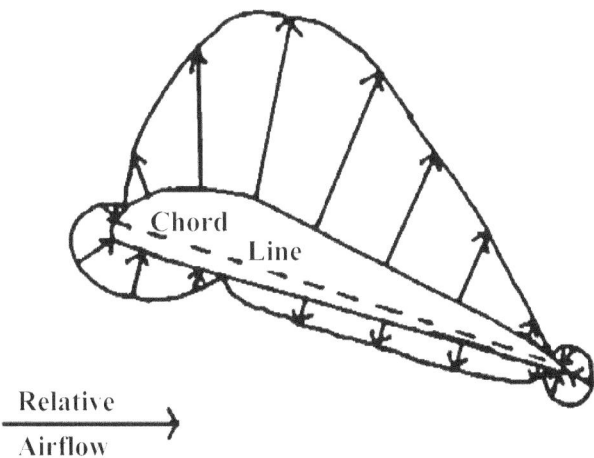

Fig 2-20. - The Effect of Increasing the AOA.

With the stagnation point moving towards the bottom of the airfoil, we also begin to get a slight increase in static pressure below the wing. This increase in pressure, combined with the larger decrease in pressure above the wing causes an increase in the lifting capacity of the wing. This increase in lifting capacity means that the airfoil will create more lift at a given airspeed, or that the airfoil is capable of creating the same amount of lift at a lower airspeed.

Lift Coefficient

The capacity of an airfoil to create lift is called the lift coefficient. The lift coefficient is essentially a measure of how well the airfoil can convert dynamic pressure into lifting pressure.

Two main factors determine the lift coefficient. They are the shape of the airfoil and the angle of attack. The reason that the angle of attack comes into play was discussed in the previous section.

The shape of the airfoil is a factor because the curvature of the top and bottom of the airfoil determines the pressure changes that will occur.

In Fig 2-21, we see a graph that represents the change in lift coefficient with a change in angle of attack. Notice that at zero AOA the lift coefficient is still greater than zero. This is because with a cambered airfoil lift is still produced at a zero angle

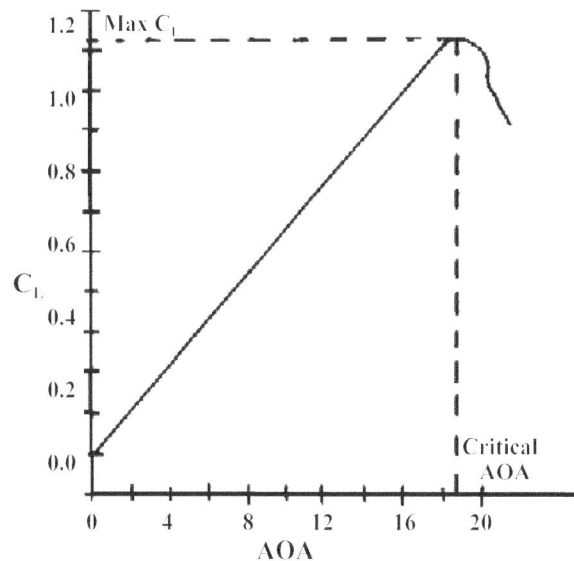

Fig 2-21. - AOA vs. Coefficient of Lift Curve - Notice that the lift coefficient increases with an increase in AOA prior to the critical angle of attack.

of attack. If we reduce the AOA even further to a negative angle, we eventually reach a point where the lift coefficient drops to zero and then to negative lift. As well, we see that the increase in lift coefficient with AOA is not indefinite. When we reach a point called the critical angle, the coefficient of lift actually begins to drop with an increase in AOA. This point is the stall and will be discussed in detail in Chapter 6.

Referring to Fig 2-22, we see a graph of AOA vs. lift coefficient for a symmetrical airfoil. Notice that with this airfoil the lift coefficient is zero when the AOA is zero. This is because the curvatures of the top and bottom of the airfoil are equal. Therefore, at a zero angle of attack there will be no difference in pressure above and below the wing.

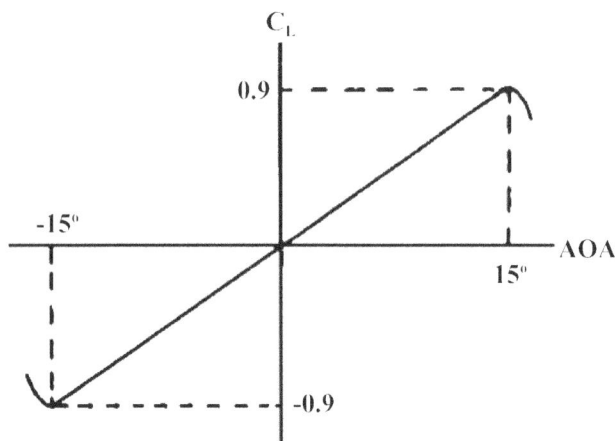

Fig 2-22. - AOA vs Coefficient of Lift for a Symmetrical Airfoil. - At zero AOA, the lift coefficient is zero. This is due to the equally curved top and bottom surfaces.

The Lift Equation

The lift equation is used to determine how much lift a wing is creating under a given set of circumstances. The equation takes into consideration factors such as the shape of the airfoil, the planform area of the wing, the AOA, the airspeed, and the density of the air:

$$L = C_L \tfrac{1}{2}\rho v^2 S \qquad (2.9).$$

$(\tfrac{1}{2}\rho v^2)$, as we have seen from equation (2.6) is the dynamic pressure of the freestream airflow. We also know that EAS is a function of dynamic pressure, while (v) is the TAS. The coefficient of lift

(C_L) is, as we know, determined by the shape of the airfoil and the angle of attack. (S) is the planform area of the wing.

Recall from (2.5) that pressure is defined as a force exerted over an area—or a force per unit area. If we multiply the dynamic pressure by the wing surface area, we will end up with the dynamic force. This is the force that would act on the wing if all of the dynamic pressure was to act directly on the planform of the wing. Once we know what the dynamic force is, we multiply it by the lift coefficient to determine the actual amount of lift that is being produced. The lift coefficient is the conversion factor which determines how much of the dynamic force is converted to lift force.

Now lets take a minute to have a look at the lift equation from a pilot's perspective. The wing surface area is a constant once the aircraft is built—the pilot has no control over wing area. As well, the cross-sectional shape of the wing (airfoil shape) is a constant once the aircraft has been built.

So we have only two variables that are controlled by the pilot. These are the dynamic pressure (equivalent airspeed) and the coefficient of lift—which the pilot controls with AOA. Therefore, to the pilot, lift is controlled with the following rule:

$$L \Longrightarrow AOA \times EAS \qquad (2.10).$$

This rule of thumb reflects what we see with experience in the aircraft. At lower airspeeds, we need a higher AOA in order to maintain equilibrium. At higher airspeeds, on the other hand, we need a lower AOA to maintain equilibrium.

It is important to remember that AOA and pitch angle are not the same thing. AOA is determined from a combination of pitch angle and flight path. The AOA and the pitch angle are the same only when the aircraft is in straight and level flight. This point will be covered in more detail in Chapter 12.

Center of Pressure

The center of pressure of an airfoil is the point at which all of the pressure can be considered to act (it is analogous to the center of gravity). It is important to differentiate between the center of pressure and the point of lowest pressure. These two points *can* coincide, but they will not necessarily be at the same point.

If we look at Fig 2-23a, we can see the pressure distribution around a symmetrical airfoil at a zero AOA. Looking, for the time being, at only the top of the airfoil, we see that the pressure change is distributed across the entire airfoil. However, the force resulting from this pressure distribution can be considered to act through a single point. This single point—the center of pressure—is the heavy arrow in the diagram.

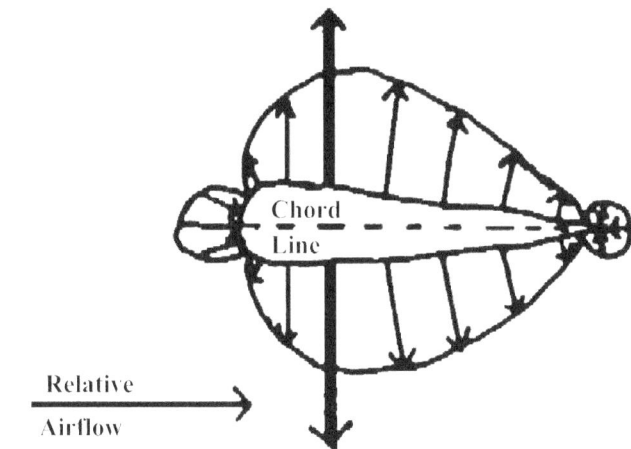

(a) A Symmetrical Airfoil at an AOA of Zero

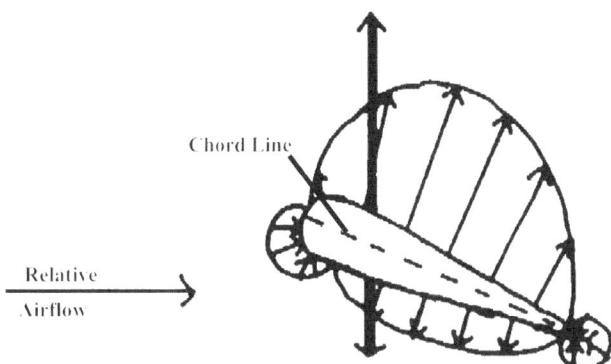

(b) A Symmetrical Airfoil at some Positive AOA

Fig 2-23. - *Symmetrical Airfoils at Various AOA.* - The C of P on top of the wing is opposite the C of P on the bottom. This results in lift with no pitching moments.

Looking at the center of pressure on the bottom of the airfoil, we see that it is exactly equal to and opposite the center of pressure on the top of the airfoil. This means that the airfoil is producing no lift. If we increase the AOA of the airfoil (Fig 2-23b), we get a larger drop in pressure over the top of the airfoil and a smaller drop in pressure underneath the airfoil. This means that the lifting force of the airfoil has increased.

Now let's look at a cambered airfoil with an AOA such that no lift is being produced (Fig 2-24). The top and bottom of the airfoil both still have centers of pressure. There is a difference, however, between a cambered and symmetrical airfoil. With a cambered airfoil at the zero-lift AOA (AKA – absolute AOA of zero, see Appendix A) , no lift is being produced, but as with the symmetrical airfoil the pressure acting on the surface is distributed along the chord. The C of P above the airfoil and the C of P below the airfoil are equal in strength and acting in opposite directions, but they do not pass through the same point. Remember from Chapter 1 that when two forces are equal and opposite, but do not pass through the same point, they form a couple. As we can see from Fig 2-24, this couple will cause a nose down pitching moment.

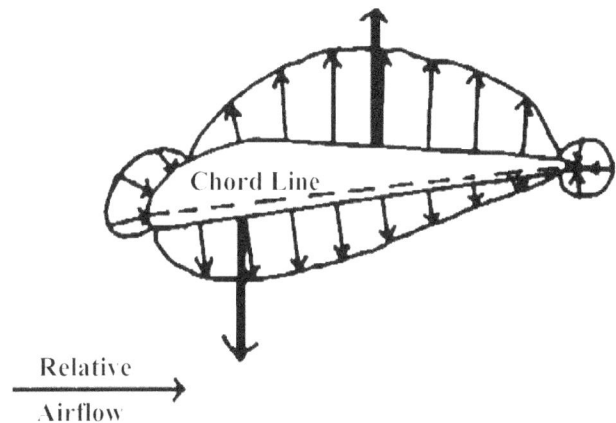

Fig 2-24. - *Pitching Moment on a Cambered Airfoil.* - The alignment of the C of P above and below the airfoil will produce a nose down pitching moment.

Aerodynamic Center

The aerodynamic center (AC) of an airfoil is the point around which the pitching moment is often measured. The AC is located at approximately the 25% point on the chord and unlike the C of P, the AC does not move with a change in AOA. The magnitude of the pitching moment is determined by multiplying the net lift by the distance between the C of P and the aerodynamic center.

Recalling equation (1.25):

$$M = FA$$

(1.25),we can determine the pitching moment on an airfoil by converting it to:

$$M_{ac} = Ld \qquad (2.11),$$

where (M_{ac}) is the pitching moment about the AC, (L) is the net lift acting on the airfoil, and (d) is the distance (arm) from the C of P to the AC.

If we increase the AOA of the cambered airfoil without changing the airspeed, the lift increases. Thus it would initially appear that the nose down pitching moment should increase. On a cambered airfoil, however, the center of pressure, moves forward with an increase in AOA. This means that even though lift is increased, the distance between the C of P and the AC is decreased. So, from equation (2.11), at a given airspeed the pitching moment will remain constant. As we change airspeed, the pitching moment will change as well.

Our depiction of the forces acting on an airfoil can now be simplified. We have an airfoil (Fig 2-25) with the lift acting at the AC, and a nose down pitching moment acting around the AC.

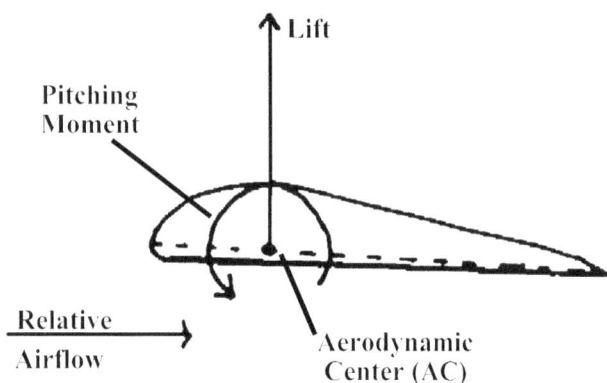

Fig 2-25. - *Lift and Moments Produced by an Airfoil.* - In reality, lift is distributed over the airfoil. However, it can be represented as a single force acting at the aerodynamic center with a pitching moment acting around the aerodynamic center.

Chapter Summary

Lift is the force that supports the weight of our aircraft in the atmosphere. It is produced by aerodynamic reactions that occur between the aircraft and the air as they move relative to one another.

The air that is moving over the aircraft can be represented graphically by streamlines. These streamlines indicate the path followed by a particle of air as it flows over an object. Streamlines cannot cross, but they can vary in their distance from one another. When Streamlines are spread widely, they indicate that the air is flowing relatively slowly, and hence the static pressure is relatively high. Conversely, if the streamlines are spaced closely, they indicate that the air is flowing at a relatively high speed, and hence the static pressure is relatively low.

In any fluid flow, the equation of continuity dictates the rate of flow. The continuity equation follows from the Conservation of Mass and indicates that fluid travelling through a restriction will be accelerated to a higher velocity or compressed to a higher density. Air is a fluid and follows the continuity equation. Since air is considered to be incompressible at low velocities, this means that air must be accelerated through restrictions.

Daniel Bernoulli lived during the 1700's and developed a theorem that relates the motion of a fluid to the pressure of the fluid. Any time a fluid is in motion, it will possess dynamic pressure as a result of the motion. An increase in dynamic pressure will result in a decrease in static pressure to maintain a constant sum.

The curvature of an airfoil causes an effective restriction to occur on top of the airfoil. From the continuity equation, this causes an acceleration of the airflow. Then from Bernoulli's Theorem, the static pressure over the top of the airfoil drops due to the increase in dynamic pressure.

At the leading edge of the airfoil there is a point where the airflow must separate. This point is called the stagnation point. The name refers to the fact that not only does the airflow separate, but at a point the flow stops—or stagnates—altogether. At the stagnation point all of the dynamic pressure of the freestream airflow is converted to static pressure. A second stagnation point forms at the trailing edge of the airfoil where the airflow is once again brought to a stop and all of the dynamic pressure is converted to static pressure.

Airspeed is the speed of the aircraft measured relative to the air. There are several types of airspeed that are of importance to us as pilots. The first is True Airspeed. TAS is the actual speed of the aircraft through the air. Next we have Equivalent Airspeed, which is a function of dynamic pressure. EAS will vary according to our TAS and the density of the air. At higher altitudes with lower densities, a given TAS will result in a lower EAS. At standard sea level, TAS and EAS will be equal. After EAS we have Calibrated Airspeed. CAS is a function of the dynamic pressure that results at high speeds when the air density varies. At low altitudes and low speeds, there is no appreciable difference between EAS and CAS. Last, but not least, we have Indicated Airspeed. IAS is the speed that we actually read off of the ASI dial. The ASI has instrument and position error inherent in it's design. These errors cause a difference between CAS and IAS. Airspeed determines how much air is flowing over the aircraft's wings, and therefore is one of the factors that determines how much lift can be produced.

The angle between the relative airflow and the chord line of the airfoil is referred to as the angle of attack. A change in AOA will change the position of the stagnation point, as well as the effective restriction that the airflow must travel through. These changes cause a change in the lifting characteristics of an airfoil. A higher AOA will mean that an airfoil will create more lift for a given amount of air flowing over it.

The ability of an airfoil to produce lift from airflow is quantified with the lift coefficient. The lift coefficient is determined by the shape of the airfoil and the AOA. The lift coefficient for a given airfoil will increase with an increase in AOA—up to a point.

The actual amount of lift being developed can be determined with the lift equation. This equation takes into consideration the coefficient of lift, air density, airspeed, and wing area. To a pilot, the lift equation reduces to the statement, "Lift is controlled by airspeed and angle of attack".

The center of pressure is the focal point for the pressure change across the chord of an airfoil. On a symmetrical airfoil, the C of P's are opposite to one another and do not move with a change in AOA. For a cambered airfoil, the top C of P is behind the bottom C of P. This forms a couple which causes a nose down pitching moment.

The nose down pitching moment created by

the C of P couple is normally calculated around a point called the aerodynamic center. The distance between the AC and the C of P is the arm, and the lift being produced is the force. Multiplying the two together gives us the magnitude of the pitching moment. The pitching moment is constant for a given airspeed because as more lift is developed at a higher AOA, the C of P moves forward, causing a decrease in the length of the arm. For calculation purposes, lift is considered to act at the AC with a pitching moment acting around it.

List of Formulae

Continuity Equation

$$\rho = m/V \qquad (2.1)$$

$$MF = \rho vA \qquad (2.2)$$

$$\rho_1 v_1 A_1 = \rho_2 v_2 A_2 \qquad (2.3)$$

$$v_1 A_1 = v_2 A_2 \qquad (2.4)$$

Bernoulli's Theorem

$$P = F/A \qquad (2.5)$$

$$P_d = {}^1/_2 \rho v^2 \qquad (2.6)$$

$$P_s + P_d = \text{Constant} \qquad (2.7)$$

$$P_s + {}^1/_2 \rho v^2 = \text{Constant} \qquad (2.8)$$

The Lift Equation

$$L = C_L {}^1/_2 \rho v^2 S \qquad (2.9)$$

$$L ==> \text{AOA} \times \text{EAS} \qquad (2.10)$$

Aerodynamic Center

$$M_{ac} = Ld \qquad (2.11)$$

Questions and Problems

1) What is a streamline? How do you interpret streamlines? Why can't they cross?

2) How does the continuity equation determine flow behavior?

3) How does Bernoulli's Theorem determine pressure patterns in a venturi?

4) Why does the air accelerate on top of a wing?

5) How does Bernoulli's Theorem determine pressure patterns around an airfoil?

6) How does Bernoulli's Theorem relate to position error?

7) What is the difference between True Airspeed and Equivalent Airspeed? Which of the two determines how much lift a wing can produce? Why?

8) Why is calibrated airspeed normally used on training aircraft instead of equivalent airspeed?

9) What is angle of attack? How does it affect the lifting properties of an airfoil?

10) What is the lift equation? From a pilot's perspective, what does the lift equation tell us?

11) What is the center of pressure? On a cambered airfoil, how is the center of pressure positioned? How does this position change with angle of attack?

12) What is the aerodynamic center? Why do we consider lift to act at the aerodynamic center?

13) How does the pitching moment of an airfoil vary with airspeed? Angle of attack?

14) Referring to Fig 2-21, an aircraft that weighs 3000 lbs is flying at 10,000 ft at a TAS of 135 knots. The aircraft has a wingspan of 42 feet and an average chord of 6 feet. What AOA must the aircraft fly at in order to maintain equilibrium? What is the aircraft's EAS?

15) After dropping skydivers, the aircraft in Q14 weighs 2000 lbs. Descending at a CAS of 140 knots, what AOA is required to maintain equilibrium? What is the aircraft's TAS passing through 4000 ft?

Chapter 3
Drag and Power Required

In order to produce lift, we need to have air flowing around the aircraft. Unfortunately, the air moving over the airframe will also create aerodynamic drag. Drag is the force that acts opposite the direction of motion and as such it will tend to slow us down.

The production of aerodynamic drag can be a complex topic. As such, we have dedicated a full chapter strictly to the development of this force. During this chapter we will investigate the various types of drag, as well as the variables that affect them.

Types of Drag

In low speed aerodynamics, we have two main types of drag. They are parasite drag and induced drag. Each is produced through separate mechanisms and each has it's own characteristics.

Parasite drag is a result of the air flowing over the airframe and will be present anytime an object is moving through the air. Parasite drag can be broken down further into skin friction, form drag and interference drag.

Induced drag is a result of the production of lift. Anytime lift is being produced the flow patterns around the wing (or, more importantly, the wingtips) will cause drag as well as lift.

Another form of drag is known as wave drag. Wave drag occurs at speeds close to or above the speed of sound (AKA – Mach 1) due to the compressibility of air and will not be discussed in this book.

Parasite Drag

As previously stated, parasite drag will occur anytime an object is moving through the air. Parasite drag is a result of the interaction of the freestream airflow and the surface of the aircraft via the boundary layer.

The Boundary Layer

The boundary layer is a thin layer of air between the surface of the aircraft and the freestream airflow (Fig 3-1). The velocity of the air that is actually touching the aircraft is zero. As we move outwards from the surface the airflow velocity gradually increases to the freestream velocity. The region of airflow with the reduced velocity is the boundary layer.

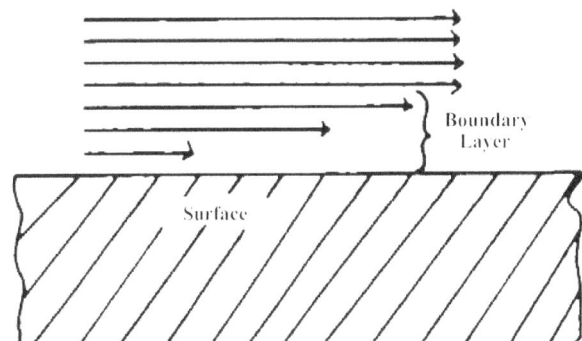

Fig 3-1. - The Boundary Layer - A thin region of air between the aircraft surface and the freestream airflow.

The velocity reduction which results in the formation of the boundary layer is caused by the viscosity of the air. Viscosity is a fluid's resistance to flow which is caused by internal friction. Viscosity also results in the air tending to "stick" to any surface passing through it.

The boundary layer can be smooth flowing or turbulent. A smooth boundary layer is known as laminar flow, and is predominant early in the flow across a surface (Fig 3-2). The boundary layer will become thicker as the air travels along the surface. Eventually, the boundary layer will form a transition region where the laminar flow becomes turbulent (Fig 3-2).

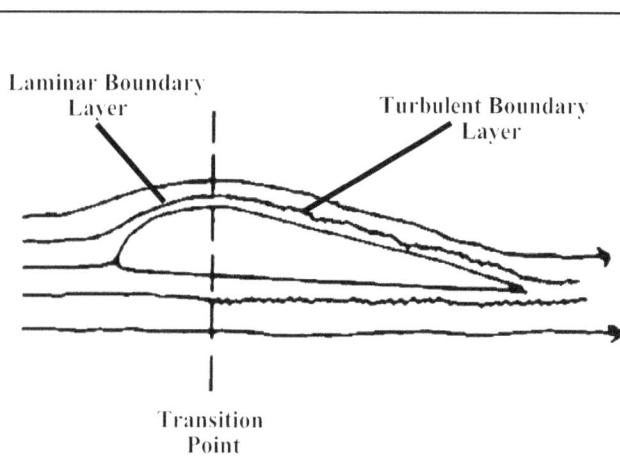

Fig 3-2. - *Laminar vs. Turbulent Boundary Layer*

Velocity profiles for both laminar and turbulent boundary layers are illustrated in (Fig 3-3). As we can see, the velocity of the air increases as it gets further from the object until the velocity matches that of the freestream.

Boundary layers have a natural tendency to transition to turbulent flow. Notice from the diagrams that a turbulent boundary layer will be significantly thicker than a laminar layer. This is a result of the mixing action occurring due to the turbulence.

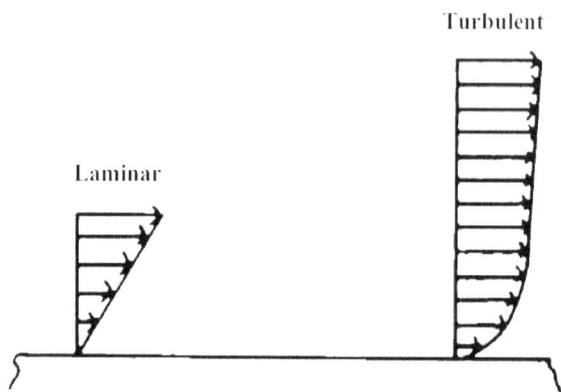

Fig 3-3. - *Boundary Layer Velocity Profiles.*

Skin Friction

Skin friction drag is exactly what the name implies—drag caused by the friction between the air and the skin of the aircraft. The mechanism through which skin friction is created is the boundary layer.

Before an aircraft passes through a location, the air at this location is stationary. This stationary air has inertia, meaning that it has a natural tendency to remain stationary. However, as our aircraft passes through this location, the friction between the air and our aircraft causes a force which accelerates the air in our direction of travel.

The problem occurs as a result of Newton's Third law. Recall from Chapter 1 that forces always occur in pairs that are equal in magnitude but opposite in direction. So if friction exerts a force on the air in our direction of travel, it must also exert a force on our aircraft in the opposite direction. This force is skin friction drag.

As we have seen from the boundary layer diagrams, a laminar boundary layer is thinner than a turbulent boundary layer. With a thinner boundary layer, skin friction drag will be less pronounced because less air is being "dragged" along with the aircraft. As the laminar boundary layer becomes thicker, and then becomes turbulent, skin friction drag will increase.

Form Drag

To begin our discussion of form drag, let's first look at a hypothetical situation in which there is no skin friction. You will see shortly that without skin friction, there will be no form drag either.

We'll begin with a cylindrical object traveling through the air (Fig 3-4). We see from the streamlines about this object that the air flows over it with two stagnation points. Remember that at a stagnation point the airflow is brought to a stop and all of the dynamic pressure is converted to static pressure. We can see this variation in pressure in (Fig 3-4b).

Since the air at the front of the cylinder is flowing at the same rate as the air at the rear of the cylinder, the stagnation pressure will be the same at both ends. This is because with equal air velocities the dynamic pressure will be the same, thus the static pressure at stagnation points will be equal.

Remember once again that pressure is a force exerted over an area. So the stagnation pressure at the nose of the cylinder is exerting a force against the direction of motion. However, the stagnation point at the tail of the cylinder will exactly counteract this rearward force with an equal forward

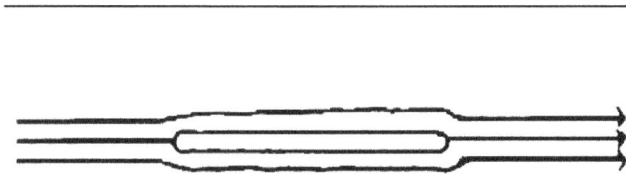

(a) Cylindrical Object with Two Stagnation Points

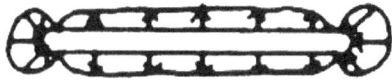

(b) Pressure distribution

Net Force = 0

(c) The Net Force

Fig 3-4. - Airflow Around a Frictionless Object.

(a) Cylindrical Object with Two Stagnation Points

(b) Pressure Distribution with Friction Present

Net Force > 0
Acting to the Right

(c) The Net Force

Fig 3-5. - The result of Skin Friction.

force (Fig 3-4c).

Now let's look at what happens when skin friction is present (Fig 3-5). With the stagnation point on the nose of our cylinder nothing changes. However, because of skin friction the velocity of the air is decreased (relative to the aircraft) towards the end of the cylinder. From equation (2.6) we can see that this will mean a decrease in dynamic pressure at the rear of our cylinder. The stagnation point at the trailing edge of the cylinder will therefore have a lower static pressure than the one at the leading edge (Fig 3-5b).

Once again, pressure is a force exerted over an area, so lower pressure means a lower force. As we can see from (Fig 3-5c), the force applied to our cylinder by the trailing edge stagnation point is smaller than the force applied by the leading edge stagnation point. The net force in this case is in the direction opposite to the direction of motion, hence this force is termed as drag.

The drag we are looking at here is called form drag because it is influenced by the shape (form) of the object. It is also commonly called pressure drag because it is caused by the pressure differences ahead of and behind the object.

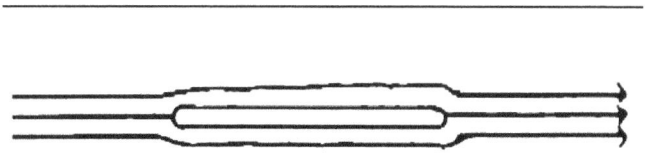

Another phenomenon can occur here which will influence form drag. As the airflow is slowed down by skin friction, it is running out of energy. Eventually a point can be reached where the boundary layer doesn't have the energy to follow the curvature of an object. When this happens, the boundary layer separates from the object and forms a low pressure "wake" behind the object (Fig 3-6).

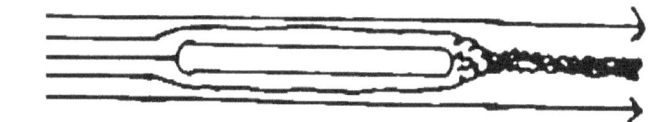

Net Force = ——————⟶

Fig 3-6. - Airflow Separation Increases Form Drag.

Form drag will develop whether or not separation takes place. However, separation will cause more significant form drag because instead of having a smaller *increase* in pressure, we will in fact get a *drop* in pressure due to the wake.

Interference Drag

Interference drag is the result of interference between the boundary layers of different components of the aircraft. Because of this interference, it is difficult to determine the total drag of an aircraft by adding up the drag from each individual component.

For example, if we were to determine that under a given set of conditions, the drag produced by the fuselage was 70 lbs. Under the same conditions, the wings are determined to produce 110 lbs of drag, the tail 40 lbs, and the landing gear 30 lbs of drag. This means that the total drag for these components is 250 lbs. However, if we put them together into a total aircraft, we would be very dismayed to find that the total drag for the aircraft under our given set of conditions is actually greater than 250 lbs (perhaps 270 lbs).

The discrepancy we see here is due to interference drag. At the junction between two components, the boundary layers will interact with one another. This causes laminar boundary layers to transition to turbulent layers earlier, and it causes turbulent boundary layers to increase in thickness. As we have seen in the section on skin friction, a turbulent boundary layer will cause more drag than a laminar one. As well, a thick boundary layer will cause more drag than a thin one.

Variables

The variables associated with parasite drag are similar to the variables associated with lift. The difference is that now we are dealing with a drag coefficient rather than a lift coefficient. The parasite drag formula is as follows,

$$D_p = C_{dp} \, \tfrac{1}{2} \rho v^2 S \qquad (3.1),$$

where (C_{dp}) is the coefficient of parasite drag. The value of (C_{dp}) will depend on how much skin friction, form, and interference drag the aircraft will produce. This in turn depends on the aircraft design characteristics which influence each type of drag.

The value of (C_{dp}) will determine how much of the dynamic pressure acting on the aircraft will be converted into parasite drag.

As we can see from equation (3.1), parasite drag will increase with the square of the velocity. Fig 3-7 is a curve which compares the velocity of an aircraft to the parasite drag being produced.

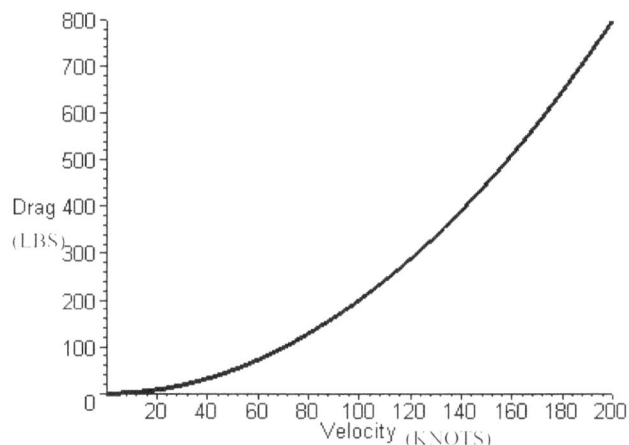

Fig 3-7. - Parasite Drag vs. Velocity

Equivalent Flat Plate Area

Instead of working with the drag coefficient and the surface area, it is sometimes more useful to work with a value known as the "*equivalent flat plate area*". The equivalent flat plate area allows a comparison of drag characteristics between two or more aircraft—whereas the drag coefficient does not since the surface area can vary.

The equivalent flat plate area (f) assumes that the drag coefficient is exactly 1. The intention is to determine the surface area of a flat plate turned face on to the airflow which would give the same amount of parasite drag as the aircraft. This is determined simply by multiplying the parasite drag coefficient by the surface area,

$$f = C_{dp} S \qquad (3.2).$$

Parasite Drag Reduction

Drag of any sort will work against us in an aircraft (except during certain maneuvers which involve a desired deceleration—such as landing). So to improve aircraft performance, we want to reduce drag as much as is possible. In order to reduce the parasite drag, we need to reduce the value of C_{dp}. This can be done via a variety of methods, but C_{dp} can unfortunately never be reduced to zero.

Skin Friction – Skin friction drag can be reduced by any method that will prevent the boundary layer from becoming turbulent, or from becoming thicker. Imperfections and rough areas on the surface of the aircraft will contribute to drag due to their effect on the boundary layer. This includes, but is not limited to, flush riveting (Fig 3-8), a smooth paint job, cleaning and waxing the airplane, or anything else that will even out the surface of the aircraft.

Regular Rivets

Flush Rivets

Fig 3-8. - *Flush Riveting vs. Regular Riveting* - The flush rivets reduce disruptions to the boundary layer and therefore reduce skin friction drag.

As well as smoothing the surface of the aircraft, reducing the surface area of the aircraft will have the effect of reducing skin friction. The amount of airplane surface that is in contact with the airflow is referred to as the wetted area. An aircraft with a smaller wetted area has less "skin" for skin friction to occur across. Thus an aircraft with a smaller wetted area will have less skin friction drag.

Form Drag – As for form drag, we need to reduce the difference in pressure ahead of and behind the aircraft. This is achieved mostly through "streamlining". As an example, let's look at the classic case of the flat plate vs. the ball vs. the streamlined shape (Fig 3-9).

The flat plate has a great deal of pressure drag. This is due to a large increase in static pressure in front of the plate, combined with a large decrease in static pressure behind the plate due to

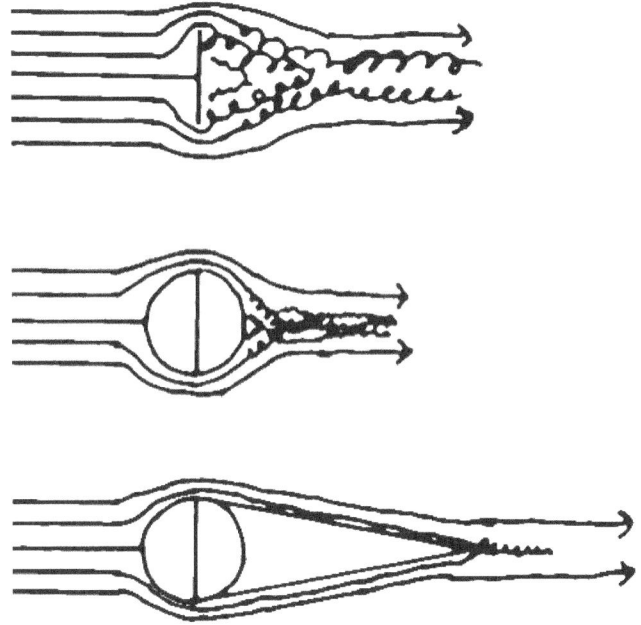

Fig 3-9. - *Shape Effect on Form Drag.*

the wake.

If we now look at a ball, things will change somewhat. Notice in Fig 3-9 that the ball has the same frontal area as the flat plate. However, the increase in pressure at the front of the ball has been decreased by the gradual curvature—as opposed to a flat surface. As well, there has been a decrease in the drop in pressure at the rear of the ball. Once again, this is due to the fact that the air can follow the curvature of the ball much more easily than the sharp directional changes experienced around a flat plate. However, we still notice a significant wake trailing behind the ball. The boundary layer is still running out of energy and separating from the surface.

So next we look at the streamlined shape. Once again the frontal area is equivalent to the frontal area of the flat plate. Like the ball, our streamlined shape experiences a smaller increase in pressure at the front due to it's curvature. As well, we can see that the wake behind the streamlined shape is reduced. This is because separation of the boundary layer is delayed as a result of the shape of the tail.

So the shape which gives us the least form drag is the streamlined shape in Fig 3-9. As we can see, this shape has a larger wetted area than the flat plate or the ball. This means that this shape will have more skin friction drag, but this disadvantage is more than compensated for by the reduction in form drag.

There is something else that can be done to reduce form drag. If we look again at the ball in Fig 3-9 we see that the boundary layer separates because it runs out of energy and can't follow the curvature of the ball. If we had some method of reenergizing the boundary layer, we could delay separation and thus reduce the size of the wake. One technique that can be used to do this is to "trip" the boundary layer so that it transitions to turbulent flow prematurely. Remember that a turbulent boundary layer has more energy than a laminar layer due to the mixing action of the turbulence, thus separation will be delayed (Fig 3-10). Once again, this will cause an increase in skin friction, but the loss of form drag will often more than compensate.

The technique of tripping the boundary layer early is the reason that golf balls have dimples and tennis balls have fuzz.

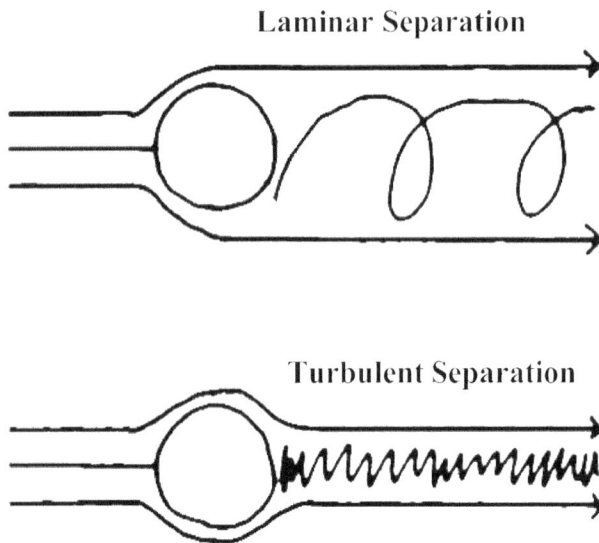

Fig 3-10. - *Separation can be Delayed by a Turbulent Boundary Layer.*

Interference Drag – As we discussed previously, interference drag results from the interaction of boundary layers from different components of the aircraft. So if we can reduce the interaction of boundary layers we can reduce interference drag. Unfortunately, reducing or eliminating the interaction of boundary layers is, for all intents and purposes, impossible.

We do, however, have another option. Instead of reducing the interaction of two boundary layers, we can merge two boundary layers into one, thus eliminating (or at least reducing) the need for interaction and interference.

The best way to merge two boundary layers is to use fairings between components (Fig 3-11). This creates a gradual junction instead of a sudden change. Acute angles between components will create larger amounts of interference, while smooth transitions will reduce the interference greatly.

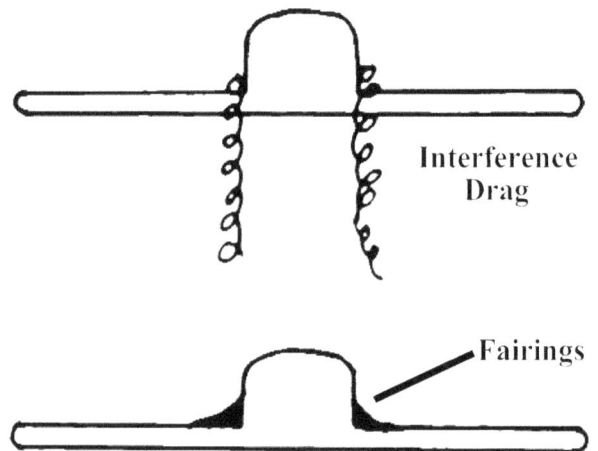

Fig 3-11. - *Fairings are Used to Smooth Out the Transition Between Two Boundary Layers.*

So to quickly sum up, we can reduce parasite drag with flush rivets, a smooth paint job, and a washed and waxed airplane. As well, we can streamline the shapes of aircraft components (i.e. wheel pants, retractable gear, etc.), and place fairings between components—especially components that are at acute angles to one another.

Induced Drag

As stated earlier, induced drag is drag created by the production of lift. If an object is travelling through the atmosphere and is not producing any lift, it will experience only parasite drag. An object that is producing lift however, will be subject to an extra decelerating force known as induced drag. This induced drag is a result of the fact that airflow over the wing is acting in three dimensions instead of just two.

The Infinite Wing

An infinite wing is a hypothetical situation in which the wing has an infinite span. This means that the wing has no wingtips, and thus the airflow will be restricted to two dimensions. Two dimensional flow occurs when there is no spanwise flow—the airflow is front to rear only, not side to side.

Pressure waves travel ahead of the wing at the speed of sound. Because of this, the air ahead of the wing is "warned" of the wings approach. In (Fig 3-12) we can see the result of these pressure waves. A small distance ahead of the wing, the air begins to move upwards to leave space for the wing to pass. This upwards flow is called upwash.

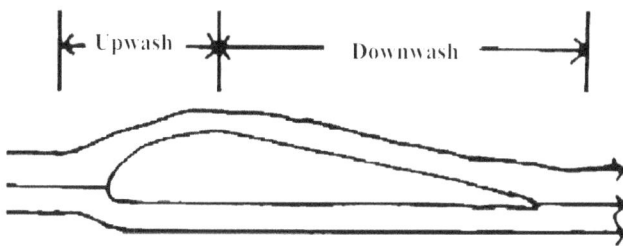

Fig 3-12. - Pressure Waves "Warn" the Air of the Wings Approach - These pressure waves cause the air to move upwards. This is upwash.

As the wing passes beyond a position, space is left for the air to fill. Since the air has travelled upwards to move out of the wings way, it must travel downwards to return to its original position. This downwards movement is termed downwash.

An important characteristic of lift is that it acts perpendicular to the relative airflow. This means that on the forward portion of the wing, lift is inclined forward, while on the rearward portion of the wing, lift is inclined rearward (Fig 3-13).

The forward and rearward components of lift will cancel each other out. So the resultant force is straight up—perpendicular to the freestream relative airflow.

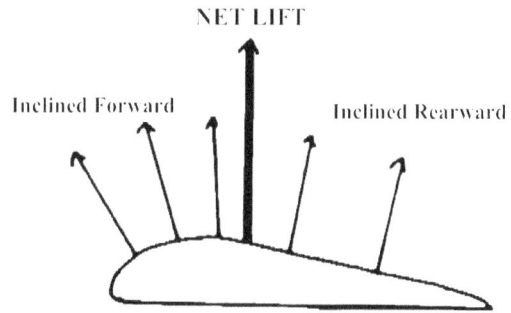

Fig 3-13. - The Lift Distribution Around an Infinite Wing - The forward component of lift cancels the aft component of lift and the net drag is zero.

The Finite Wing

If we now consider a wing that ends at some finite distance from the root—that is, it has wing tips—we will find that things change a little bit. The airflow patterns around the wing tips cause induced drag.

Upwash still occurs as a result of pressure waves travelling ahead of the wing. As well, downwash will occur behind the wing. However, the downwash for a finite wing will exceed the upwash.

Let's look at the rear view of an aircraft that is producing lift (Fig 3-14). As we can see here, there is a difference in static pressure above and below the wing—this we discovered in the previous chapter. We also know that air pressure has a natural tendency to equalize. This equalization is achieved by air from a high pressure region flowing into the low pressure region.

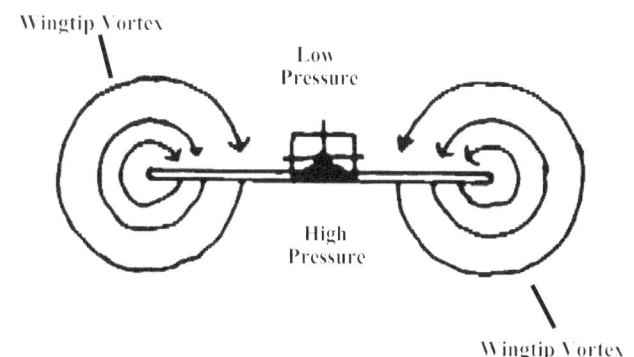

Fig 3-14. - The Formation of Wingtip Vortices.

We can see from the diagram that this flow causes the air to follow a circular path around the wing tips with the direction being down and out—clockwise on the left and counter-clockwise on the right. This circular flow is called "wing tip vortices". Wing tip vortices are strongest at the wing tips, but they have an effect along the entire span of the wing.

At first this doesn't look to be causing any drag. However, if we now look at the side view of one of the wings, we can see that the rotational flow is causing a downflow above the wing. This downflow forces the air travelling over the wing to be deflected downwards. This is downwash and as a result of the wing tip vortices, it will be greater than the upwash. The downwash behind a wing can be seen in Fig 3-15.

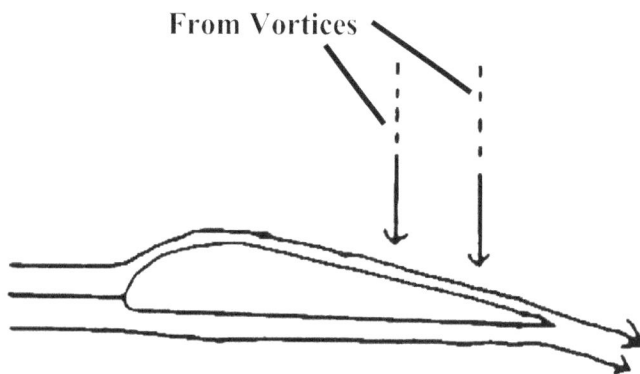

Fig 3-15. - *Downwash on a Finite Wing.*

Downwash and the Lift Vector

We can see from the discussion and diagrams on wing tip vortices and downwash that the relative airflow actually varies in direction across the chord of the wing. That is, towards the trailing edge of the wing downwash effects cause the relative airflow to be tilted downwards.

Since downwash now exceeds the upwash, the rearward component of lift will also exceed the forward component of lift. This means that the net lift force is now going to be tilted backwards—that is, it will no longer be perpendicular to the freestream relative airflow (Fig 3-16). We can now divide the actual net force lift into perpendicular components. Then we will see a vertical component which is the effective lift, and a horizontal component which is induced drag.

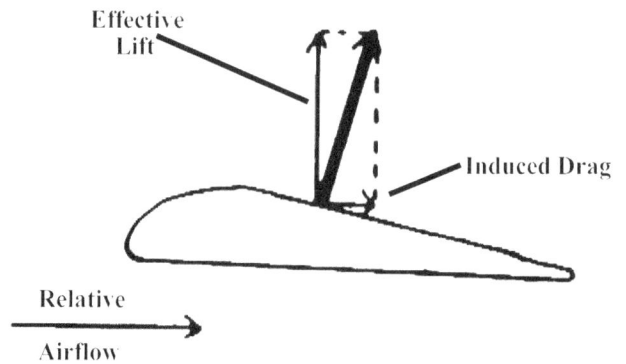

Fig 3-16. - *Downwash Causes a Rearward Tilt of the Lift.* - This rearward lift can be divided into components. The component that is perpendicular to the freestream relative airflow is the effective lift. The aft component, which is parallel to the freestream relative airflow is induced drag.

At higher angles of attack, the pressure difference above and below the wings becomes more pronounced. This causes the wing tip vortices to become stronger, thus increasing downwash. The increase in downwash will result in more induced drag since the lift vector will be tilted back further.

Loss of lift

Looking back to figure 3-16, we can see that the vertical component of lift is slightly smaller than the net lift force being developed. This loss of lift is a result of downwash, just as the induced drag is a result of downwash.

In order to produce the same amount of lift that we would produce without the presence of downwash, we need to maintain a higher angle of attack. The difference between the AOA that we would need without the downwash and the AOA we need with downwash is called the induced angle of attack (Fig 3-17).

Notice that at higher angles of attack the induced angle of attack is larger as well. This is because downwash is greater at higher angles of attack.

Variables

One of the main variables that affects induced drag has already been mentioned—lift. If induced drag is caused by lift, then it makes sense to say that if more lift is being generated, more induced drag will be created. This is indeed the case—induced drag will increase with an increase in the lift coefficient.

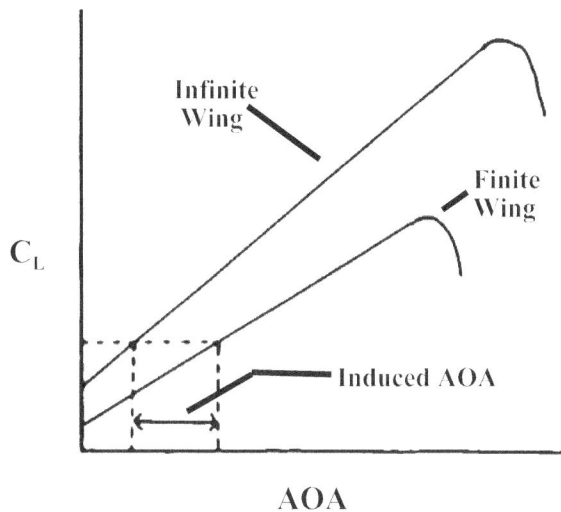

C_L

AOA

Fig 3-17. - *Induced AOA.* - Due to downwash, some lift is lost. The difference between the AOA required to produce a given amount of lift with and without downwash is called the induced AOA.

Another primary factor influencing induced drag is the span of the wing. Remember that wing tip vortices are most powerful at the wing tips, and they become less pronounced towards the wing root (although they never disappear altogether). If we could have a wing with a smaller portion of that wing at the tip, then the effect of the vortices would be reduced. So an increase in wing span will decrease the effects of induced drag.

What we in fact want to do is to increase the aspect ratio of the wing. The aspect ratio is the ratio between the wing span and the wing chord:

$$AR = {}^b/_c \qquad (3.3),$$

where (b) is the wing span, and (c) is the wing chord. For wings that do not have regular shapes, we can multiply (3.3) by the span over the span (in other words we multiply by 1), and we get another method for determining aspect ratio:

$$AR = {}^b/_c \times {}^b/_b = b^2/S$$

therefore:

$$AR = b^2/S, \qquad (3.4),$$

where (S) is the wing planform area.

The advantage of a high aspect ratio is the reason that gliders have long thin wings.

So now that we've seen the factors that affect induced drag, let's look at the induced drag formula. Initially the formula looks just like the lift equation or the parasite drag equation:

$$D_i = C_{di} {}^1/_2 \rho v^2 S \qquad (3.5).$$

At first glance, it looks like induced drag will behave just like parasite drag and increase with the square of our velocity. However, we need to look a little bit more at this equation. The value of (C_{di}) will vary with the AOA as follows:

$$C_{di} = \frac{k C_L^2}{AR} \qquad (3.6).$$

Equation (3.6) still doesn't tell us much, but we know that if we are producing more lift, the lift coefficient will be higher. As for the value (k), it is a constant which depends on the planform shape of the wing. For elliptical wings with no fuselage and no parasite drag, the value of k is $1/\pi$.

Using the lift equation (2.9), we can now solve for (C_L). We can then figure out what (C_{di}) is, and using it we can solve for induced drag:

$$L = C_L {}^1/_2 \rho v^2 S \qquad (2.9),$$

therefore:

$$C_L = \frac{L}{{}^1/_2 \rho v^2 S}$$

and using (3.6):

$$C_{di} = \frac{k C_L^2}{AR},$$

therefore:

$$C_{di} = \frac{L^2}{\frac{({}^1/_2 \rho v^2 S)^2}{\pi AR}}.$$

From this we get:

$$C_{di} = \frac{L^2}{\pi AR {}^1/_4 \rho^2 v^4 S^2},$$

and substituting this back into (3.5),

$$D_i = \frac{L^2}{\pi AR \, \frac{1}{4}\rho^2 v^4 S^2} \left(\frac{1}{2}\rho v^2 S\right),$$

therefore,

$$D_i = \frac{L^2}{\pi AR \, \frac{1}{2}\rho v^2 S}.$$

Recalling from (3.4) that

$$AR = b^2/S,$$

we finally get

$$D_i = \frac{L^2}{\pi \, \frac{1}{2}\rho v^2 b^2} \qquad (3.7).$$

As the induced drag equation (3.7) shows us, induced drag will increase as more lift is being produced. For an aircraft in equilibrium lift is equal to weight, therefore heavier aircraft will produce more induced drag. Induced drag increases with the square of the lift being produced. So if we increase the weight of an aircraft by 10%, the induced drag will be increased by 21%.

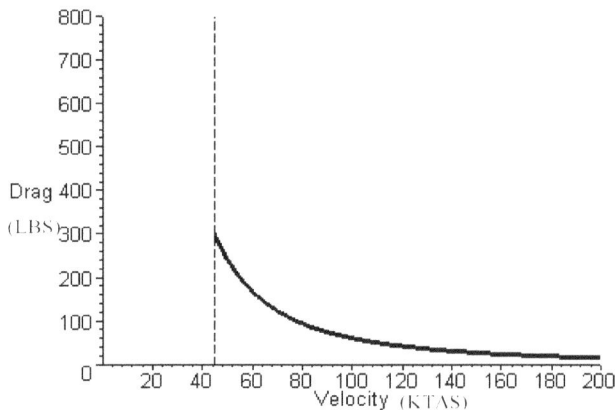

Fig 3-18. - *Induced Drag vs. Velocity* - Induced drag will decrease as the airspeed is increased.

Anything that is in the denominator of the equation will cause a reduction in induced drag. The most significant of these would be the wingspan and the velocity since they are squared. As well, the density of the air will have an effect on induced drag. Higher density (lower altitude) will

result in less induced drag—all other things being equal.

Notice that $\left(\frac{1}{2}\rho v^2\right)$ is in the denominator. We know from the previous chapter that this is our dynamic pressure, and our EAS is a function of dynamic pressure. So with a low EAS, induced drag will be higher, while at a higher EAS induced drag will be lower.

Figure 3-18 is a curve which represents induced drag vs. velocity. Notice that the curve is not plotted down to a velocity of zero. This is because of the stall—which is discussed in detail in Chapter 6.

Induced Drag Reduction

There are a number of factors that can affect induced drag. Most of them are apparent from the induced drag equation, such as the weight of the aircraft, the airspeed, and the wingspan. There is also something else which is not so obvious from the equation.

We know that induced drag is caused by the wing tip vortices. So if we could somehow reduce wing tip vortices we could therefore reduce induced

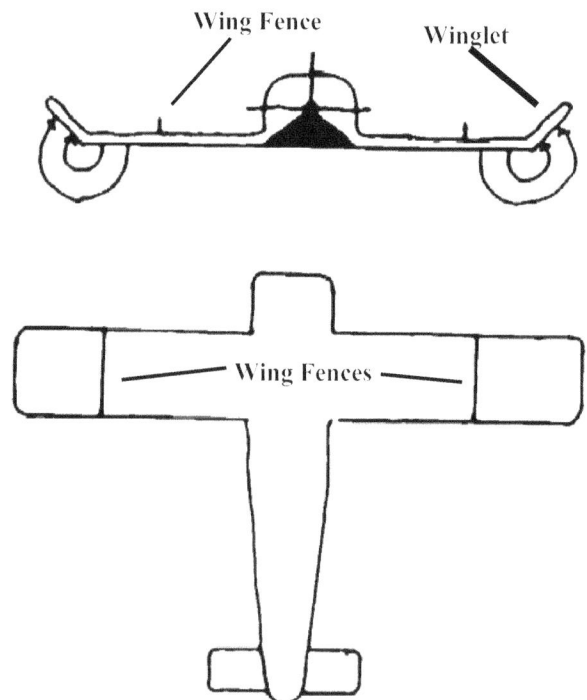

Fig 3-19. - *Wing Fences and Winglets.* - The disruption of wingtip vortices provides a reduction in induced drag and an increase in lift.

drag.

Theoretically, an infinitely long wing would have no induced drag because there would be no wing tips for the vortices to form around. Unfortunately this can't happen in the real world due to a number of practical considerations.

Another way, however, to reduce wing tip vortices is to create some sort of barrier to the airflow. This is the basic idea behind wing fences and winglets (Fig 3-19). As well, we have another phenomenon which will reduce induced drag for us. This is ground effect.

When a wing is flying in close proximity to the ground (less than one spanlength), the presence of the ground acts as a barrier which prevents the formation of wing tip vortices (Fig 3-20). The closer the wing is to the ground, the more significant the result of ground effect.

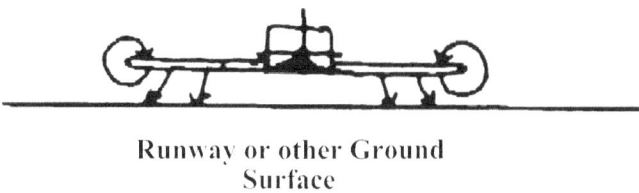

Runway or other Ground Surface

Fig 3-20. - *Ground Effect* - The presence of the ground prevents or reduces the formation of wingtip vortices. This decreases induced drag and increases lift while flying in close proximity (less than one spanlength) to the ground.

Ground effect decreases rapidly as the wing gets further from the ground. At 10% of the span length, about 50% of the normal induced drag is present. At 50% of the span length, about 90% of the normal induced drag is present. For an aircraft with a wingspan of 30 ft (light trainers are usually in this range) this means wing heights of 3 ft and 15 ft, respectively. As we can see from these heights, the aircraft needs to be extremely low for ground effect to be significant. As well, low wing aircraft will obviously benefit more from ground effect than will high wing aircraft (for a given wingspan).

Ground effect will have a large influence on heavy aircraft at low airspeeds. This is because these aircraft have the highest amounts of induced drag. Lighter aircraft at higher speeds will gain less because more of their total drag consists of parasite drag which is not affected by ground effect.

Total Drag

The total drag produced by an aircraft consists of both parasite and induced drag. The combination of both of these types of drag makes for an interesting pattern of behavior at various airspeeds. We will see in the next section that at very low airspeeds, drag will actually decrease as we speed up. Then, as we pass a minimum point, the drag will increase again with a gain in airspeed.

Drag Curve

The total drag curve looks like a "U" (Fig 3-21). This is because parasite drag increases with the square of the velocity, while induced drag decreases with an increase in the square of the velocity. If we add these two curves together, we get the total drag curve.

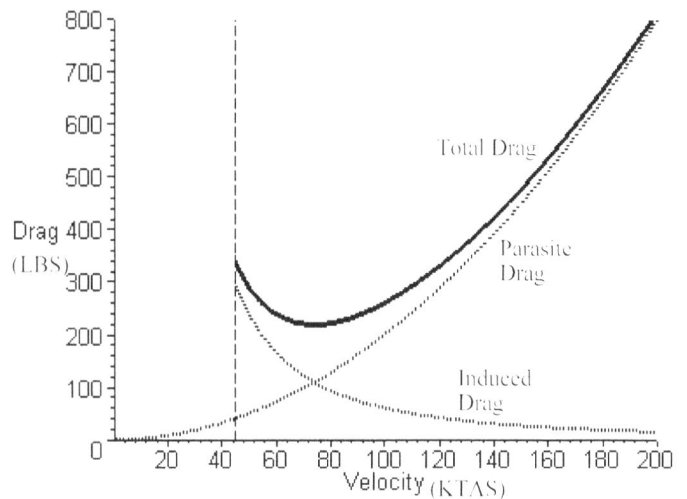

Fig 3-21. - *The Drag Curve* - This is the total drag created by an aircraft vs. the airspeed. Notice that the total drag starts high, decreases to a minimum value, and then increases again. This pattern is due to the combined effects of parasite and induced drag. Parasite drag increases with the square of the speed, while induced drag decreases with the square of the speed.

Variables

The variables that affect the total drag of an aircraft are the same variables that affect the individual types of drag. Fig 3-22a is a total drag curve for a given aircraft at high altitude compared to a drag curve for the same aircraft at sea level. Notice that the curve is shifted to the right. This is because a higher TAS is required to experience the same dynamic pressure at a higher altitude. Fig 3-22b is a comparison of drag curves at different weights.   Notice that the amount of drag has increased.   This is because of the increase in induced drag caused by the increase in weight (notice that the weight effect is more pronounced at lower airspeeds).

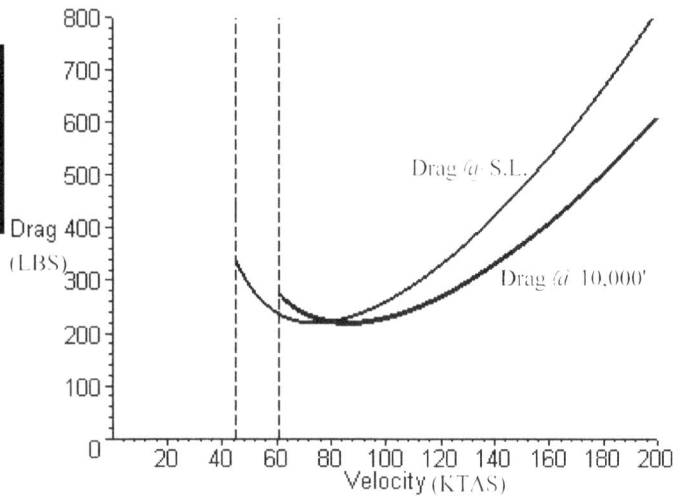

(a) The Effect of Altitude on the Drag Curve - This curve is in terms of TAS. The drag produced at a given EAS does not change with altitude.

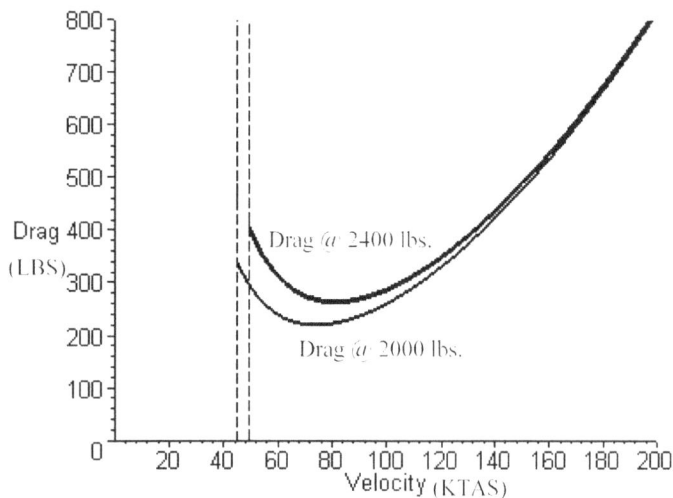

(b) The Effect of Weight on the Drag Curve

Fig 3-22. - *The Effect of Altitude and Weight on the Total Drag Curve. - An increase in altitude shifts the curve to the right. An increase in weight will increase the drag being produced.*

Lift/Drag Ratio

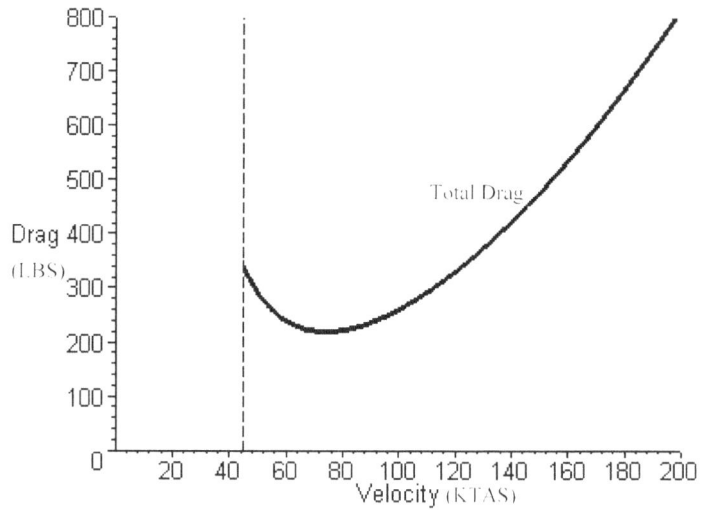

(a) Drag Curve

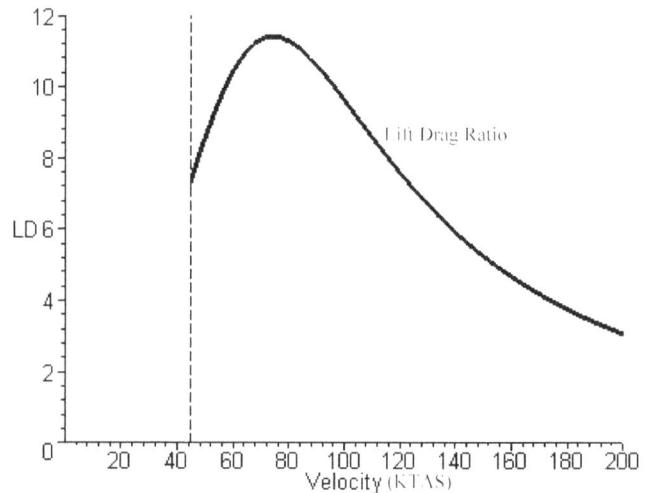

(b) Lift/Drag Ratio Curve

Fig 3-23. - *Drag vs. Lift/Drag Ratio - The best lift to drag ratio coincides with the minimum drag point. This is also the point where parasite drag is equal to induced drag.*

The ratio between the lift being produced and the drag being produced is an important consideration in determining aircraft performance. It is essentially a measurement of how efficiently the aircraft is operating. That is, it determines how well the aircraft is converting dynamic pressure into lift versus how much of that dynamic pressure becomes drag. It will be shown in Chapter 8 that for an aircraft in a glide the lift to drag ratio is equal to the aircrafts glide ratio.

Figure 3-23a is a drag curve for an airplane.

Since the weight of the aircraft is a constant, it can be used to determine the L/D ratio which is displayed in figure 3-23b. As we can see, the maximum L/D ratio occurs at the minimum drag point. It can be shown that at the best L/D (minimum drag) speed, parasite drag is equal to induced drag. It can also be shown that the maximum L/D ratio always occurs at the same angle of attack—thus a change in weight will dictate a change in the airspeed required. This is a result of the induced drag curve moving with a weight change—that is lowering with a weight decrease and becoming higher with a weight increase.

Power Required

Recall from Chapter 1 that power is the rate of doing work, and work is a force applied over a distance—the transfer of energy. Drag is a force which must be counteracted by thrust. However, propeller aircraft do not produce thrust. They produce power. Because of this, we must consider the power required by the aircraft as opposed to the drag being produced.

Thrust Required

Thrust required is very straightforward. In order to maintain level flight at a constant airspeed, the thrust provided by the engine must equal the drag being produced by the aircraft.

If the thrust available is less than the drag being produced, the aircraft will either decelerate, or descend to maintain airspeed. If the aircraft descends, the weight of the aircraft is providing the thrust required that is not provided by the engine. For example, in figure 3-24a, the engine is providing no thrust, so all of the thrust is being provided by the weight of the aircraft. If, on the other hand, the thrust available is greater than the drag being produced, the aircraft will either accelerate or climb. If the aircraft climbs, the weight of the aircraft is providing the "drag" that is not created by the aircraft (Fig 3-24b).

So essentially, thrust required is equal to the drag produced by the aircraft plus the component of the weight that is acting opposite the flight path (in level flight, this component is zero). As a result of this, thrust required will vary with airspeed just as total drag does. Therefore, a total drag curve is often referred to a thrust required curve—meaning

that it represents the thrust required to maintain equilibrium in level flight.

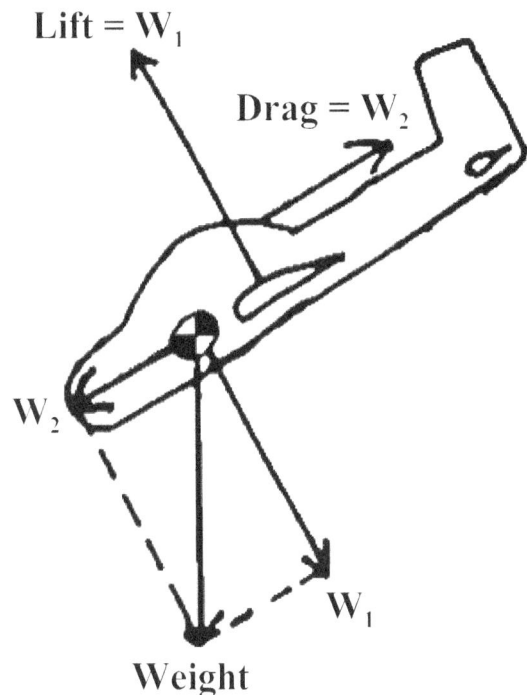

(a) Aircraft in a Power Off Descent - All Thrust Comes from the Forward Component of Weight.

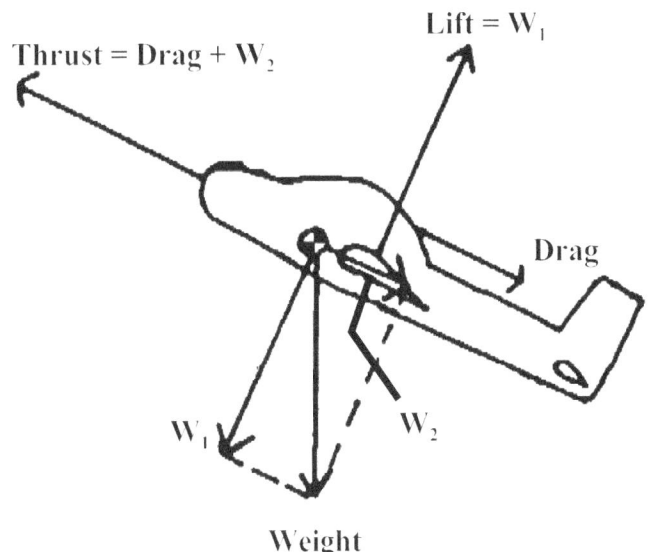

(b) Aircraft in a Climb - Excess Thrust is Countered by the Aft Component of Weight.

Fig 3-24. - *Forces Acting on an Aircraft.*

Force, Work, and Power

Once again, a quick review from Chapter 1. A force is a phenomenon that causes an object to accelerate. Work is a force applied over some distance and is a transfer of energy. Power is the rate of doing work.

The purpose of this section is to take the information presented in Chapter 1 and perform a unit conversion. This is so that we can work with knots instead of fps.

To start with, we need an equation from Chapter 1 and a unit definition:

$$Pwr(HP) = Fv/550 \qquad (1.14)$$

for v in fps.

$$1 \text{ knot} = 6080 \text{ ft/h}.$$

Now we need to convert (1.14) into an equation that we can use with knots instead of fps. Initially, we will convert knots into fps:

$$\frac{6080 \text{ ft/h}}{(60 \text{ min/h})(60 \text{ s/min})} = 1.689 \text{ fps}$$

then,

$$Pwr(HP) = \frac{Fv\ (1.689)}{550}$$

$$Pwr(HP) = Fv/325 \qquad (3.8).$$

Power Required

Equation (3.8) gives us a method for converting drag in pounds and velocity in knots into power in units of horsepower. Since we are dealing in particular with an aircraft, we can change the equation once again. We know that the force being applied is equal to the drag being produced by the aircraft, therefore (3.8) changes to:

$$Pwr = Dv/325 \qquad (3.9).$$

The concept of power required is essentially the same as that of thrust required. The power required is the power that must be supplied by the aircrafts propeller in order to maintain level flight at a constant airspeed. If the prop is not providing enough power, the aircraft will either decelerate or descend. If the prop is providing excess power, the aircraft will either accelerate or climb.

Power required can be divided up in a manner similar to that of drag. So we end up with parasite power required and induced power required. The sum of the two gives us the total power required.

For parasite power required, we combine equations (3.1) and (3.9) to get:

$$P_{req(P)} = \frac{(C_{dp}\,{}^1\!/_2\rho v^2)vS}{325} ,$$

therefore:

$$P_{req(P)} = \frac{C_{dp}\,{}^1\!/_2\rho v^3 S}{325} \qquad (3.10).$$

So we can see from equation (3.10) that, parasite power required increases with the *cube* of the velocity (Fig 3-25). This poses a problem to aircraft designers who want to build faster airplanes. A very large increase in power will often yield only a small increase in speed. Along with this, the extra weight of a larger engine is often prohibitive. Much more success is found with reducing the drag than increasing the power.

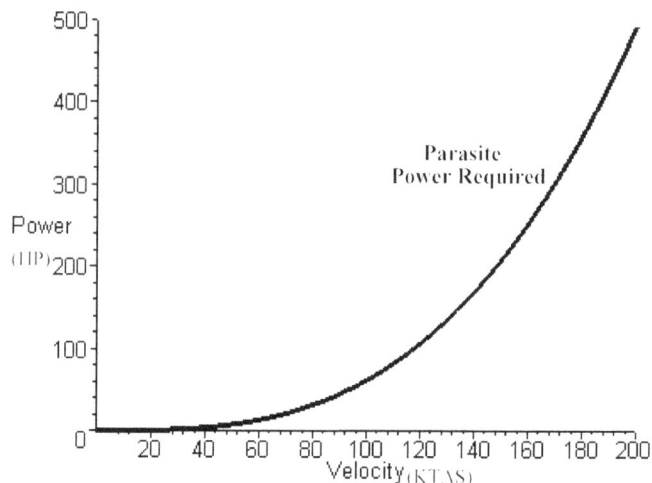

Fig 3-25. - *Parasite Power Required vs. Airspeed.*

For induced power required, we combine equations (3.7) and (3.9) to get:

$$P_{req(I)} = \frac{L^2\ (v)}{\pi\,{}^1\!/_2\rho v^2 b^2\ (325)} ,$$

therefore:

$$P_{req(I)} = \frac{L^2}{(325)\pi^{1/2}\rho vb^2} \qquad (3.11).$$

According to (3.11), induced power required varies *inversely* with velocity (Fig 3-26). As we can see from the induced power curve, induced power decreases as we speed up.

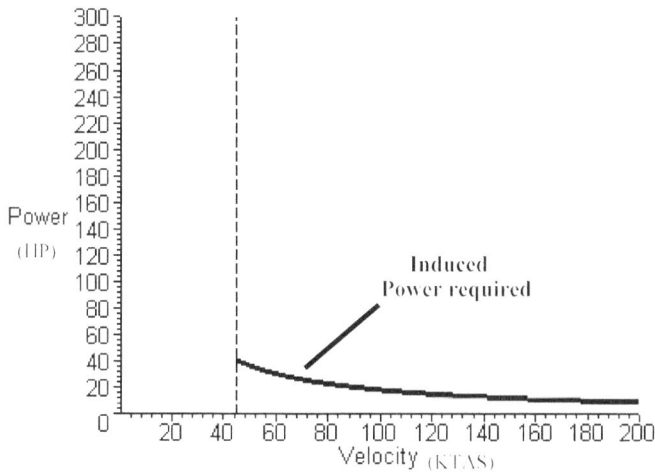

Fig 3-26. - Induced Power required vs. Airspeed.

The total power required is the sum of the parasite and induced power required (Fig 3-27b). As we can see, the power required curve follows a "U" shape similar to the thrust required curve. The power curve however, is flatter on the left side and steeper on the right side than the thrust curve. This is due to the velocity effects on each type of power required.

Fig 3-27 illustrates the difference between the thrust required curve and the power required curve for the same aircraft.

Chapter Summary

Drag is the force that opposes the motion of an aircraft. For low speed aerodynamics, it can be subdivided into two main types, parasite drag and induced drag.

Parasite drag is present any time an object moves through the air and can be further divided into skin friction, form drag, and interference drag. The formation of a boundary layer between the object's surface and the freestream airflow is the

mechanism that creates parasite drag. This boundary layer can be laminar or turbulent. A laminar boundary layer produces less skin friction drag than a turbulent boundary layer, but is more prone to separation which increases form drag. Parasite drag is proportional to the square of the airspeed. As such, it increases rapidly with an increase in airspeed.

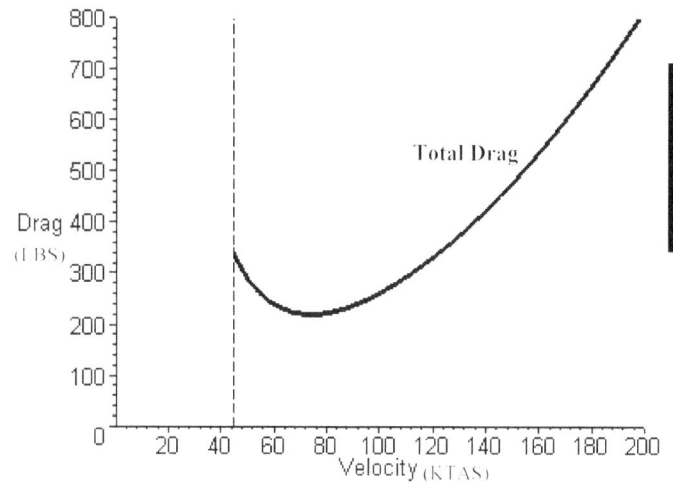

(a) Drag vs. Airspeed

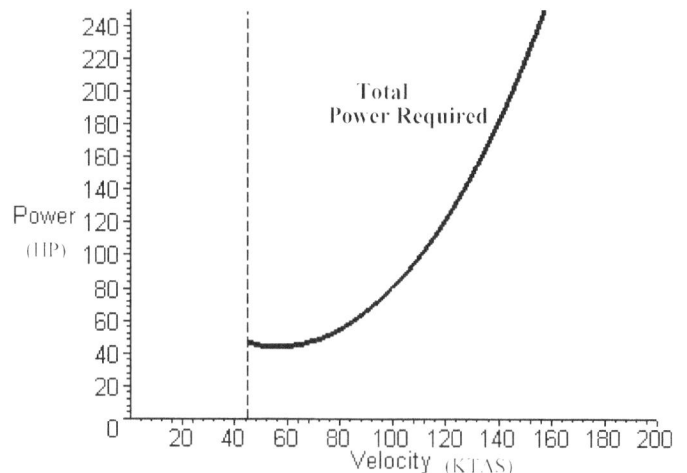

(b) Power Required vs. Airspeed.

Fig 3-27. - Drag Curve vs, Power Curve.

Induced drag is caused by the production of lift. The wing tip vortices induced by pressure differences above and below the wing cause downwash behind the wing. This downwash is effectively a change in the relative airflow. Since lift acts perpendicular to the relative airflow, the downward tilting of the airflow causes a rearward tilting of the lift. The aft component of the lift is induced drag. Induced drag is greater at higher lift coefficients, thus it is high at low airspeeds and decreases as the airspeed of the aircraft increases.

The total drag of an aircraft is the sum of the parasite drag and induced drag. As a result, drag will be high at low airspeeds, decreasing to a minimum and then increasing as speed is increased further. Since lift is constant, the best lift/drag ratio will be realized at the minimum drag point.

Power required is the rate that the aircraft loses energy to the atmosphere and is a function of drag and velocity. Due to velocity effects, the power required curve is shaped slightly differently than the thrust required (drag) curve, although the general trends are the same. On the left (low speed) side of the curve, power required decreases gradually with an increase in speed. On the right (high speed) side of the curve, power required increases very rapidly with an increase in speed. Because of this rapid increase in power required, large increases in power result only in small increases in speed.

List of Formulae

Parasite Drag

Variables

$$D_p = C_{dp}{}^1/_2\rho v^2 S \qquad (3.1)$$

Equivalent Flat Plate Area

$$f = C_{dp}S \qquad (3.2)$$

Induced Drag

Variables

$$AR = {}^b/_c \qquad (3.3)$$

$$AR = b^2/S \qquad (3.4)$$

$$D_i = C_{di}{}^1/_2\rho v^2 S \qquad (3.5)$$

$$C_{di} = \frac{kC_L{}^2}{AR} \qquad (3.6)$$

$$D_i = \frac{L^2}{\pi\,{}^1/_2\rho v^2 b^2} \qquad (3.7)$$

Power Required

Force, Work, and Power

$$Pwr(HP) = Fv/325 \qquad (3.8)$$

Power Required

$$Pwr = Dv/325 \qquad (3.9)$$

$$P_{req(p)} = \frac{C_{dp}{}^1/_2\rho v^3 S}{325} \qquad (3.10)$$

$$P_{req(l)} = \frac{L^2}{(325)\pi\,{}^1/_2\rho v b^2} \qquad (3.11)$$

Questions and Problems

1) What are the two main types of drag? How do they behave in relation to airspeed?

2) What is the difference between skin friction and form drag? Which of the two main types of drag do they form part of?

3) How would an aircraft designer try to reduce interference drag? Why is this method used?

4) How does boundary layer separation affect form drag?

5) Why does downwash cause induced drag? Why does it cause a reduction in lift?

6) How does total drag behave in relation to speed? Why?

7) How do variations in aircraft weight affect the power required curve?

8) How does the drag curve relate to the power required curve?

9) Why doesn't a significantly larger engine necessarily mean a significantly higher cruise speed?

10) A 3100 lb aircraft produces 240 lbs of drag at it's best L/D ratio speed of 90 kts. How much drag will this aircraft produce at 180 kts.? How much horsepower is required at each of these speeds and at sea level?

Chapter 4
Thrust and Power Available

As we've seen from the previous chapter, an aircraft in flight will always be producing drag. This drag is unavoidable, although we do try to reduce it. Drag, left to it's own devices, will cause the aircraft to decelerate. In order to maintain equilibrium, there must be a force supplied to cancel out the negative effects of drag. This force is thrust.

For an aircraft in level flight to maintain equilibrium, thrust must equal drag. These two forces cancel each other out to develop a net force of zero (Fig 4-1).

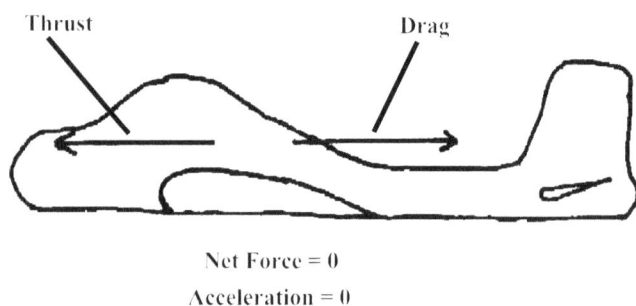

Net Force = 0
Acceleration = 0

Fig 4-1. - *Thrust and Drag* - The net force in equilibrium must be zero. This is accomplished when thrust and drag exactly counter one another.

Thrust is provided by the aircraft engine. For propeller driven aircraft it would be more accurate to say that the engine provides power to the prop, and the prop provides power for the aircraft.

Notice that for a prop aircraft the term power is used instead of thrust. This is because the engines that turn propellers don't produce thrust, they produce power. Therefore aircraft performance is dependant on power available as opposed to thrust available.

Production of Thrust

The fundamental concept behind the production of thrust is Newton's third law. Recall from Chapter 1 that forces occur in pairs which are equal in magnitude and opposite in direction. This concept of "action/reaction" is Newton's third law.

Let's begin with a parcel of air that is stationary in the atmosphere . Due to Newton's first law (inertia), this parcel of air will tend to remain stationary. But then along comes an aircraft engine (prop or jet) which applies a force to the parcel of air. The force causes the air to be accelerated in the direction opposite the aircrafts flight path according to Newton's second law (F = ma). Then, due to Newton's third law an equal and opposite force is applied to the aircraft engine (Fig 4-2).

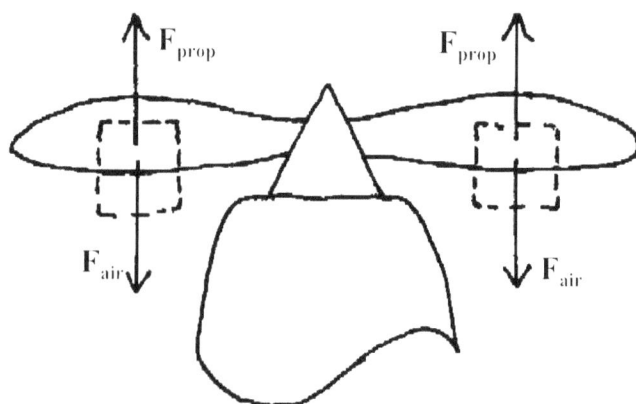

Fig 4-2. - *Thrust is a result of Newton's Third Law* - The force applied to an air particle results in an equal and opposite force being applied to the propeller.

This equal and opposite force which is applied to the aircraft engine is thrust. If thrust is equal to drag, the aircraft will maintain a constant airspeed and altitude. This is due to the fact that thrust and drag add up to give a net force of zero.

To determine how much thrust is being produced, we need to know how much air is being accelerated and how large the acceleration is. The mass flow of air through the propeller is determined by dividing the mass of air through the prop by the amount of time taken for this mass to flow:

$$Q = m/t \qquad (4.1).$$

Once we know the mass flow, we can multiply it by the change in velocity which occurs through the prop disc:

$$T = Q(v_2 - v_1) \qquad (4.2).$$

Equation (4.2) tells us how much thrust is being produced by an aircraft engine. With a slight manipulation, equation (4.2) reduces to Newton's second law:

$$F = ma \qquad (1.1).$$

Prop vs. Jet

Thrust is a force, and force is proportional to mass and acceleration. Because of this, we can produce the same force with a large mass and small acceleration or a small mass and a large acceleration.

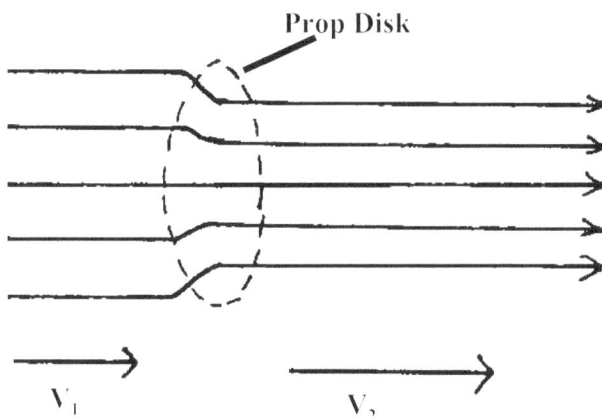

Fig 4-3. - *Propeller Action* - A propeller provides a relatively small acceleration to a relatively large mass of air.

A propeller acts on a relatively large mass of air and provides it with a relatively small acceleration (Fig 4-3). A jet on the other hand, acts on a relatively small mass of air and provides it with a relatively large acceleration (Fig 4-4).

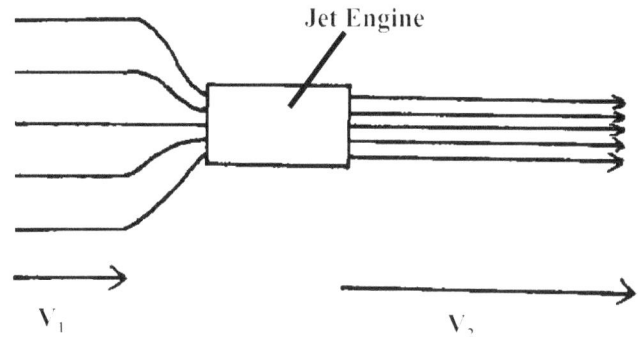

Fig 4-4. - *Jet Action* - A jet engine provide a relatively large acceleration to a relatively small mass of air.

The main reason that a reciprocating engine with a propeller would be used over a jet engine (aside from the obvious problem of price), is because propellers are much more efficient at lower airspeeds—while jets are better suited to higher speeds. The reason for this lies in the source of inefficiencies for the prop and the jet. In both cases the air is disrupted and put in motion. The energy put into the airflow by the engine is essentially wasted energy. With a jet, high exhaust velocities mean that large amounts of energy are left behind in the airflow. At higher speeds, a prop loses efficiency due to compressibility effects.

Engines *generally* flow from least powerful to most powerful as follows:

– *Reciprocating and Propeller* – Used mostly by general aviation and small charter operations.

– *Turbine and Propeller (Turboprop)* – Used mostly by small to medium airline operations as well as some government and military applications.

– *Turbofan* – Used mostly by medium to large airlines as well as some government and military applications.

– *Turbojet* – With a handful of exceptions (e.g. – the Concord and the DC-9), turbojets are usually used for combat applications.

– *Rocket* – Outside of spacecraft and experimental research vehicles, rocket engines are not normally used for manned aircraft.

Unfortunately, each time we step up to a more powerful engine, we do so at the cost of efficiency.

The Reciprocating Engine

The reciprocating engine is the most common engine currently in use in aviation. It is also the most efficient engine developed to date—a fact which contributes greatly to the engines popularity.

The most common type of reciprocating engine is the four stroke. The name comes from the fact that this type of engine has four sequential stages (strokes) which are used to produce the power. The concepts behind the four stroke are fairly straightforward, however, putting these concepts into practice is quite complex. In this section we will look at the principles behind the production of power with a four stroke engine.

The basic structure of the engine is a group of cylinders, each of which produces power independently of the other. The cylinders are hollow and they have valves for intake and exhaust. As well, inside each cylinder there is a piston which is free to move up and down in the cylinder. The piston from each cylinder is connected by a connecting rod to the crankshaft. This crankshaft is where the engine connects to external components which use the power it produces. In the case of airplanes, the main external component is the propeller—along with engine accessories (alternator/generator, fuel pump, vacuum pump, etc).

The four strokes of a four stroke engine are 1) intake, 2) compression, 3) combustion, and 4) exhaust.

During the intake stroke, the intake valve is opened and the exhaust valve is closed. The piston is moving down which leaves space inside the cylinder. This space is then filled by the fuel/air mixture which comes from the carburetor.

For the compression stroke, both of the valves are closed and the piston is moving up. The fuel/air mixture is sealed inside the cylinder and is being compressed. At the end of the compression stroke, the spark plug fires and ignites the fuel/air mixture.

Once the mixture is ignited, the combustion heats the gases rapidly. This heating increases the pressure of the gases and pushes the piston down. This stroke is where the engine actually gets it's power, and hence it is sometimes called the power stroke.

The burned gases left in the cylinder after the power stroke need to be ejected before the cycle can begin again. During the exhaust stroke, the intake valve is closed and the exhaust valve is opened. As the piston moves up through the cylinder, the exhaust gases are pushed out through the exhaust valve.

The power produced by a four stroke engine is dependant on a number of factors. They include the mean (average) pressure (P) in the cylinder during the power stroke, the length (L) of the stroke, the area (A) of the piston head, and the number (N) of power strokes per minute. These variables are related according to the following formula:

$$P(HP) = \frac{PLAN}{33,000} \qquad (4.3)$$

Remember that pressure is a force per unit area. So the pressure multiplied by the piston head area will give us a force. If we multiply the total force being applied to the piston head by the distance it is applied over, we find out how much work is done during each power stroke.

Once we know how much work is done with each power stroke, it is simply multiplied by the number of power strokes per minute. This gives us the amount of work done per minute. As we discussed in Chapter 1, work per unit of time is power. So the PLAN part of (4.3) will give us the power in foot pounds per minute.

Since one horsepower is defined as 33,000 foot pound per minute, we divide PLAN by 33,000 to determine the horsepower of the engine.

The number of power strokes (N) is determined by multiplying half of the engine RPM by the number of cylinders:

$$N = (RPM)(c)/2 \qquad (4.4).$$

The reason we only use half of the RPM is because the crankshaft rotates twice for every power stroke. The piston moves up and down once for the first two strokes. This rotates the crankshaft once. Then the last two strokes move the piston up and down again, which rotates the crankshaft once more.

Types of Power

We have several types of power to consider when we talk about aircraft. The first is Indicated Horsepower (IHP). IHP is the power being produced inside the cylinders of the engine. This is determined with the use of equation (4.3).

As we know, one of the important variables that affects IHP is the mean pressure inside the cylinder during the power stroke. This pressure is in turn affected directly by the density of the fuel and air mixture taken into the cylinder. With a more dense mixture entering the cylinder (don't confuse this with a more *rich* mixture) a larger mass of fuel and air is available to burn. With more fuel and air burning in the cylinder more energy is available in the form of heat to increase the pressure pushing down on the piston.

Unfortunately, at higher altitudes the air is less dense (see Appendix II). Therefore the amount of fuel being burned—and thus the power available—will be reduced at higher altitudes.

More important to us than IHP is Brake Horsepower (BHP). BHP is the power that we actually get from the engine. The name comes from the device that was once used to measure horsepower—the Prony Brake. BHP will always be less than IHP. The reason for the difference is that some of the power produced by the engine must be used to run the engine. The three strokes that do not produce power will use up some of the power produced by the power stroke. As well, friction between moving parts of the engine will use up some of the power produced.

Since IHP decreases with an increase in altitude, so will BHP. (Fig 4-5) is a chart of BHP vs density altitude for a normally aspirated (non-turbo-charged) reciprocating engine.

Next on the list is Shaft Horsepower (SHP). SHP is the power received by the propeller. BHP and SHP will be different if reduction gearing is used between the crankshaft and the prop. This would be done if the engine needed to run at high RPM to produce it's maximum rated power. Unfortunately, propellers lose efficiency at high speeds due to compressibility and shockwave formations. Reduction gearing will allow the engine to run at high RPM while the propeller still operates at relatively low RPM. However, some power is lost to the gears due to friction.

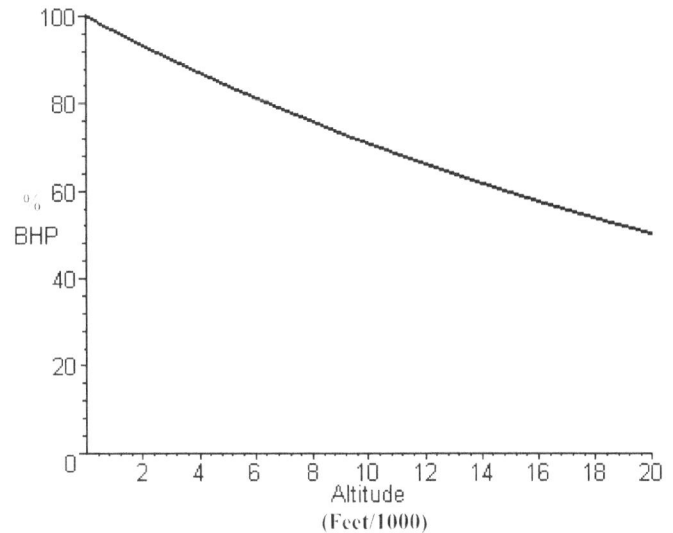

Fig 4-5. - *%BHP vs. Density Altitude for a Normally Aspirated Engine.*

The most important type of power in regard to aircraft performance is Thrust Horsepower (THP). THP is the power that we can use. The engine provides power to the propeller in the form of torque, and the propeller provides the power that we use to operate the aircraft. The propeller, however, is an airfoil and will lose some power to aerodynamic inefficiencies. These losses will vary with airspeed and RPM as will be discussed in the next two sections.

- - - - - - - - - - - - - -

So to sum up, the path followed by power from the engine to the airmass around the aircraft is as follows:

– *Equation (4.3)* gives us <u>Indicated Horsepower</u>.

– *Internal Power Losses* – <u>Brake Horsepower</u>.

– *Reduction Gearing Losses* – <u>Shaft Horsepower</u>.

– *Aerodynamic Losses* – <u>Thrust Horsepower</u>.

The Propeller

Reciprocating and turbine engines produces power in the form of torque (power applied in a circular form). This torque is directed through the crankshaft for it's application. Unfortunately, this rotational power is not of much use to us directly, so we need to convert it to some useful form. The device that we use for this conversion is the propeller.

The propeller is essentially a wing which is designed to rotate about a horizontal axis (Fig 4-6). Because of the propeller's direction of travel, it will produce "lift" in the forward direction. This "lift" is termed thrust, as it provides us with the forward force which we need to counteract drag.

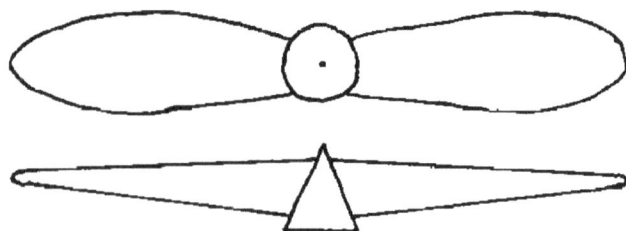

Fig 4-6. - *The Propeller* - The prop is essentially a rotating wing oriented to provide thrust instead of lift.

Just like a wing, a prop will also produce drag. In the case of a prop, however, this drag is termed torque (Fig 4-7). With a wing, the lift to drag ratio determines the efficiency of the wing. Likewise, the thrust to torque ratio determines the efficiency of a prop. The thrust to torque ratio will be determined by the angle of attack of the prop—just like that of a wing. So we have thrust—which is useful to us—and torque—which works against us. In order for the prop to maintain a constant RPM, the torque must be canceled out by the engine.

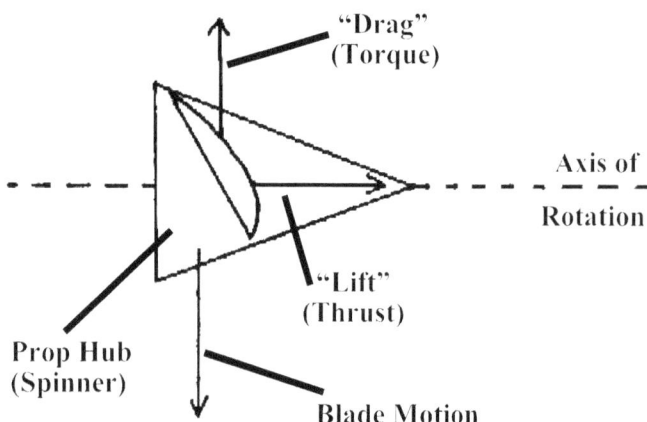

Fig 4-7. - *Forces Produced by a Prop.*

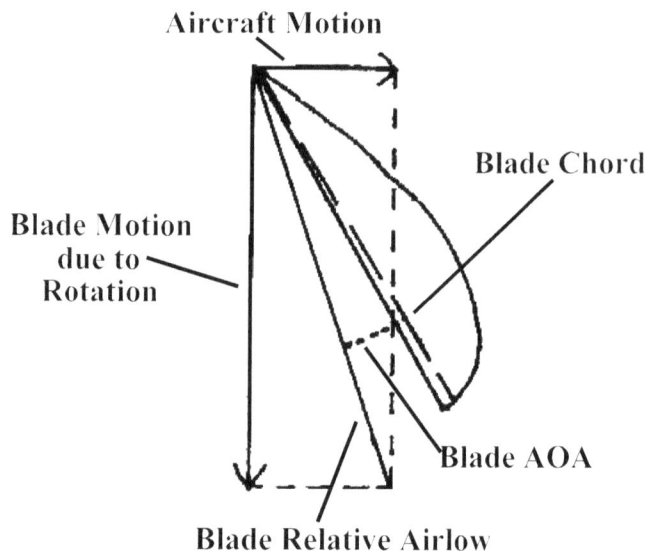

Fig 4-8. - *Prop Blade AOA* - The relative airflow of the prop is determined by both the rotational velocity of the prop and the forward speed of the aircraft.

Since a propeller is a wing, it will be affected by the same variables as a wing. This means that the speed the prop travels, as well as the angle of attack of the prop will determine how much thrust is being produced—as well as how much torque. The speed of the prop is determined largely by the prop RPM.

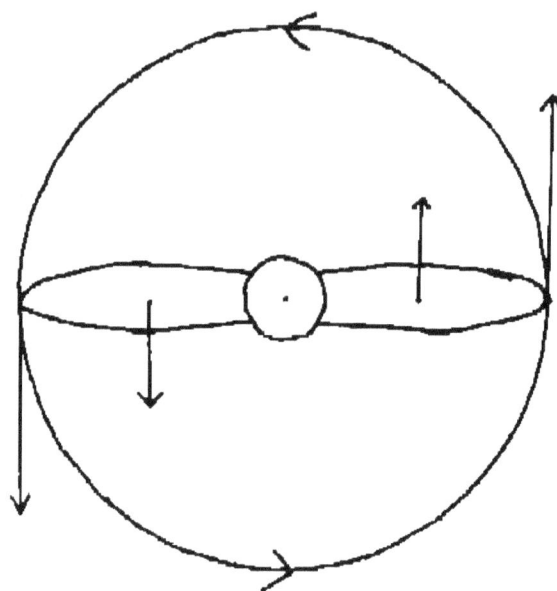

Fig 4-9. - *Velocity of Prop Sections* - As the distance increases from the center of the prop, a given RPM equates to a higher velocity for the blade section.

Recall that angle of attack is the angle between the chord line and the relative airflow. The relative airflow of a prop section is determined by the rotational velocity of that section and the forward velocity of the aircraft (Fig 4-8). With a high RPM and a low airspeed, the prop AOA will be high. With a low RPM and a high airspeed, the prop AOA will be low.

The chord line orientation of a given prop section will vary along the span of the prop. The reason for this variation is to compensate for the change in rotational velocity along the prop. The entire prop is rotating at the same angular velocity (RPM). So as we get further from the prop hub, the actual speed of the prop section will increase (Fig 4-9). Because of this, props are twisted to maintain a constant angle of attack along the prop span.

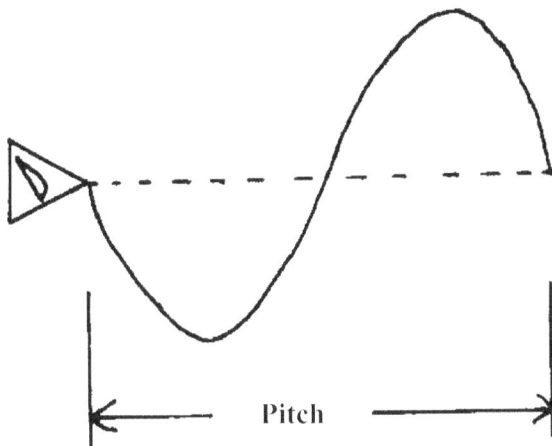

(a) Coarse Pitch

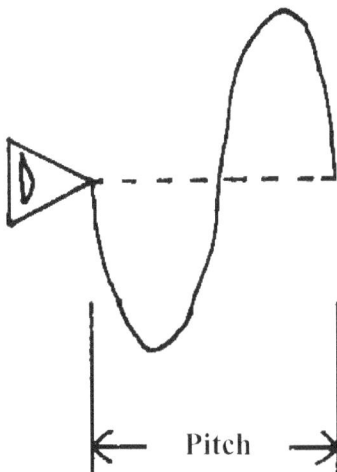

(b) Fine Pitch

Fig 4-10. - *Propeller Pitch* - The distance forward the prop travels in one revolution.

The orientation of the prop chord is referred to as the pitch of the propeller. Pitch is measured in inches, it is the distance that the prop would travel forward in one rotation if the angle of attack were zero. The term coarse pitch refers to propellers that move forward a large distance (Fig 4-10a). The term fine pitch refers to propellers that move forward a short distance (Fig 4-10b).

Thrust Horsepower Available

Referring to (Fig 4-11), we can see that the prop angle of attack for a given RPM will vary with airspeed. This means that the thrust to torque ratio will also vary with airspeed—but not proportionally. The lift to drag ratio of a wing increases until it reaches a maximum, and then decreases. A prop works the same way—a maximum point will be reached, and then efficiency will drop (Fig 4-12). The peak efficiency for most propellers is about 80%.

Fig 4-11. - *Variation of Prop AOA with Airspeed* - An increase in airspeed leads to a decrease in prop AOA, and vice versa.

Because of this efficiency change with the prop, the power actually available for us to use will change with airspeed. So even though the engine may produce 200 BHP at a given altitude, the useful power we actually get out of the prop will vary with speed. The power developed by the prop at a maximum throttle setting is referred to as thrust horsepower available (THP$_A$). THP$_A$ will follow a curve similar to the prop efficiency curve (Fig 4-13).

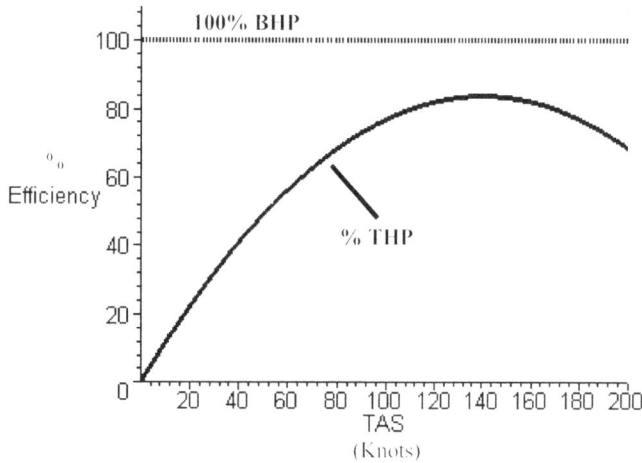

Fig 4-12 - *Prop Efficiency vs. Airspeed.*

THP is determined by the following formula,

$$THP = BHP(e) \qquad (4.5),$$

where (e) is the prop efficiency. (e) will always be less than one.

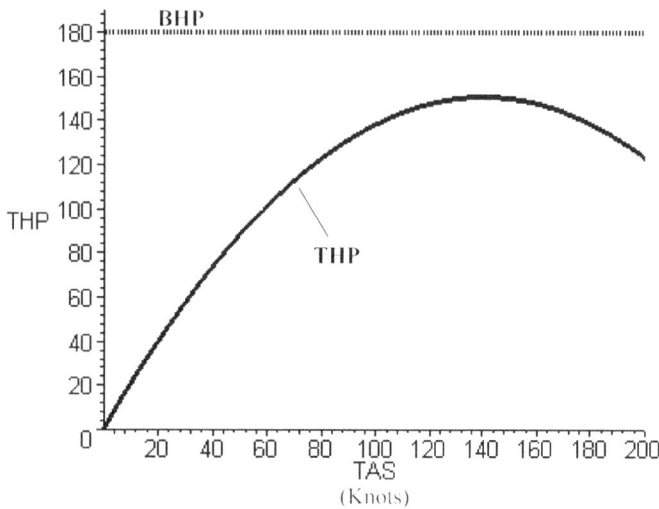

Fig 4-13. - *THP vs. Airspeed.*

We can see from (Fig 4-13 & 4-14) that the peak efficiency of a fixed pitch prop can only be achieved at one particular airspeed. The prop selected for an aircraft will depend on the role of the aircraft, and therefore what speed requires better prop efficiency.

For example, an aircraft which is meant to take off in a short distance, and climb well, needs to

have high prop efficiency at relatively low speeds. This type of aircraft would be better suited to a fine pitch prop - which performs better at low airspeeds. A fine pitch prop is sometimes referred to as a climb prop for this reason.

On the other hand, an aircraft which is meant to cruise fast will need a prop that is more efficient at higher airspeeds. This means the aircraft needs a coarse pitch prop—also referred to as a cruise prop.

Most fixed pitch propellers are a fair compromise between a climb prop and a cruise prop—unless the aircraft is meant for a specific purpose (e.g. – carrying skydivers would normally require a climb prop).

Variable Pitch Propeller

We run into trouble if we want an aircraft that performs well for the take off, climb, *and* in cruise. For this, we need a propeller that can change it's pitch. A variable pitch prop can maintain high efficiency at a variety of airspeeds (Fig 4-14).

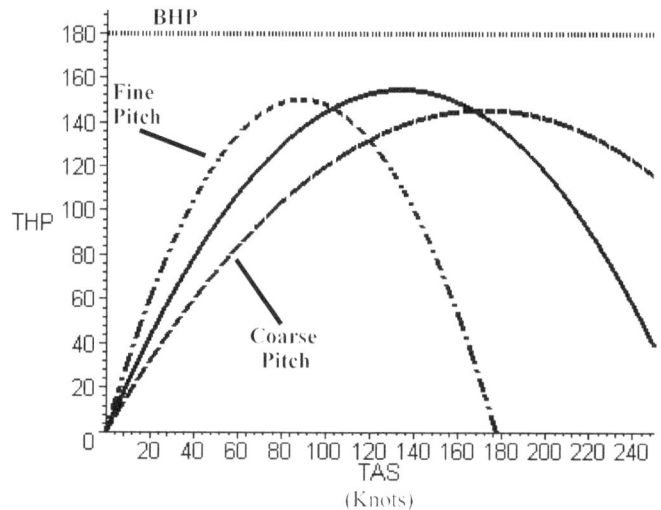

Fig 4-14. - *A Variable Pitch Prop's Efficiency Curve* - A variable pitch prop can maintain high efficiency over a wide range of airspeeds due to it's ability to maintain an efficient AOA.

In an aircraft with a variable pitch prop, the pitch would be set to fine for take off and climb, and then set to coarse for cruise. This would allow a maximum utilization of the engine power for both the climb and cruise portions of the flight.

The most common type of variable pitch prop is the constant speed prop. This type of propeller has a governor in the hub which regulates the pitch to maintain a constant RPM. The pilot sets the RPM with the prop controls in the cockpit. If the aircraft then changes speed, the drag (torque) on the prop changes, causing a change in RPM. This change in RPM is detected and rectified by a prop governor system. The governor system rotates the prop in order to change it's pitch so that the RPM will return to the preset value.

If the aircraft speeds up, the drag on the prop will be reduced, thus causing the prop to speed up. The governor senses this and increases the pitch of the prop which causes it to slow down. On the other hand, if the aircraft slows down, the prop will tend to slow down as well. When the governor senses this, it causes the prop to decrease pitch and thus speed up.

Through these processes, the angle of attack of the prop is held constant for a given power setting. If the power setting is changed, the RPM will still remain constant. This means that the pitch must change in order to create the amount of torque necessary to counteract the power. For this reason, each power setting in an aircraft with a constant speed prop will also have an optimum RPM setting.

Power is set with the throttle which controls the intake manifold pressure in the engine. With a fixed pitch propeller, we measure power with RPM because a change in manifold pressure will cause a change in RPM. With a constant speed propeller, however, we need a manifold pressure gauge to determine the power being produced—the prop governor will prevent RPM changes.

Left Turning Tendencies

There are some phenomenon that occur as a result of a propeller's rotation that we need to be aware of. These phenomenon do not contribute to thrust, but they are an unavoidable consequence of using propellers. As a group, they are often referred to as the left turning tendencies.

Slipstream – The first effect to consider is slipstream. Slipstream is a result of the rotation of the air as it moves back over the aircraft. The prop rotates clockwise as seen from the cockpit, so the air travelling back over the aircraft will rotate the same way. This means that on top of the aircraft, the airflow is moving to the right (Fig 4-15). When this airflow strikes the vertical stabilizer, it pushes the tail to the right, which in turn pushes the nose to the left.

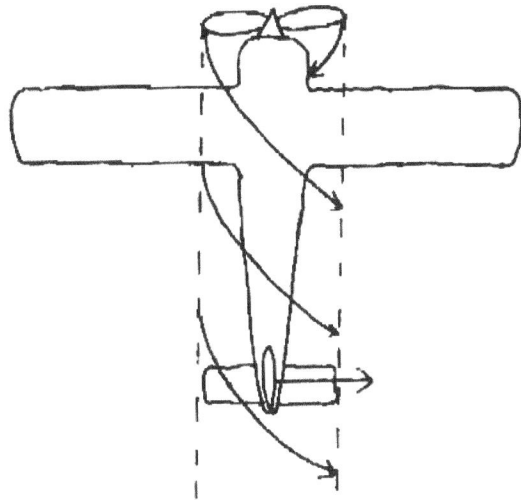

Fig 4-15. - *Slipstream Effect.* - This effect is increased by conditions of a high power setting and a low airspeed.

Slipstream is most pronounced at high power settings and low airspeeds. This is because at high power settings, the air that has flowed through the prop is rotating faster. As well, since the aircraft is moving slower, the rotating air has more time to affect the vertical stabilizer.

Asymmetric Thrust – Combined with slipstream, we have asymmetric thrust. Asymmetric thrust is a result of the right side of the prop disc producing more thrust than the left side. We can see from (Fig 4-16) that this imbalance will tend to pull the nose of the aircraft to the left.

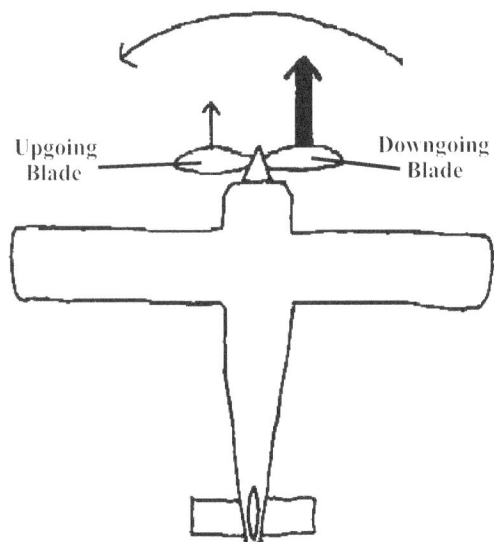

Upgoing Blade

Downgoing Blade

Fig 4-16. - *Asymmetric Thrust.*

The difference in thrust on each side of the prop is caused by the angle of attack of the wings. To explain, let's look first at an aircraft that is flying at a very low angle of attack. This means that the axis of the prop disc is aligned, or nearly aligned, with the flight path. (Fig 4-17a) indicates that under this condition, the blades of the prop are both at the same angle of attack.

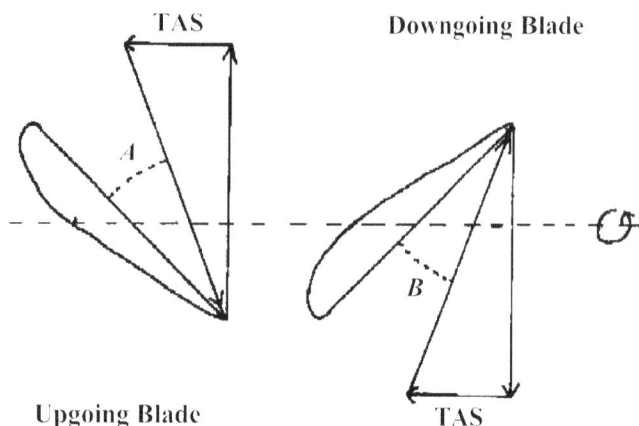

(a)   The Effect of Low AOA Flight. - 'A' is the AOA for the upgoing blade, and 'B' is the AOA for the downgoing blade. The two angles are equal, so both blades are producing the same amount of thrust.

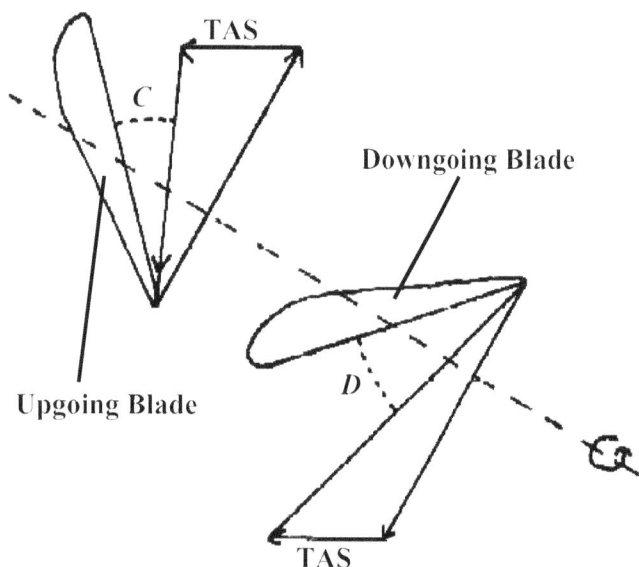

(b)   The Effect of High AOA Flight - 'C' is the AOA of the upgoing blade and 'D' is the AOA of the downgoing blade. Not only does the downgoing blade have a higher AOA, but it also has a slightly higher airspeed than the upgoing blade. As a result, the downgoing blade will produce more thrust than the upgoing blade during high AOA flight.

Fig 4-17. - *The Cause of Asymmetric Thrust.*

Now if we look at that same aircraft at a high angle of attack, we see something different. The axis of the prop disc is now tilted upwards relative to

the flight path. So the angles of attack for the prop blades are now indicated by (Fig 4-17b). As we can see, the right (downgoing) blade has a higher angle of attack. Because both blades are moving at the same speed, this means that the right blade will produce more thrust.

Like slipstream, asymmetric thrust is also more pronounced at low airspeeds and high power settings. Low airspeeds because this is where we usually experience a high angle of attack. High power settings because if the prop is producing a large amount of thrust to begin with, the difference between the two blades will be that much greater.

Torque – Torque is another left turning tendency. As the prop is being forced to the right (as seen from the cockpit), there is an equal and opposite reaction occurring as a result of Newton's third law. This equal and opposite reaction tends to roll the aircraft to the left.

Once again, this tendency is more pronounced at low airspeeds and high power settings. This is because at high power settings, there is more torque being produced by the engine. As well, at low airspeeds there will be more drag (torque) acting on the prop due to the props higher angle of attack.

To counteract the rolling tendency created by torque, some aircraft manufacturers will mount the left wing at a slightly higher angle of incidence so that a canceling rolling moment to the right will be created. This solution, however, creates it's own problem. The extra lift created on the left wing causes more induced drag. This drag imbalance will create a yaw to the left.

Gyroscopic Effect – Last on our list of left turning tendencies is gyroscopic effect. A gyroscope is a spinning mass. Gyroscopes have two characteristics that are of great interest to pilots (not only for aerodynamics, but for the study of the instruments as well).

The first is rigidity in space. This means that a gyroscope will tend to remain in the same plane of rotation—this is a result of the inertia of the spinning mass.

The second property of gyroscopes that is of interest to us here is precession. When a force is applied to a gyroscope which tends to displace it from it's plane of rotation, that force will actually manifest itself 90° later in the direction of rotation (Fig 4-18). We won't concern ourselves with the

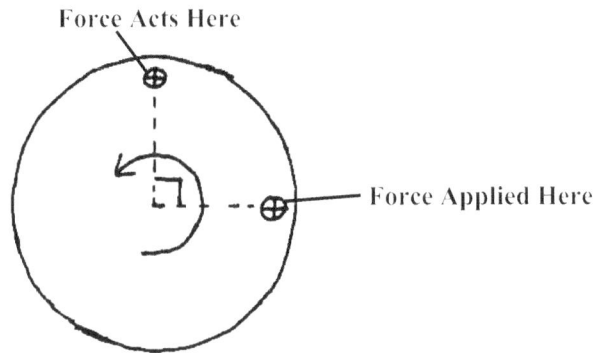

Fig 4-18. - *Gyroscopic Precession* - A force applied to a gyroscope will act 90° later in the direction of rotation.

details here, but the reason for precession lies in the conservation of angular momentum.

A propeller, since it is spinning, is a gyroscope. This means that a propeller will tend to remain in the same plane of rotation. However, if we apply a force to it, that force will manifest itself 90° later in the direction of rotation (clockwise).

So if we raise the tail of the aircraft, we are effectively applying a forward force to the top of the prop disc. This force manifests itself 90° later—which is on the right of the prop. What we end up with due to precession is a tendency for the nose of the aircraft to swing to the left. If we lower the tail, the nose will swing to the right.

This effect is most pronounced at high RPM, since this is where gyroscopic effect is the strongest. The most noticeable place for precession is during the take off roll in a tail wheel aircraft. Many tail wheel aircraft require the pilot to raise the tail (lower the nose) during the take off roll in order to achieve flying speed. This will cause gyroscopic precession to take effect, and the nose of the aircraft will tend to move to the left.

Most aircraft these days have some correction built in to the aircraft design for the left turning tendencies. There are several different types of corrections that can be used. They include offsetting the vertical fin, offsetting the engine, offsetting a wing, or some other imbalance to counteract the engines left turning tendencies.

Most of these corrections are adjusted for normal cruise flight. So in a climb (high power, low speed) the aircraft would still have left turning tendencies. As well, in a descent (low power, high speed), the aircraft will actually have right turning tendencies due to overcorrection.

The methods for controlling these left and right turning tendencies will be discussed in the next chapter.

Chapter Summary

Thrust is the force provided by the engine that opposes aerodynamic drag. In the production of thrust the engine (or propellor) exerts a force on a parcel of air. As a result of Newton's Third Law, the parcel of air exerts an equal and opposite force on the engine (or prop). This equal and opposite force is thrust. This fundamental concept applies to all types of atmospheric engines (i.e. – piston, turbine, jet), however the details vary somewhat from one type of engine to the next.

In the case of the reciprocating engine, it is more convenient to work in terms of power (energy transferred per unit of time) than thrust (force). The reciprocating engine produces power by burning a fuel/air mixture in a cylinder where the increased pressure can be applied to a piston. Several of these cylinders are connected to a crankshaft which carries the power to external destinations.

When power is produced in a piston engine, several sets of losses lead to some inefficiencies. Internal friction and the three non-power strokes reduce the amount of power we can get from the engine. If reduction gearing is used, further frictional losses occur. Last, but certainly not least, aerodynamic losses at the prop reduce the useful power to the "thrust horsepower" value.

For a given BHP, the amount of thrust horsepower available will vary with airspeed due to changes in prop efficiency. These changes result from the change in prop blade AOA as the relative airflow shifts with the airspeed. Aircraft manufacturers will sometimes use variable pitch props so that the blade angle can change to suit the relative airflow. This allows a much higher efficiency than a fixed pitch prop.

All single engine prop powered aircraft are plagued by the "left turning tendencies". These tendencies include slipstream, asymmetric thrust, torque, and gyroscopic precession. As a result of these, the aircraft will tend to yaw and/or roll to the left—especially at high power settings and low airspeeds. Built in corrections for these tendencies are normally adjusted for cruise flight. So in a climb, manual corrections to the right will be needed while in a descent manual corrections will be needed to the left.

List of Formulae

The Production of Thrust

$$Q = m/t \qquad (4.1)$$

$$T = Q(v_2 - v_1) \qquad (4.2)$$

The Reciprocating Engine

$$P(HP) = \frac{PLAN}{33,000} \qquad (4.3)$$

$$N = (RPM)(c)/2 \qquad (4.4)$$

Thrust Horsepower Available

$$THP = BHP(e) \qquad (4.5)$$

Questions and Problems

1) Based on Newton's laws, how is thrust produced?

2) What are the advantages of a piston engine over a jet? Vice versa?

3) What are the strokes of a four stroke engine?

4) What are the four types of power? What is the difference between them?

5) Why does thrust horsepower vary with speed?

6) What is the advantage of a variable pitch propeller?

7) What are the four left turning tendencies of the propeller?

8) Why are the four *left* turning tendencies not right turning tendencies?

9) When are the four left turning tendencies most pronounced? Why?

10) What are some of the built-in corrections used to combat the engines left turning tendencies?

11) What is the difference between a climb prop and a cruise prop? Why is this difference important?

Chapter 5
Flight Controls

So far in this book, we've discussed some of the phenomenon that affect aircraft. However, we have not looked at how we as pilots control these phenomenon. In this chapter, we will consider the control surfaces we use to manipulate the aircraft and how they affect the behavior of the aircraft.

Without flight controls, an aircraft would be pretty useless. It would simply be hurtling through the atmosphere with no way to control speed, height, direction, touchdown point and time, etc. With flight controls, an aircraft becomes an effective means of transportation.

Axis of the Aircraft

Because we exist in a three dimensional world, all objects have three axis' to rotate about. Movement about these axis', or combinations of these axis' allow us to control the orientation of an object relative to some fixed reference. In flying, the reference that we use is the horizon.

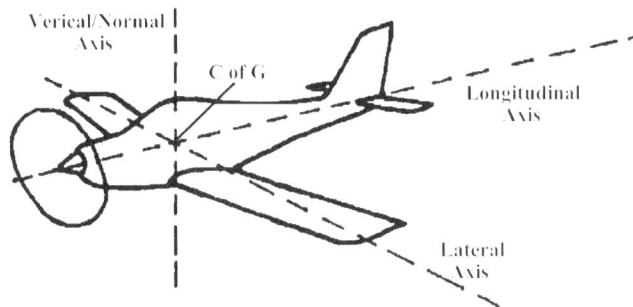

(a) The Rolling Movement – Movement about the longitudinal axis.

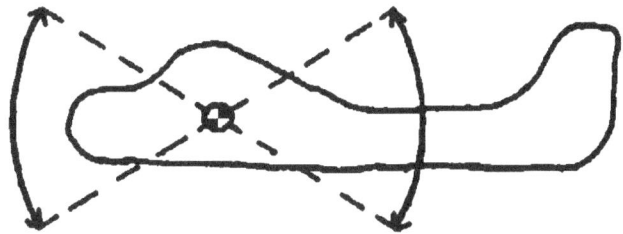

(b) The Pitching Movement – Movement about the lateral axis.

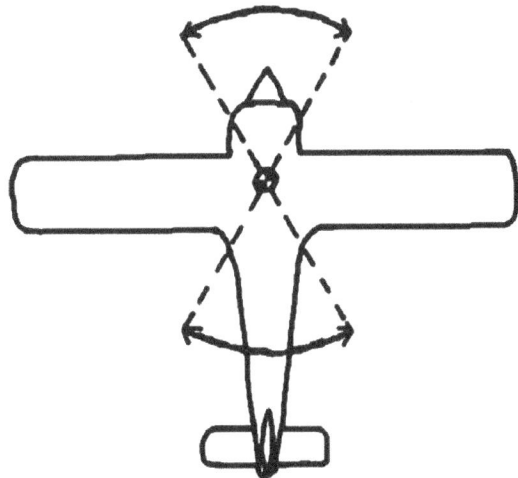

Fig 5-1. - *The Three Axis of the Aircraft* - The aircraft rotates around these three axis', which are perpendicular to each other. All three axis' pass through the G of G.

The three axis' (Fig 5-1) that we must consider for an aircraft are the *longitudinal* axis (nose to tail), the *lateral* axis (wing tip to wing tip),

(c) The Yawing Movement – Movement about the normal/vertical axis.

Fig 5-2. - *The Movements of and Aircraft* - A movement is a rotation about one of the three axis'

and the *vertical* or *normal* axis (top to bottom). Each of these three axis' intersect at the aircraft's center of gravity, and they are perpendicular to one another.

Movement about the longitudinal axis (Fig 5-2a) is called "roll". When the aircraft is rolling, one wing is moving up and one wing is moving down.

Movement about the lateral axis is called "pitch". When the aircraft is pitching (Fig 5-2b), the nose will be moving up or down.

Movement about the vertical or normal axis is known as "yaw". When an aircraft is yawing (Fig 5-2c), the nose will be moving from side to side.

Primary Flight Controls

The primary flight controls are the controls that have a direct and intentional influence on the attitude of the aircraft. The attitude is the position of the aircraft relative to the horizon. The elevator controls pitch, the ailerons control roll, and the rudder controls yaw.

Elevator and Pitch

The elevator is the control surface on the trailing edge of the horizontal stabilizer (Fig 5-3). There are in fact usually two separate elevators, but they act as a single control surface. Both elevators move in the same direction—up or down.

Under normal circumstances, all of the pitching moments on the aircraft cancel out. The

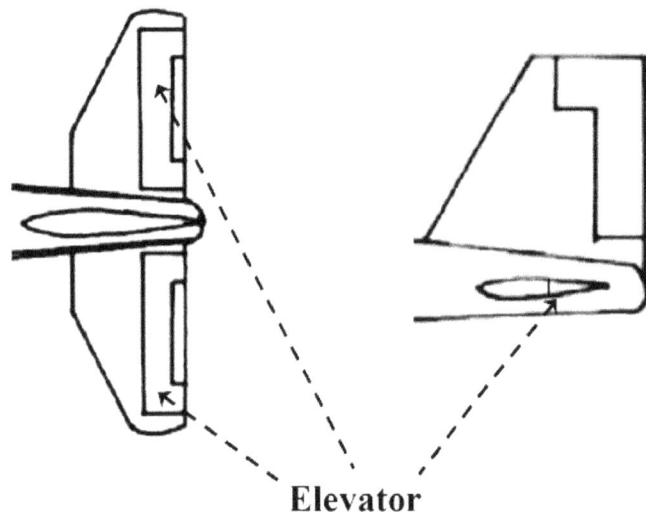

Elevator

Fig 5-3. - The Elevator.

pitching moment of the wing and the weight/lift couple combine to produce a nose down moment (Fig 5-4). In order for the aircraft to maintain equilibrium, this moment needs to be countered by a nose up moment. The nose up moment is provided by the thrust/drag couple and a down force on the tail.

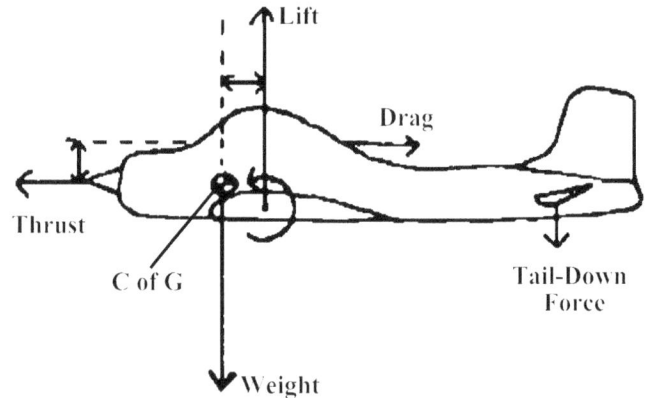

Fig 5-4. - Pitching Moments on an Aircraft - The wing moment and the lift/weight couple create a nose down moment. The tail and the thrust/drag couple create a nose up moment.

In order to pitch the aircraft, we need to temporarily upset the equilibrium of the aircraft. We do this by varying the down force on the tail—thus varying the nose up moment. When the nose up and nose down moments are not equal, the aircraft will rotate about the lateral axis.

Varying the down force on the tail is where the elevator comes in. The horizontal stabilizer is essentially an inverted wing—it produces lift in the downwards direction. The amount of lift being produced will depend on a number of factors—each of which were discussed in Chapter 2. The elevator is used to change the effective camber of the stabilizer surface (Fig 5-5). This change in

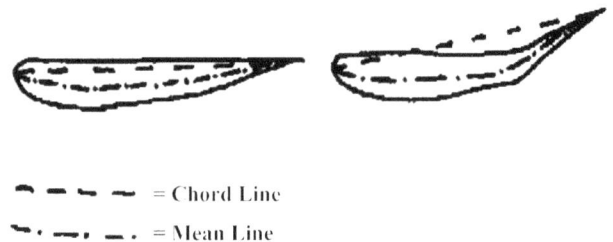

= Chord Line
= Mean Line

Fig 5-5. - The Elevator's Effects on Tail Surface Camber.

camber will change the lift coefficient of the stabilizer, thus changing the amount of lift being produced.

When the pilot pulls back on the control column, the elevator pivots upward. This increases the camber of the stabilizer thus increasing the down force. With an increase in down force, the nose up pitching moment overcomes the nose down pitching moment and the aircraft pitches up.

When the pilot pushes forward on the control column, the exact opposite occurs. The effective camber of the stabilizer is decreased and as a result the nose down moment overcomes the nose up moment.

Stabilator

The stabilator is an arrangement where the horizontal stabilizer and the elevator are combined into a single component (Fig 5-6). The concept was initially developed to overcome problems associated with flight at speeds near that of sound. As it turns out, however, the stabilator design comes in handy for slower aircraft as well—mainly as a way to reduce weight and drag.

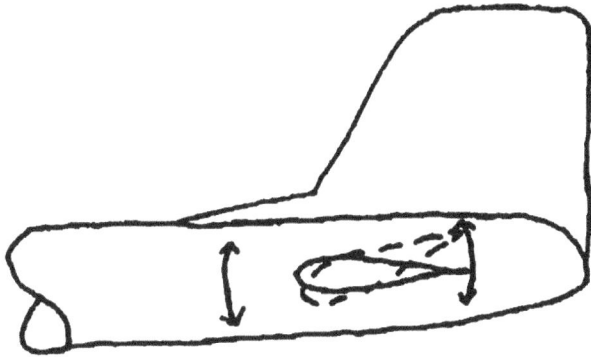

Fig 5-6. - The Stabilator.

With a stabilator, instead of varying the camber of the control surface, The entire stabilator rotates with control inputs. The angle of attack of the surface is thus varied, and pitching moments controlled in this manner.

Downwash and Ground Effect

In Chapter 3, we discussed how the relative airflow varies along the chord of the wing, and how

this variation results in downwash. Because downwash occurs behind the wing, it will have a direct influence on the operation of the horizontal stabilizer.

The downwards direction of the airflow behind the wing will increase the angle of attack of the stabilizer thus increasing the tail-down force (Fig 5-7a). However, if we were to reduce or eliminate downwash, we would effectively be reducing the angle of attack of the stabilizer (Fig 5-8b). This means that for the same control surface position, the tail-down force would be decreased, causing the nose to pitch down. In order to prevent this decrease in tail-down force, the elevator must be deflected upwards further.

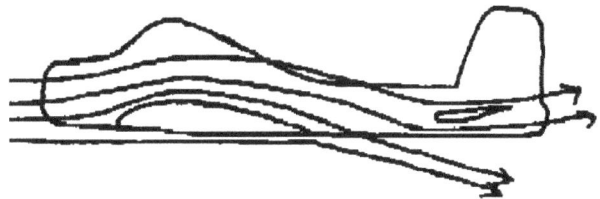

(a) Tail Surface Out of Ground Effect

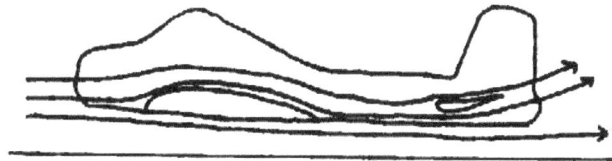

(b) Tail Surface In Ground Effect

Fig 5-7. - The Effect of Ground Effect on the Tail Surface - Decreased downwash will decrease the AOA of the horizontal stabilizer and therefore reduce the effect of a given elevator deflection.

Ground effect is the mechanism which will cause this decrease in downwash. So when an aircraft is flying in ground effect, a larger rearward deflection of the control column will be required than when the aircraft is out of ground effect.

Canards

A canard is another type of horizontal stabilizer, but it is mounted to the airframe forward of the wings instead of aft (Fig 5-8). With a canard, positive lift is produced instead of negative lift—like the more common tail arrangement.

Once again, the pitching moment of the wing and the lift/weight couple combine to produce a nose down moment. Instead of balancing this

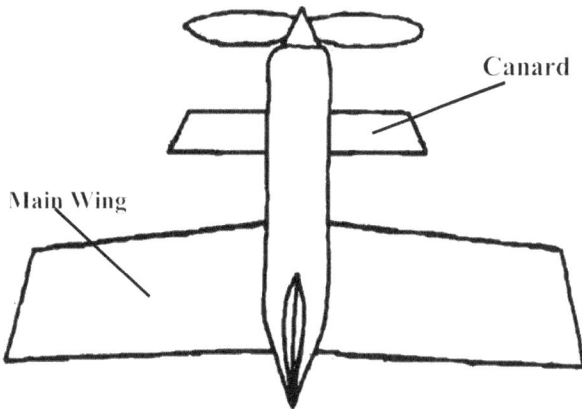

Fig 5-8. - *A Canard can Replace the Tail Surface.*

moment with a tail-down force, though, a canard produces an upwards force on the nose. A canard can have a stabilizer/elevator combination or it can have a stabilator. The function of the canard for pitch control is much the same as a rearward horizontal stabilizer. Control inputs on the canard, however, will vary the amount of nose-up force as opposed to tail-down force.

Ailerons, Roll, and Bank

Ailerons are the control surfaces that we use to control the movement about the longitudinal axis (roll). They move up and down opposite to one another, and are normally located on the trailing edge of the wing tips (Fig 5-9), although this location can vary somewhat. The movement of the

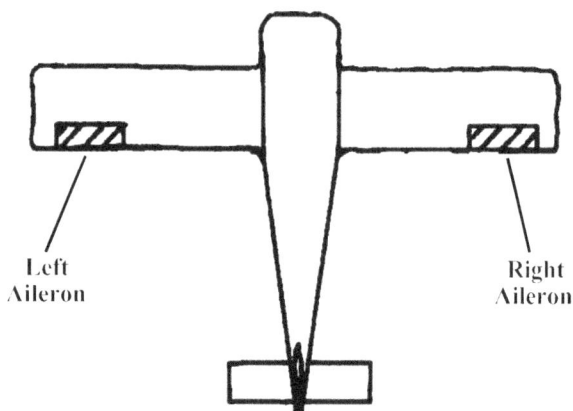

Fig 5-9. - *Ailerons* - Ailerons are used to control the rolling movement.

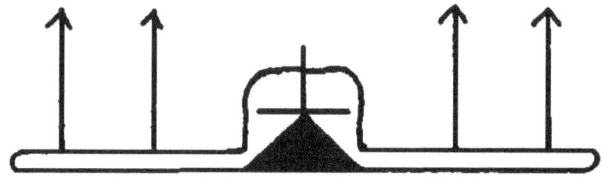

Fig 5-10. - *Ailerons Neutral* - The lift produced by both wings is equal and no rolling moment is produced.

ailerons is normally controlled with the control column by turning it side to side (or by the stick by moving it side to side).

As long as the ailerons remain in the neutral position (flush with the wings), both wings will be producing the same amount of lift (Fig 5-10). This means that there is no imbalance of forces across the wingspan and the aircraft will maintain it's position relative to the horizon (that is, the aircraft will not roll).

If, however, we were to deflect the control column to the left, the left aileron would move up and the right aileron would move down (Fig 5-11). The up going aileron (the left one in this case) would effectively decrease the camber and AOA of the wing, thus reducing lift. The down going aileron, on the other hand, effectively increases the camber and AOA of the wing, thus increasing lift.

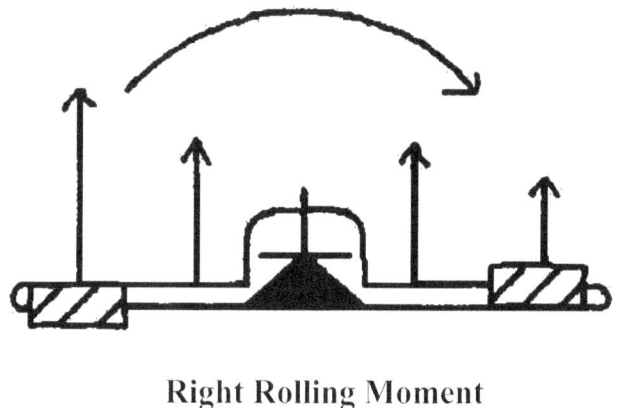

Right Rolling Moment

Fig 5-11. - *The Effect of Aileron Deflection.* - The downgoing aileron increases lift while the upgoing aileron decreases lift. The net effect is to create a rolling moment toward the upward aileron.

So with the ailerons deflected, we have an imbalance of lift across the wingspan. The total lift will still equal weight, but will be distributed differently across the wing. This imbalance of lift will cause a rolling moment to form, and the aircraft will roll in the direction of the aileron deflection.

It is important to note here that the ailerons do not control the bank attitude of the aircraft. Ailerons control the rolling movement, which is used to transition the aircraft from one banked attitude to another. Once the aircraft is established in a banked attitude, the ailerons must be neutralized in order to maintain it.

Rudder and Yaw

Yaw, as previously mentioned, is the movement about the vertical or normal axis. We control yaw with the rudder, which is mounted to the trailing edge of the vertical stabilizer (Fig 5-12).

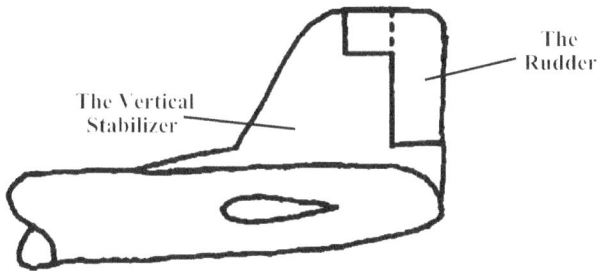

Fig 5-12. - *The Rudder* - This control surface controls yaw.

The main idea behind the rudder is the same as the elevator or the ailerons. It pivots on it's hinge line, which changes the camber of the stabilizer (Fig 5-13). This change in camber will cause an aerodynamic force to act on the stabilizer, pushing the tail to the side opposite the rudder deflection. This force acting on the tail will rotate the aircraft about it's normal axis—thus producing yaw.

In practice, however, the rudder is not normally used to *produce* yaw. It is used instead to *control* yaw. The difference is subtle, but important, and will be discussed further in the next section and in future chapters.

Coordinated Flight

The key to understanding the proper function of the rudder is to understand the concept of coordinated flight. If an aircraft is in coordinated flight, the freestream airflow is parallel to the longitudinal axis as seen from above (Fig 5-14a).

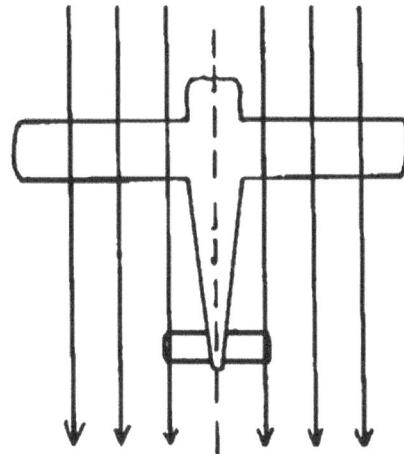

(a) Coordinated Flight – Zero Sideslip

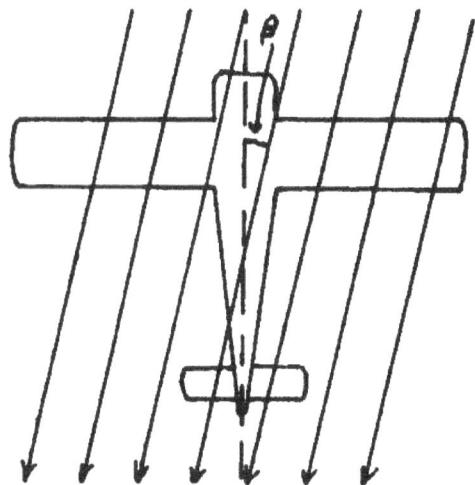

(b) Uncoordinated Flight – Sideslip Angle

Fig 5-14. - *Coordinated and Uncoordinated Flight.*

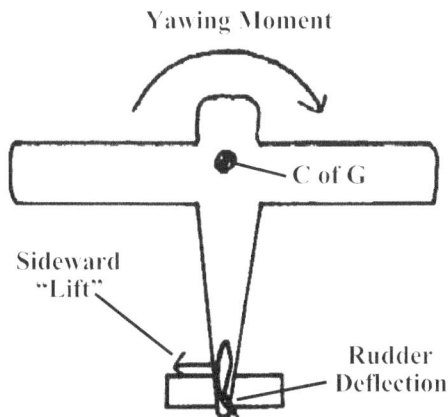

Fig 5-13. - *The Rudder* - When the Rudder Pivots, the Camber of the Vertical Stabilizer Changes and Produces an Aerodynamic Force.

If, on the other hand, the airflow is at an angle to the longitudinal axis (Fig 5-14b), the aircraft is said to be uncoordinated.

This uncoordinated flight is also referred to as a sideslip. Normally, a sideslip is undesirable for reasons which will be discussed in more detail at various points in this book. In brief, it comes down to the drag produced by the fuselage travelling sideways, the roll induced by the yaw (discussed in two sections time), and the effect a sideslip has on the stall characteristics of the aircraft (discussed in the next chapter).

Note that uncoordinated flight—or a sideslip—doesn't simply mean that yaw is present. It means that the amount of yaw present is too much or too little for the aircraft's angle of bank. When the wings are level, no yaw should be present, but when the aircraft is banked, there should be an amount of yaw corresponding to the bank angle (more about this in Chapter 8).

So uncoordinated flight doesn't mean that yaw is present—it means that *unwanted* yaw is present.

Causes of Unwanted Yaw

Since uncoordinated yaw is usually an unwanted characteristic of flight, we normally eliminate it through coordinated use of the rudder. Why, you may ask, do we need a rudder to *eliminate* yaw when it is the rudder that *produces* yaw? In fact, even though the rudder can produce yaw, there are several other sources of unwanted yaw which we need to consider.

The first and most obvious causes of unwanted yaw are the four left turning tendencies of the propeller. These tendencies were discussed in detail in the previous chapter. However, as a quick review, we have:

1) *Torque*
– Action/Reaction as a result of the torque applied to the prop by the engine.

2) *Slipstream*
– A result of the rotation imparted on the airflow by the propeller.

3) *Asymmetric Thrust*
– At high angles of attack (generally meaning low airspeeds), the down going prop blade

has a higher angle of attack than the up going prop blade, thus it produces more thrust. This imbalance in thrust yaws the aircraft left.

4) *Gyroscopic Precession*
– Due to the gyroscopic properties of the propeller, as the tail is raised, the aircraft will yaw to the left.

As mentioned in the previous chapter, most aircraft have some sort of built in correction for these left turning tendencies. The problem is, these corrections are designed for one combination of power and speed (usually the design cruise speed). If the aircraft is being flown at a speed and/or power setting that does not coincide with design cruise, corrections must be made by the pilot. These corrections are made with the rudder.

In a flight condition where the aircraft has a high power setting and a low airspeed (e.g. – climbing), the built in correction will be insufficient and it will be necessary to use *right* rudder to keep the aircraft coordinated. On the other hand, if the aircraft has a low power setting and a high airspeed (e.g. – descending), the built in correction will be over compensating, and *left* rudder will be necessary.

The next source of unwanted yaw to look at is turbulence. Turbulence and gusts can cause the aircraft to pitch, roll, and yaw. The control of these movements is fairly straightforward, simply using the appropriate control surfaces to return the aircraft to the position we want will do the trick ("*sure* it will", you say).

Our last source of unwanted yaw is called aileron drag. Aileron drag produces a phenomenon known as adverse yaw. With adverse yaw, the aircraft is rolling in one direction while yawing in the other direction. This brings the aircraft into uncoordinated flight, and must be corrected with the rudder.

When we deflect the ailerons with the control column, they travel in directions opposite to one another—that is, one goes up and the other goes down.

The wing with the down going aileron is now producing more lift than the other wing. The problem is, as we discussed in Chapter 3, an increase in lift also means an increase in induced drag. On the other hand, the wing with the up going aileron is now producing less lift—meaning less induced drag.

So as the aircraft rolls to the left, for example, the right wing is producing more drag—due to more lift—than the left wing (Fig 5-15). This excess drag on the right will cause the aircraft to yaw to the right even as it is rolling left.

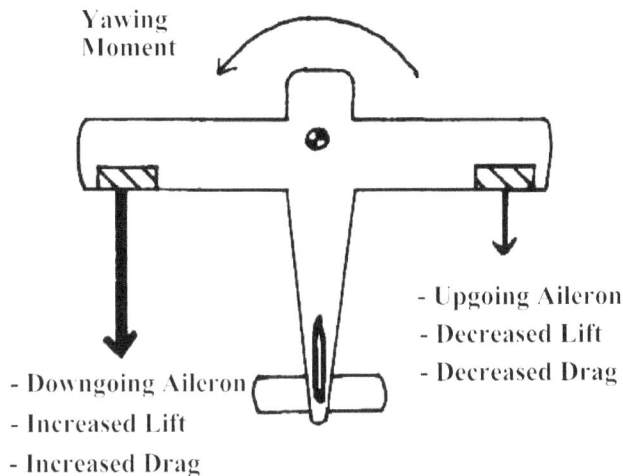

Fig 5-15. - *The Cause of Adverse Yaw.*

Most modern aircraft have some sort of built in correction for aileron drag. The first option to consider is linking the ailerons and rudder. When this is done, an aileron input will automatically include the corresponding rudder input due to mechanical linkages. This rudder input will cancel the adverse yaw that is present as a result of the roll.

(a) Differential Ailerons

(b) Frise Ailerons

Fig 5-16. - *Some Solutions for Adverse Yaw* - Differential and frise ailerons created parasite drag on the downgoing wing to balance the induced drag produced on the upgoing wing. This reduces and sometimes eliminates the imbalance in aileron drag that produces adverse yaw.

Other options include differential ailerons or frise (pronounced "freeze") ailerons (Fig 5-16). Differential ailerons don't pivot equal amounts to one another. The down going aileron produces more lift—thus producing more induced drag. The up going aileron moves up further. Because of it's projection into the airflow, the up going aileron will now produce more form drag. So one wing is producing extra induced drag, while the other wing is producing extra form drag. The two counter one another to reduce the yawing moment (Fig 5-17).

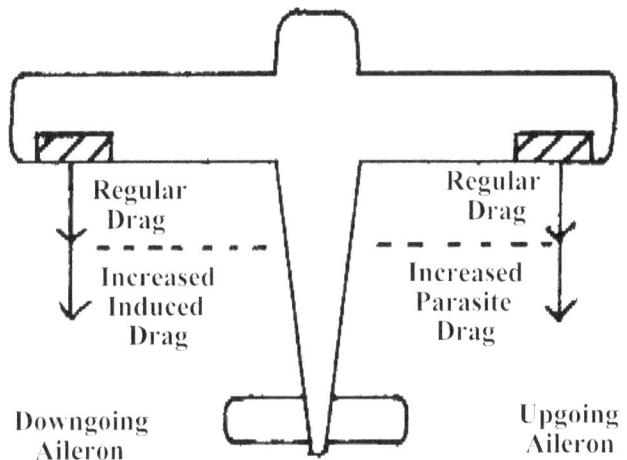

Fig 5-17. - *The Balance of Drag Across the Wingspan.*

Frise ailerons work on a similar premise. The down going aileron still increases the lift on the wing—thus increasing induced drag. The up going aileron is designed, however, so that part of the aileron projects beneath the wing—thus causing form drag. Once again, the two drags counter one another and the net yawing moment is reduced.

Secondary Effect of Yaw

One of the reasons that yaw is undesirable under normal flight conditions is because of it's secondary effect. As the aircraft yaws about the normal axis, the two wings are no longer travelling at the same speed (Fig 5-18). As we've seen in Chapter 2, the lift being produced is a function of speed. Because of this, the faster wing will be producing more lift, and the aircraft will roll in the direction of the yaw. This roll will also be aggravated by the stability characteristics of the

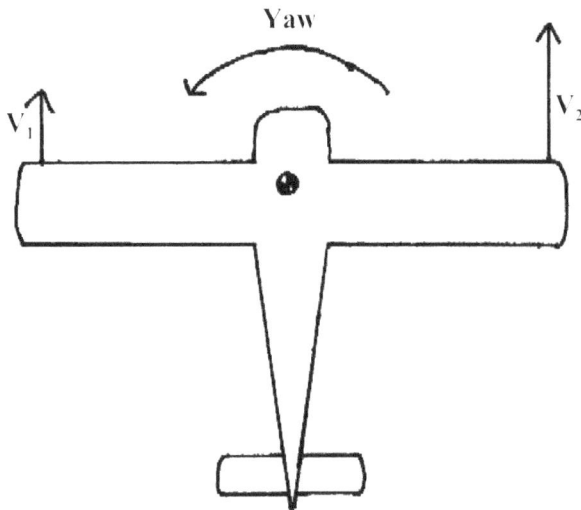

Fig 5-18. - *Speed Differential Between Two Wings During Yaw* - The wing on the outside of the yaw will be travelling faster through the air and will therefore produce more lift. This leads to a roll into the yaw, i.e., a left yaw will lead to a left roll.

aircraft (discussed in Chapter 7).

This roll can lead to problems if it is not kept in check. The use of rudder to prevent the unwanted yaw will eliminate the roll as well. On the other hand, the use of rudder to produce a yaw will also produce a roll.

Control Balancing

The flight controls on an aircraft will have aerodynamic forces acting on them that tend to return the surface to neutral (Fig 5-19). This means that at larger deflections it becomes progressively more difficult to utilize the controls. Up to a point, this is a good thing because it prevents us from overstressing the aircraft—a lack of resistance would lead to frequent over application of the controls. However, it can also prevent us from getting maximum safe deflection out of the controls.

Fig 5-19. - *Air Loads Acting on a Control Surface.*

To solve this problem, many aircraft have some type of control balance incorporated into their design. For aerodynamic balancing, part of the control surface extends to the other side of the hinge line (Fig 5-20). Aerodynamic forces on that side of the surface tend to cause a greater deflection. These forces will balance or partly balance the forces which tend to neutralize the surface.

Fig 5-20. - *Air Loads on a Control Surface With a Dynamic Balance.*

Another type of balancing is known as "mass balancing". This is done for a separate set of reasons—the prevention of flutter and the prevention of gust load imbalance. Flutter will be discussed further in Chapter 9. For now, we will look at the effect of gust loads on a control surface.

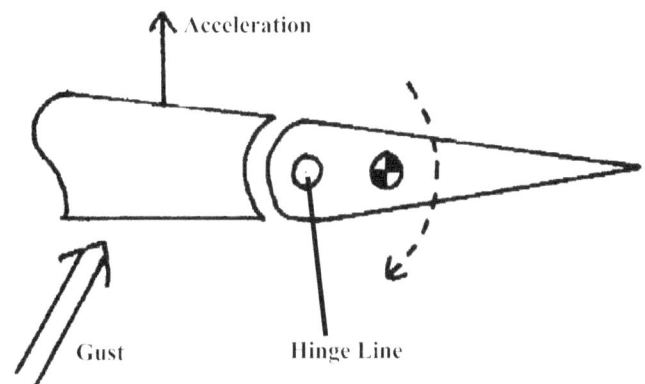

Fig 5-21. - *Gust load Effect on an Unbalanced Control Surface.* - These loads can deflect the control surface in the direction that aggravates the gust condition.

Using an elevator as an example (Fig 5-21), consider what would happen to an unbalanced surface experiencing a gust induced acceleration. The material that the elevator is constructed out of extends only to one side of the hinge line. This means that the center of gravity of the surface is also on that side. If a gust was to force the tail up under these circumstances, the force applied to the control surface would be applied at the hinge line. Since the hinge line and the C of G do not line up, a moment would be created which would force the deflection of the surface. In this example, the elevator would be forced down. We know that a downwards deflection of the elevator will reduce the down force on the tail, thus causing it to rise even further. So we see here that a gust applied to an unbalanced surface will cause the surface to aggravate the displacement caused by the gust.

To solve this problem, a mass balance is attached to the control surface (Fig 5-22). This balance extends to the other side of the hinge line and thus moves the C of G of the control surface to or beyond the hinge line. The couple now produced by a gust load will tend to move the surface in such a way as to negate the gust.

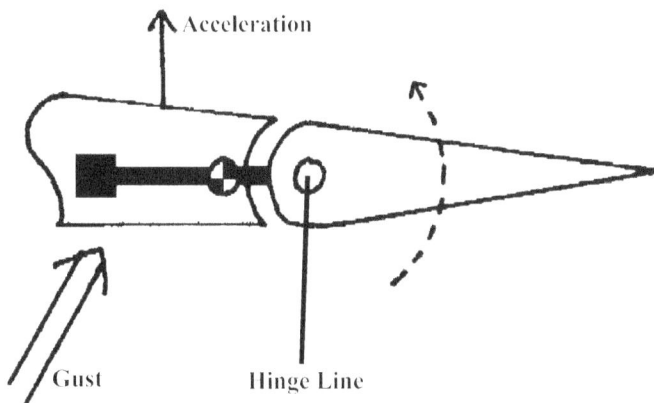

Fig 5-22. - *Gust Load Effect on a Balanced Control.*

Secondary Flight Controls

The secondary flight controls are the controls which do not have a direct or intended effect on the *attitude* of the aircraft, but they do have an immediate bearing on the behavior and flight characteristics of the aircraft.

Trim Tabs

Trim of an aircraft is the state or position that the aircraft will tend to remain in when left to it's own devices. This idea is closely related to aircraft stability, which will be discussed in Chapter 7. The concept of trim can be applied around all three axis', but in pilot circles—especially single-engine pilots—the term trim is usually directed towards pitch trim.

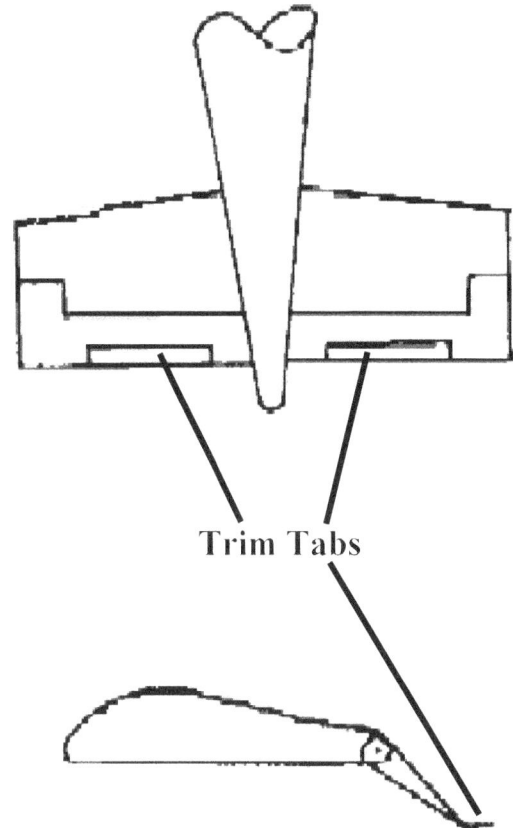

Fig 5-23. - *An Elevator Trim Tab* - Trim tabs are used to reduce control forces for the pilot. They can also be placed on the rudder and the ailerons. Trim tabs work by cambering the control surface and causing an aerodynamic force ("lift") to hold the surface in place.

Trim tabs (Fig 5-23) are what we use to control the trim of the aircraft. Without trim tabs flying would be a very physically demanding activity.

In the case of pitch trim, the trimmed position is measured as an angle of attack. This angle of attack is the angle at which no pitching moments are present. If we were to decide to fly an aircraft at an angle of attack other than the trimmed angle, a pitching moment would be present and we would have to counteract it with constant pressure on the controls. Instead of allowing this to continue, however, we could readjust the trimmed angle via the trim tab(s).

The trim tab(s) will move in the direction opposite to the control surface. As a result, it will create a camber on the surface which causes a "lift" force to move it in the direction we desire.

So when we want to change the angle of attack in the aircraft, we can adjust the trim tab(s) to relieve pressure on the control column.

Servo and Anti-Servo Tabs

Servo and anti-servo tabs are often confused with trim tabs. This is mostly due to the fact that the functions of the two (trim/servo or trim/anti-servo) are often combined into a single control surface. The servo or anti-servo tab is intended to vary the trimmed angle of attack just like the trim tab does. As such they will be located in the same place as a trim tab (the trailing edge of a control surface). The difference lies in the reason for the trim change.

In the case of a trim tab, the trim change is used to reduce the pilot workload—the trim tab eliminates the need for constant pushing or pulling on the control column. A servo tab is almost the exact same thing, except that it is used to move the control surface directly—not just to reduce loads.

Anti-servo tabs are the exact opposite. They are used to induce artificial loads on the control surface so as to prevent overstressing of the aircraft due to over-enthusiastic control application. On some control surfaces, there is no natural resistance to the movement of the surface. This must be countered by some sort of "built-in" resistance. The anti-servo tab moves in the same direction as the control surface and produces an aerodynamic moment which tends to return the control surface to it's neutral or trimmed position.

A good example of where the anti-servo tab would be used is on a stabilator. On a stabilator, the pivot point must be placed at the aerodynamic center in order to prevent the "lift" being produced from forcing a control deflection. For this reason, symmetrical airfoils are usually used for stabilators since they do not create a pitching moment around the aerodynamic center. So a deflection of a stabilator will not create a force tending to return it to neutral. This force must be supplied by an anti-servo tab which is linked mechanically to the control surface and deflects automatically as the surface is deflected. A greater deflection of the control surface means a greater deflection of the anti-servo tab, and thus a greater returning tendency for the control surface.

Flaps

Flaps are normally located on the inboard trailing edge of the wings (Fig 5-24). They both will move in the same direction to increase the camber of the wings.

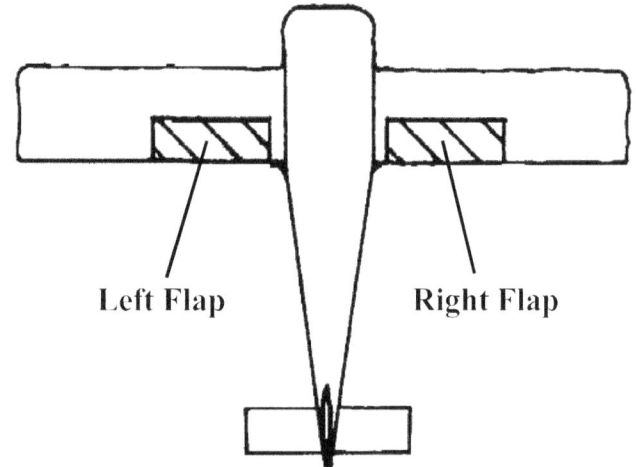

Fig 5-24. - *Flaps.*

Flaps are considered to be high lift devices. This is because they allow the wing to produce a higher coefficient of lift than the same wing without flaps. (Fig 5-25) is a typical graph of angle of attack vs coefficient of lift for a flapless wing, and the same wing with flaps. Notice that the maximum coefficient of lift is higher with the flapped wing, but that it occurs at a lower angle of attack.

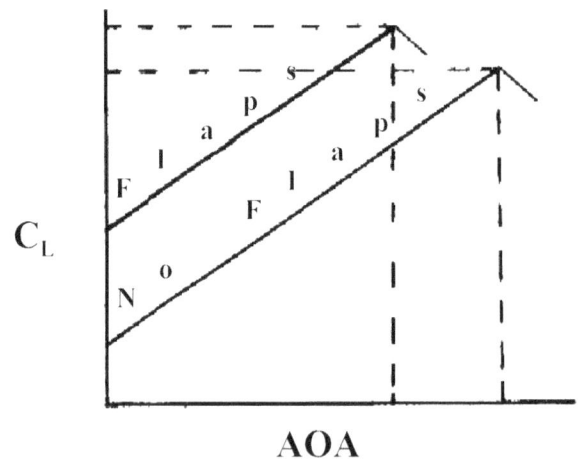

Fig 5-25. - *AOA vs. Lift Coefficient With and Without Flaps.*

Don't be confused by the term "high lift device", an aircraft with flaps extended will not necessarily be producing more lift than an aircraft with flaps retracted. It will simply be producing the same amount of lift at a lower airspeed and/or angle of attack. The term "high lift device" simply means that the wing *can* produce more lift for a given dynamic pressure—meaning the coefficient of lift will be higher.

Flaps will also increase the amount of drag produced by the wing, particularly when extended a large amount. This explains why they are not permanent fixtures. We need to extend flaps to operate at lower airspeeds, and retract them in order to operate at higher airspeeds. Most flap designs have multiple settings to work with (e.g. – 10, 20, 30, and 40 degrees).

We have several different types of flaps to mention here. Each has pros and cons in terms of lift, drag, and complexity of design.

1) Plain Flap

– A section of the trailing edge is hinged so that it can pivot downwards. This is a more common type on general aviation aircraft.

The Plain Flap

2) Split Flap

– A section of the underside of the trailing edge is hinged to pivot downwards. The split flap produces a great deal of drag.

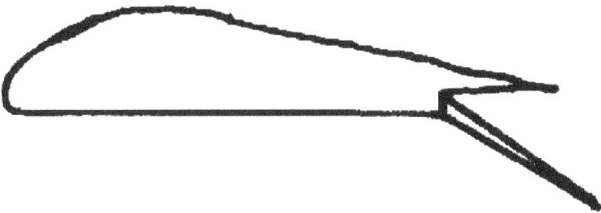

The Split Flap

3) Slotted Flap

– As the flap pivots down, a slot opens up between the flap and the wing. The airflow passing through this slot energizes the airflow over the wing to delay separation (more

about separation in the next chapter). This allows for a very large increase in the lifting capacity of the wing.

The Slotted Flap

4) Fowler Flap

– The fowler flap slides back on rails as it pivots downward. This increases the surface area of the wing as well as the camber. As we can recall from the lift equation, an increase in wing area will increase the lift of the wing—all other things being equal. The fowler flap adds complexity to the wing since it requires rails instead of a hinge.

The Fowler Flap

5) Flaperon

– On some aircraft, the ailerons will both move downward as the flaps are extended (or in place of the flaps). This configuration allows the ailerons to contribute to the lift as well as the roll control and is known as flaperons.

6) Leading Edge Flap

– Another type of flap (which is more common on larger aircraft) is the leading edge flap—also called the droop snoot. Leading edge flaps allow the camber of the wing to be increased even more. This allows the aircraft to fly at even lower airspeeds due to the increased lifting capacity. Once again, this type of flap will need to be retractable in order for the aircraft to operate at higher airspeeds.

The "Droop Snoot" - Leading Edge Flap

Aircraft can have any one or any combination of these flaps. Which type(s) are used depends on how much extra lift potential the aircraft needs and how complex the designers allow the aircraft to get.

Spoilers

Spoilers are devices which derive their name from the fact that they "spoil" lift. The surface projects above the wing to disrupt the airflow over the wing—causing premature separation of the boundary layer (Fig 5-26). This separation causes a significant reduction in the lift being produced by the wing. Spoilers will also produce some drag, but their primary purpose is to reduce lift.

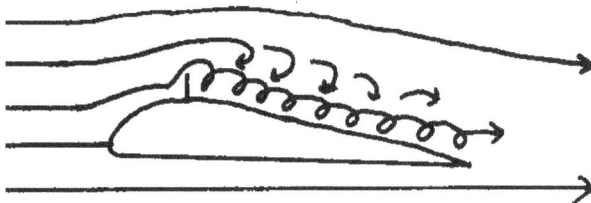

Fig 5-26. - *Spoilers* - Airflow over the top of the wing is disrupted to reduce the lift being produced.

The most common use of spoilers is to control the rate of descent (and hence the angle of descent).

Another common use of spoilers is to reduce the landing roll. Once an aircraft has landed, we usually want to stop in the shortest distance consistent with safety. However, the wings are still producing lift. Referring back to equations (1.18) and (1.19), we see that friction is reduced if the normal force is reduced. In this case, the normal force is the amount of weight resting on the wheels, so if the wings are producing lift, rolling friction will be reduced and our landing roll will be increased

due to the reduced effectiveness of our brakes. We can solve this problem by extending spoilers and reducing the lift, thus increasing the weight on the wheels.

On some aircraft, differential spoilers are used to control roll. When control column inputs are made, one spoiler stays retracted while the other is extended. This means that each wing is producing a different amount of lift, and the aircraft will roll. Obviously, the roll will be in the direction of the extended spoiler, since this wing experiences the loss of lift.

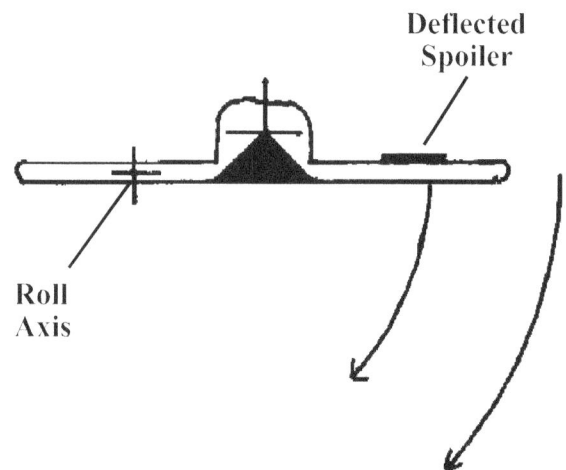

Fig 5-27. - *Spoiler Roll* - The roll axis shifts towards the high wing due to the loss of lift that accompanies the spoiler deflection.

Unfortunately, when a spoiler is used to control roll instead of ailerons, the aircraft doesn't roll about the longitudinal axis. The roll axis is instead shifted towards the up going wing (Fig 5-27). This can cause problems when maneuvering down low (i.e. – takeoff or landing) because the likelihood of a wing striking the ground is increased.

Speed Brakes

Speed brakes are intended to produce drag. Normally we don't want drag, so aircraft are designed to have the lowest amount of drag possible. However, in some scenarios (e.g. approach and landing) we actually want to increase the drag of the aircraft. This is especially true of aircraft that are extremely clean and fast.

As well, aircraft with turbofan engines often employ speed brakes (and other features) to increase drag on the approach. This is because high bypass turbofan engines are often very slow to power up. So during the approach, instead of reducing thrust like we would for other aircraft, we can increase drag to achieve the same effect. If an overshoot becomes necessary, we can reduce the drag much more quickly than we can increase thrust.

The idea behind speed brakes is pretty straightforward. A surface from some position on the aircraft is projected into the airstream to produce form drag. This form drag will either slow the aircraft down or increase the descent rate.

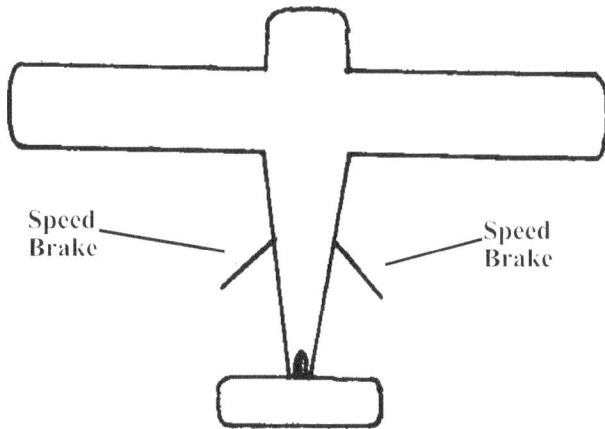

Fig 5-28. - *Speed Brakes* - Surfaces that extend into the airflow to produce large amounts of form drag. Speed brakes can be located in a variety of places.

Speed brakes (Fig 5-28) can be located in a variety of places. Generally, they are placed so that the disrupted airflow won't interfere with the operation of other surfaces such as control surfaces, stabilizers, wings, etc. This will not be the case if we have a speed brake/spoiler combination—in which case the brake will spoil lift as well.

Chapter Summary

The flight controls are the means by which we control the operation of an aircraft. They are generally divided into two main groups—primary and secondary controls. The primary flight controls are our method of manipulating the aircrafts orientation relative to the horizon (attitude). The secondary flight controls contribute to other aspects of control, such as variations in lift and drag potential.

The primary controls include the elevator or stabilator to control the pitching movement, the ailerons to control the rolling movement, and the rudder to control the yawing movement. The elevator or stabilator can be tail mounted or on the forward portion of the aircraft, in which case it would be referred to as a canard. Each of the primary controls acts as a surface that modifies the camber of an airfoil (either the wing, the horizontal stabilizer, or the vertical stabilizer).

The secondary controls include trim tabs, flaps, spoilers, and speed brakes, among others.

Trim tabs are intended to reduce loads on the control column or rudder pedals. The most common trim tab location is on the elevator, although some aircraft have trim tabs on the rudder and/or ailerons as well. Rudder trim is very common on multi-engine aircraft.

Flaps are used to increase the lift potential of the wings, as well as the drag of the wings. They will both move downward the same amount to increase the camber of the wings—thus increasing lift and drag. Their are several different types of trailing edge flaps, as well as leading edge flaps and flaperons.

Spoilers project into the airflow above the wing to disrupt it. This causes a reduction in the lift available from the wings. Spoilers can be used to control descent rate. As well, they are often used to reduce the landing roll by allowing all of the weight of the aircraft to rest on the wheels and thus improve braking.

Speed brakes are drag producers. They will project into the airflow at some point where the disrupted airflow will not interfere with some other aircraft component. When the surface is extended, it will produce form drag. This form drag allows the pilot to control speed and/or rate of descent.

Questions

1) What are the three axis' of the aircraft? Where are they on the aircraft? What are the movements about these axis'? What is the common point of the three axis'?

2) What is a primary control? Secondary control?

3) What are the three primary controls? What movements do they control?

4) How does an aileron vary the amount of lift produced by a wing?

5) What is the difference between an elevator, stabilator, and canard?

6) Explain the production of adverse yaw. What can we do to prevent or counteract adverse yaw?

7) How does ground effect affect the operation of the tail mounted elevator/stabilator? Why? Would this phenomenon occur with a canard? Why/why not?

8) What does it mean to be in coordinated flight?

9) The presence of yaw will make the aircraft roll. Why? Will this roll be in the direction of the yaw, or in the opposite direction? Why?

10) What are some causes of unwanted yaw?

11) What is the purpose of a trim tab? How does a trim tab accomplish it's function?

12) What are some different types of flaps? Explain the function of flaps.

13) What do spoilers do? Why are they used?

14) What do speed brakes do? Why are they used?

Chapter 6
Stalls and Spins

As we've seen from Chapter 2, lift produced by a wing varies according to EAS and angle of attack. As airspeed decreases, we need to increase the lift coefficient in order to maintain a constant amount of lift. This is done by increasing the angle of attack. As well, when we increase the amount of lift required without increasing the airspeed, we have to increase the angle of attack. Unfortunately, this can't go on indefinitely. We'll eventually reach a limit to the amount of lift potential we can get from a wing.

When we reach this limit, the lift coefficient will actually drop with an increase in angle of attack (Fig 6-1). This will always occur at the same angle of attack for a given wing and is known as a stall.

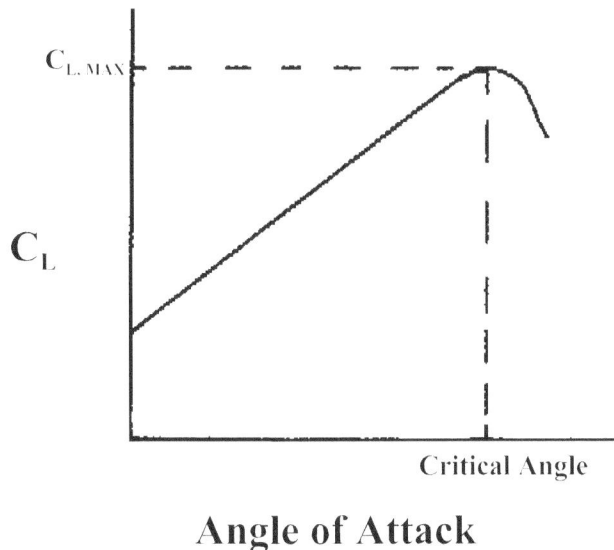

Angle of Attack

Fig 6-1. - *Lift Coefficient vs. AOA* - The lift coefficient will increase to a maximum point, and then decrease as the AOA is increased. The point at which this occurs is the critical AOA and the condition is known as a stall.

Lift

Let's quickly review how lift is produced. As we have seen from equation (2.9), lift is determined by the coefficient of lift (C_L), the dynamic pressure ($\frac{1}{2}\rho v^2$), and the wing surface area (S):

$$L = C_L \tfrac{1}{2}\rho v^2 S \qquad (2.9).$$

Dynamic pressure is determined by our equivalent airspeed, and the lift coefficient is determined by our angle of attack (for a given wing). As we change the value of variables in the equation, we have to change other variables as well to maintain a balance. For example, a decrease in (v) dictates an increase in (C_L) or a decrease in lift. If we want to maintain a constant amount of lift (enough to counteract weight for example), a decrease in (v) means that we must increase (C_L). Increasing the coefficient of lift means that we must increase the angle of attack.

Pressure Gradients

Recalling the continuity equation, we know that air flowing through a venturi tube experiences a drop in static pressure. Let's take a look at the pressure distribution inside a venturi tube while air is flowing through it (Fig 6-2). As we can see, there are areas of relatively high static pressure at the beginning and end of the tube, with the area of relatively low pressure in the middle. The point of lowest static pressure is at the narrowest point in the tube (cross section *b*).

The air in Fig 6-2 is flowing from left to right, so from *a* to *b*, the air is flowing from a region of high pressure to a region of low pressure. This is known as a favorable pressure gradient because air left to it's own devices will always flow from a high a low. From *b* to *c*, however, the air is flowing from a region of low pressure into a region of high pressure. This is known as an adverse pressure gradient since the air is flowing against the direction dictated by the basic laws of fluid motion.

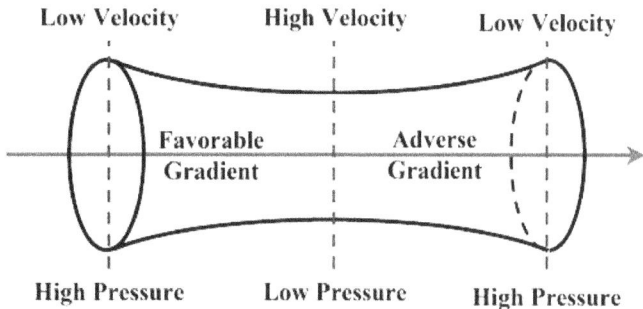

Fig 6-2. - Pressure Gradients in a Venturi Tube - Air flowing from a to b is in a favorable pressure gradient. Air flowing from b to c is in an unfavorable, or adverse, pressure gradient.

Now let's look at how this applies to a wing. For the moment we will ignore the airflow below the wing and concentrate on what is happening on top of the wing. For a wing at a low angle of attack (Fig 6-3), we have higher than atmospheric static pressure near the leading edge (*a*). As we move back towards the minimum pressure point on the wing, the static pressure is dropping (*b*). Moving beyond that point towards the trailing edge, the static pressure now increases back to the stagnation pressure (*c*). So, forward of the minimum pressure point on the wing we have a favorable pressure gradient, while aft of the minimum pressure point, we have an adverse

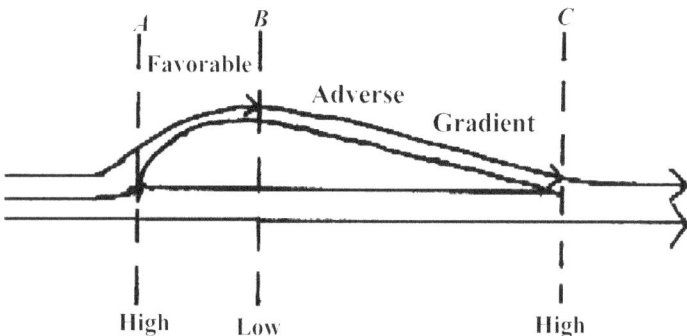

Fig 6-3. - Pressure Gradients on a Wing - Past the minimum pressure point, the air is flowing through an adverse pressure gradient—meaning that it is flowing from a region of low pressure into a region of high pressure.

pressure gradient. At low angles of attack, this is not a big problem, but as we will see, things get worse at higher angles of attack.

As we increase the wing's angle of attack, the stagnation point moves down on the leading edge and the minimum pressure point on the wing moves forward. Once again, the static pressure at the leading edge (*a*) is higher than atmospheric. But this time, the minimum pressure is lower than the minimum in Fig 6-3. As well, this minimum pressure point is further forward on the wing than with the lower angle of attack. At the trailing edge, the pressure has once again increased.

So we have a favorable pressure gradient on the forward portion of the wing again, with an adverse pressure gradient on the aft portion of the wing. As we can see from the diagrams, the minimum pressure at high AOA is lower than the minimum pressure at a low AOA. Also, the air has to travel further with an adverse pressure gradient at a higher angle of attack.

As the AOA is increased, this trend continues and the adverse pressure gradient gets stronger and longer. The high pressure at the trailing edge eventually overcomes the kinetic energy of the air and causes a flow reversal towards the rear of the wing (Fig 6-4). This flow reversal is known as boundary layer separation, and as the angle of attack is increased it moves further forward on the wing.

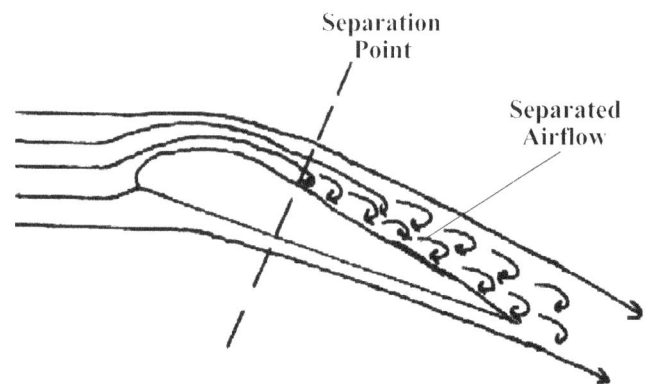

Fig 6-4. - Flow Reversal Due to the Adverse Pressure Gradient - This reversal of the airflow is known as boundary layer separation. The separation point moves forward as the AOA is increased.

Beyond the separation point, the boundary layer no longer follows the contours of the wing. Due to this, the lift produced after the separation

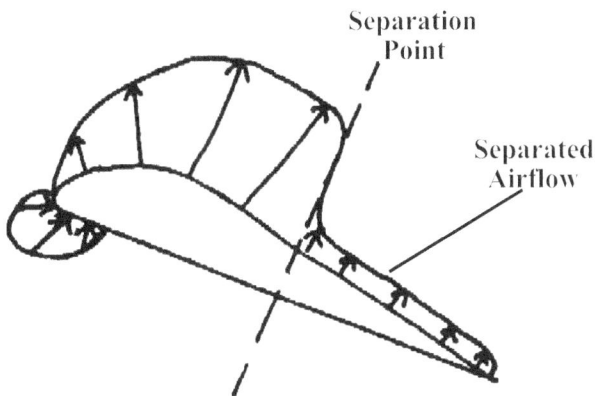

Fig 6-5. - *Lift Distribution Forward and Aft of the Separation Point.*

point is significantly reduced (Fig 6-5).

As the angle of attack is increased, the separation point moves forward and more of the aft section of the wing is affected by separated airflow. Eventually, so much of the wing is engulfed by separated and turbulent airflow that the wing can no longer produce sufficient lift to support the weight of the aircraft. This condition is known as the stall and will always occur at the same angle of attack for a given wing.

Due to the effects of the separated airflow, the wing forms a large wake at high angles of attack. This wake causes an increase in form drag. So at the stall, we get both a large decrease in lift and a large increase in drag.

Center of Pressure

In Chapter 2 we discussed the concept of the center of pressure and it's movement. We know that with an increase in angle of attack, the center of pressure moves forward and decreases the arm between itself and the aerodynamic center. We also know that for a given airspeed, an increase in angle of attack will result in an increase in lift. The combination of increased lift and decreased arm means that for a given airspeed, the pitching moment of the wing remains constant.

With a change in airspeed however, the pitching moment of the wing will change as well. A decrease in airspeed will cause a decrease in nose down moment and an increase in airspeed will cause an increase in the nose down moment.

This is all well and good until we reach the stall. At the stall, we experience a sudden decrease in lift combined with a sudden afterwards shift in the center of pressure. The net effect is that the nose down moment will increase as we enter the stall.

We can see in Fig 6-6a the center of pressure for a wing at a very low angle of attack.

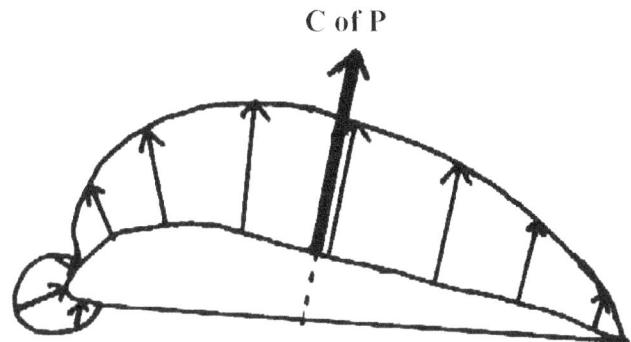

(a) Low Angle of Attack

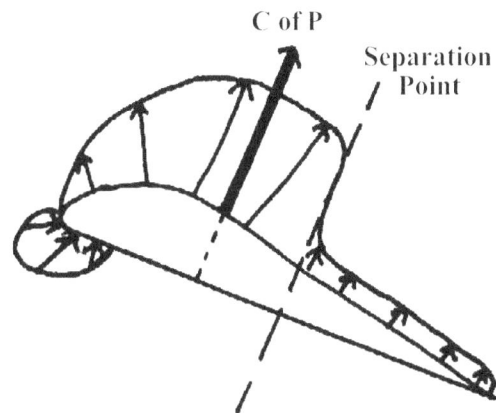

(b) High Angle of Attack

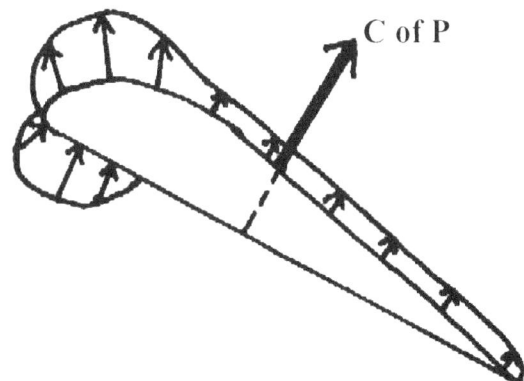

(c) Stalled

Fig 6-6. - *The Effect of AOA and the Stall on the Center of Pressure.*

Fig 6-6b indicates what we already know about the lift increase and center of pressure movement with an increase in angle of attack. Now, however, we can see why this movement occurs. The lift at the forward portion of the wing has increased due to the angle of attack, while the lift on the aft portion of the wing has decreased due to airflow separation. The net effect is an increase in lift with a forward shift in the center of pressure.

As we increase the angle of attack further, even more lift is produced on the forward portion of the wing and separation causes a reduction of lift on more of the aft portion of the wing. Eventually we get to the stall angle (Fig 6-6b). At this point, most if not all of the wing is in the separated flow. This means that very little lift is being produced by the wing, and the lift which *is* being produced is spread more or less evenly along the chord of the wing. The more even lift distribution causes the center of pressure to shift back at the stall.

Aircraft Behavior

There are several points of aircraft behavior which are applicable to stalls. The first and most critical effect is the aircrafts inability to maintain altitude. Loss of lift means that initially the aircraft will not be in equilibrium and a downwards acceleration will result. Eventually, equilibrium will be regained as the change in flight path inclines drag upward. Unfortunately this equilibrium cannot be attained until a descent is established, so loss of altitude is inevitable.

As we approach the stall, particularly if we approach it gradually, the aircraft will usually provide us with some warning of the impending stall. Airflow separation on the wing causes turbulent air to flow over the rear fuselage and tail section (Fig 6-7). This turbulent air creates a noise as well as a shaking of the aircraft known as "buffeting". On most aircraft buffeting will precede the stall slightly and will intensify in the stall.

In the previous section, we discussed the movement of the center of pressure during a stall and the effect that this movement has on the pitching moment of the wing. As the aircraft stalls, the nose down moment increases. This means that unless something happens to increase the nose up moment, the nose of the aircraft will drop at the stall. This nose drop will cause a reduction of angle of attack which will allow many general aviation

aircraft to recover from the stall spontaneously if allowed. This nose drop is increased by the effect that the horizontal stabilizer has as the aircraft begins to descend.

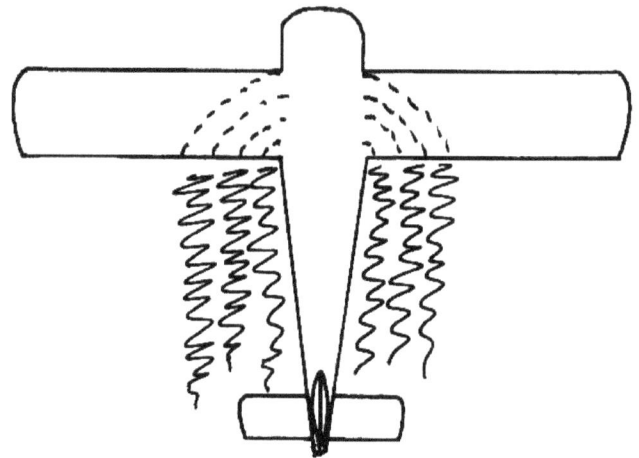

Fig 6-7. - Separated Airflow will Cause Airframe Buffeting.

So to sum up, we have three items of aircraft behavior that concern us during regular (non-aggravated) stalls. First and foremost is the loss of altitude in a stall. Next we have airframe buffeting which we can use as a warning or as a symptom of a stall. And last we have the pitching down of the nose which is associated with the stall. The details of each of these stall characteristics will vary from one aircraft type to another, but generally we will see all three of them with a stall.

Stall "Speed" Variables

The expression "stall *speed*" is very misleading. Unfortunately, we often relate the likelihood of a stall to the speed of the aircraft. In actual fact, the aircraft will always stall at the same *angle of attack*. This means that the stall speed will only remain constant for a *given set of conditions*.

For example, the published stall speeds (V_s) for most general aviation aircraft are based on the following conditions:

1) Maximum Gross Weight
2) No Vertical Accelerations (1g)
3) Power set to Idle or Zero Thrust
4) C of G in a Specified Location
5) Flaps Retracted
6) Landing Gear Extended

Any variations from the conditions which the speed is based on will cause a variation in the stall speed.

The stall speed in landing configuration (V_{so}) will be based on similar conditions, except the configuration will be that of landing (obviously) – i.e. flaps extended.

Weight – To tackle the conditions one at a time, let's begin with weight. If the aircraft is at a higher weight, it must be producing more lift in order to maintain equilibrium. Likewise, if the aircraft is operating at a lower weight, it must be producing less lift in order to maintain equilibrium.

Looking back to the lift equation,

$$L = C_L \tfrac{1}{2}\rho v^2 S \qquad (2.9),$$

we can see that a change in lift (L) dictates a proportional change in one of the variables on the right of the equation. The question is, which variable is it that we must change.

A careful look at the equation will show you that everything except v^2 is a constant. The aircraft will always stall at the same angle of attack, thus the lift coefficient at the stall will always be the same. The surface area of the wing (S) is a constant, and the air density (ρ) is a constant for a given altitude. So if we change the weight of the aircraft, we must change v^2 by an equal proportion. Considering this, we come up with the following formula,

$$\text{Stall Speed} = V_s \sqrt{(GW/MGW)} \qquad (6.1).$$

In this equation, (GW) is the current gross weight of the aircraft, and (MGW) is the maximum gross weight of the aircraft. So as we can see, a reduction in weight will result in a corresponding reduction in stall speed.

Load Factor – Load factor is a similar issue. Load factor is the lift being produced by the aircraft divided by the weight of the aircraft. We normally measure load factor in "g's", or multiples of gravitational acceleration. If an aircraft is "pulling g's", it is producing lift greater than the weight. Referring again to the lift equation, this increase in lift will require an increase in one of the other variables.

Once again, the aircraft will always stall at the same angle of attack—and thus at the same lift coefficient. So the only item remaining as a

variable is again v^2.

Manipulating the lift equation again, we come up with an equation similar to equation (6.1).

$$\text{Stall Speed} = V_s(\sqrt{n}) \qquad (6.2),$$

where (n) is the load factor in g's,

$$n = L/W \qquad (6.3).$$

We can see from equation (6.3) that for 1 g flight lift is equal to weight.

Load factors can be imposed on the airframe by a number of factors. They include pulling out of a dive or into a climb, turning, and turbulence. Each of these items will have an effect on the stall speed of the aircraft due to the loads they impart on the aircraft.

Power – The power that is being used during a stall will have an effect on the stall speed for a couple of reasons. First of all, the slipstream from the propeller induces airflow over the wing root (Fig 6-8). This reduces the angle of attack of the root, and it increases the effective airspeed of the root.

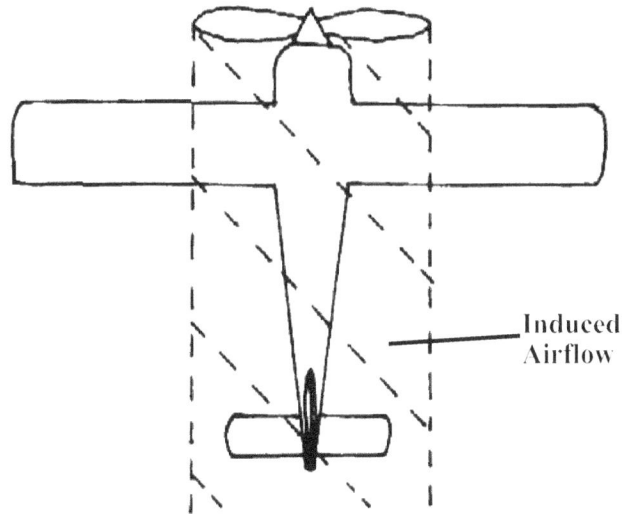

Fig 6-8. - *Slipstream Effect on the Wing Root* - Induced airflow over the root delays the stall and reduces the stall speed.

Combined with this effect, the thrust produced by the propeller will be inclined upwards relative to the flight path (Fig 6-9). This is due to the high angle of attack. So the propeller will actually take some of the load off of the wings—effectively reducing the lift required to be produced by the

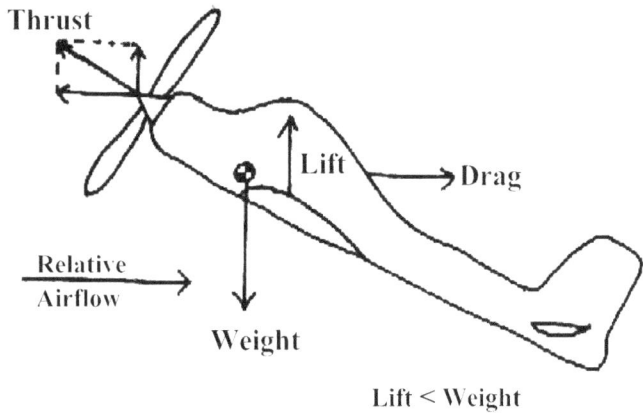

Lift < Weight

Fig 6-9. - *Inclined Thrust Line* - This inclination causes a component of thrust to contribute to the overall lift of the aircraft. This reduces the load on the wings and therefore reduces the stall speed.

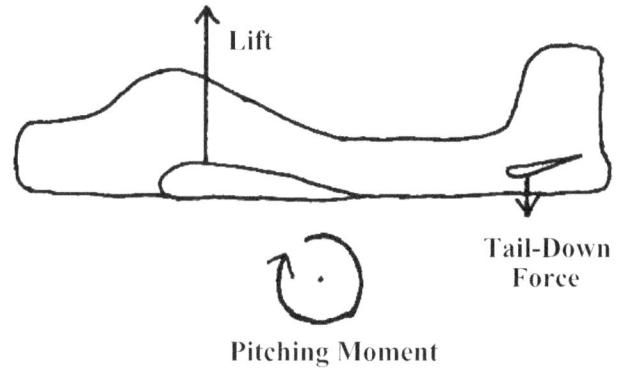

Fig 6-10. - *Pitching Moment Produced by the Tail.*

wings.

These two effects will reduce the stalling speed of the aircraft, but they also have undesirable effects. The slipstream delays the stall at the wing root, causing the wingtips to stall at the same time as the roots, or possibly even prior to the roots. It is actually desirable to have the wing root stall first so that lateral (roll) control can be maintained in the stall. Combined with this problem are the left turning tendencies of the propeller. If power is carried into a stall, the aircraft could yaw to the left. The problem with this will be discussed in more detail later, but yaw in a stall could lead to the development of an autorotation.

Center of Gravity – The next variable to consider is the position of the C of G. Recall from Chapter 1 our discussion of the weight and lift couple. This couple causes a nose down pitching moment which is balanced by the thrust and drag couple. Unfortunately, the thrust/drag moment will vary with speed and power setting. As well, the lift/weight moment will vary with weight and C of G location. So there must be another moment produced to maintain a proper balance. This moment is formed between the lift of the wing and the down force acting on the tail (Fig 6-10).

Normally, we consider the lift required to be equal to the weight. This is conveniently simple, and for most discussions it is accurate enough. However, the lift required is in fact slightly greater than the weight of the aircraft due to the down force on the tail. This down force is added to the weight

of the aircraft in order to determine the actual lift required.

Remember from Chapter 1 that a moment is equal to the force applied multiplied by the arm. In the case of the lift/weight moment, the force is the weight, and the arm is the distance between the C of G and the wing's AC (Fig 6-11). As for the tail down moment, the force is the tail down force, and the arm is the distance between the wings AC and the tail's AC.

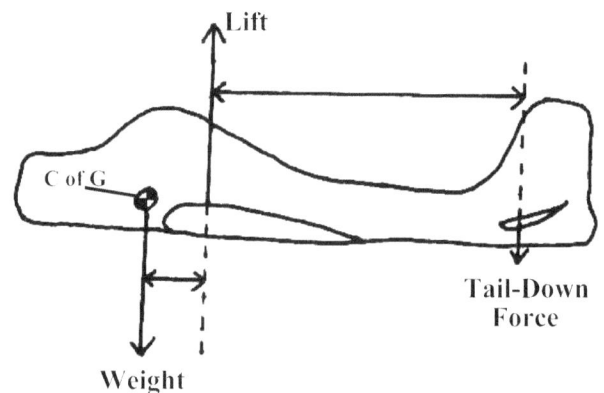

Fig 6-11. - *Moment-Arms on an Aircraft.*

We can consider for the time being, an aircraft that is kept at a constant weight. If the C of G is moved forward, the nose down moment due to the lift/weight couple is increased due to the increased arm. The arm for the tail down moment, however, cannot be increased—the wings and tail (and thus the two AC's) are fixed in place. So in order to increase the tail down moment, we have to increase the tail down force. Since lift required is the sum of the weight and the tail down force, shifting the C of G forward will therefore cause an increase in lift required. As we have already seen, an increase in lift required will result in an increase in stall speed.

You may take notice that if the C of G was to be placed *behind* the AC, the tail would have to produce positive lift—thus *reducing* the load on the wing and reducing the stall speed. This is in fact true. As well, placing the C of G aft will improve other performance figures (rate of climb, range, etc.) as we will see in Chapter 8. There are problems, however, associated with an aft C of G. The main problem is one of stability, and will be discussed in detail in the next chapter.

Notice that for the sake of simplicity, the wing's pitching moment about the AC was omitted from the above discussion. This will have an effect on the stall speed as well, but the trend resulting from C of G movement is what we wish to see here.

Configuration – The aircraft configuration will have a direct bearing on the stall speed. The most common method for changing configuration is the use of flaps. As we've seen from Chapter 5, flaps increase the camber of the wing, thus increasing the maximum coefficient of lift. Looking to the lift equation again, we can see that an increase in lift coefficient will result in a decrease in velocity. Thus extending flaps will decrease the stall speed for a fixed amount of lift.

Spoilers, on the other hand will increase the stall speed since with them extended less of the wing is contributing to lift.

Landing gear can have varying effects on the stall speed. On some aircraft, the gear will have no effect at all. While on others the gear may increase or decrease the stall speed.

Altitude Effect

As we climb higher in the atmosphere, the air gets less dense. As a result of this, we must travel faster to experience the same dynamic pressure ($\frac{1}{2}\rho v^2$). So it would appear initially that if all other things are held constant, our stall speed increases with altitude. This is true if we are referring to our *TAS*. If on the other hand we are referring to our *EAS*, this statement is not true.

Recall from Chapter 2 that our EAS is a function of our dynamic pressure—which is a function of density as well as speed. Therefore,

even though our *true* stalling speed is increasing, our *equivalent* stalling speed is not changing with altitude.

The difference is important since airspeed control is usually accomplished with reference to the EAS/CAS. This is done due to the fact that aircraft performance speeds (including the stall) depend on dynamic pressure. To a pilot the use of TAS is usually restricted to navigation problems.

Once again, most low speed general aviation aircraft will deal with CAS instead of EAS. This is because the difference between the two at low speed and low altitude is minute.

Stall Warning Devices

When stalls occur inadvertently, they generally occur at low altitudes. Reasons for this will be discussed in the *Critical Areas* section of Chapter 19. Because of the low altitude usually associated with "surprise" stalls, the loss of altitude can be critical. This fact can be reiterated simply by reading the *many* accident reports in which a pilot found his/her self in an inadvertent stall at low altitude. Unfortunately, many of these accidents end with fatalities.

Fairly early in the history of aviation it was recognized that low altitude stalls were not good for ones flying career. This fact, combined with the commonality of stall/spin accidents, lead to the development of devices which would warn the pilot of an impending stall. These devices invariably took advantage of the fact that a wing will always stall at the same angle of attack. Remember that with a change in angle of attack, we also get a change in the position of the stagnation point. It is this movement of the stagnation point that allows us to detect a stall before it occurs.

One common type of stall warning is the switch or vane type. At low angles of attack, the stagnation point is above a vane on the leading edge of the wing. The flow of air downward over the vane holds it down, which in turn keeps an electrical circuit open. Once the stagnation point moves down below the vane (due to an increase in angle of attack), the airflow pushes the vane up. This closes the electrical circuit and activates an alarm in the cockpit.

Another type of stall warning is set up with a hole in the leading edge of the wing that has or two reeds behind it. With the stagnation point above the hole, air simply flows over the opening with no effect. As the stagnation point moves onto the opening, the increase in pressure forces air through the reeds which then buzz. The buzzing noise is then carried to the cockpit via tubes and connections.

Yet another type of stall warning device is an Angle of Attack Indicator. Aircraft that have some relatively undisturbed airflow over some portion of the airframe (usually on the fuselage) can have a probe placed in the relative airflow. This probe rotates to align itself with the relative airflow, and sends a signal to an instrument in the aircraft. This instrument displays the angle of attack and can be used for a variety of functions, including a stall warning.

Many larger aircraft have angle of attack indicators that automatically activate a "stick shaker" and/or a "stick pusher" at some critical point approaching a stall. The stick shaker is a more prominent warning than a light or buzzer, and the stick pusher will physically force the pilot to recover from the situation.

Planform Effects

The effect of the planform shape on the stall characteristics of a wing are very important. Every stall starts at some point on the planform and then spreads out from there. Where the stall begins and how it spreads has a large effect on the behavior of the aircraft in the stall. Planform effects are due to the three dimensional characteristics of the airflow

over the wing.

Ideally, we want a wing which stalls at the root first with the stall spreading from there (Fig 6-12). The reasons for this are warning and controllability. When the root stalls, the buffeting gives us advanced warning to the stall of the entire wing. As well, as we enter the stall, the ailerons are not immersed in separated airflow. This means that lateral (roll) control is still effective in the early stages of the stall.

If the wing tips were to stall first (Fig 6-13), noticeable buffeting would probably not occur. Combined with this lack of warning, any aileron input would be ineffective since the ailerons themselves would be stalled. Aileron deflection may actually induce a roll in the direction opposite the input. The reason for this will be discussed later in the chapter.

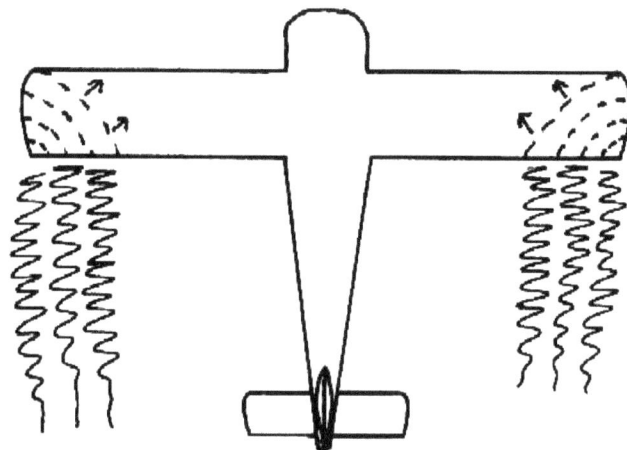

Fig 6-13. - *Unfavorable Stalling Pattern* - If the wingtips stall first, there will be no warning, and control difficulties may occur.

Wings that have a rectangular planform tend to stall at the wing root first, with the stall progressing outward from there (as in Fig 6-12). Unfortunately, this cannot be said of all wing planform shapes. Figures 6-14 illustrates the stall patterns associated with different planforms. These stall patterns are based on the assumption that no other design feature has been incorporated to alter the pattern to a more favorable one. Some of the design features that *could* be used to make corrections are discussed in the next section of the chapter. In most cases, manufacturers strive to produce a favorable stall pattern in order to improve the safety and reliability of the aircraft.

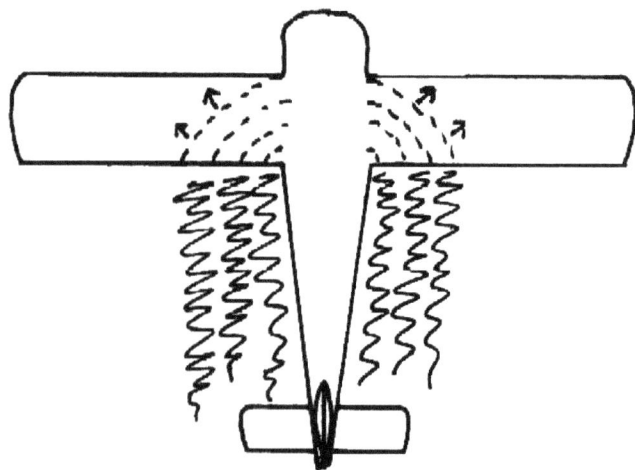

Fig 6-12. - *Favorable Stall Pattern* - When stalls begin at the wing root, they provide some warning and some controllability in the early stages.

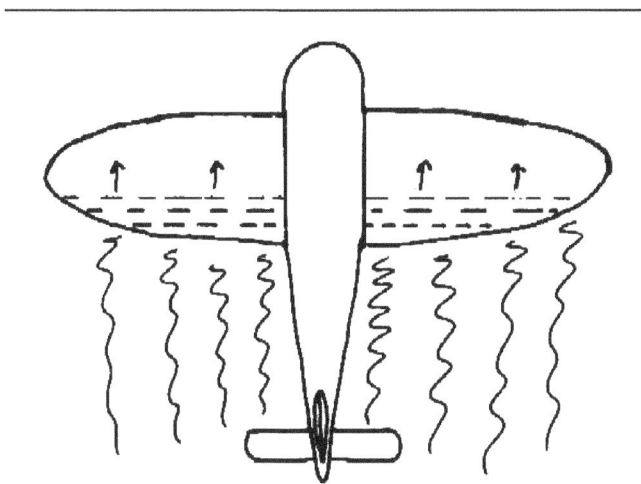

(a) An elliptical wing stalls at the root with the stall pattern expanding forward.

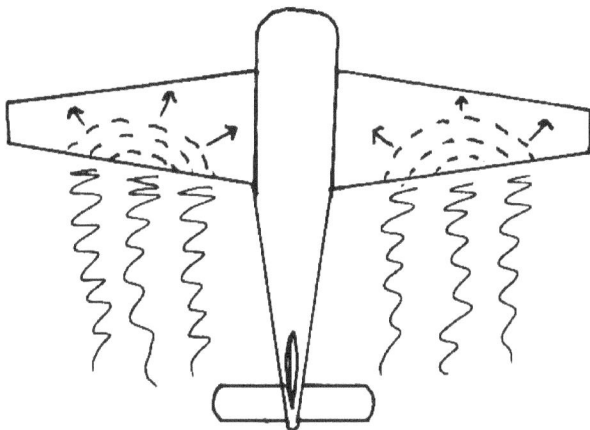

(b) A tapered wing stalls at approximately mid-span with the stall pattern expanding outward.

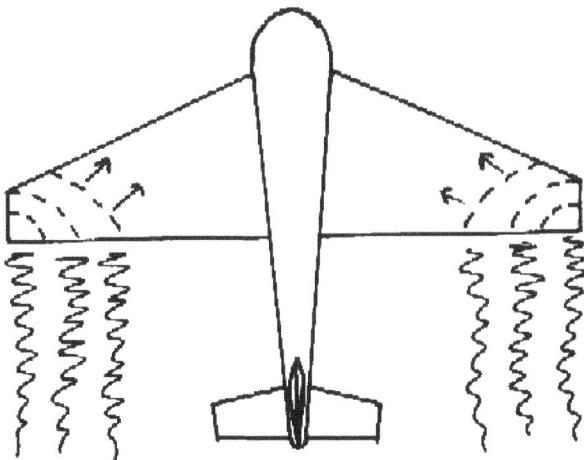

(c) A delta wing stalls at the wingtips first with the stall progressing inward.

Fig 6-14. - *Planform Effects on Wing Stalling Patterns.*

Washout and Stall Strips

If the stall characteristics of the chosen planform are not favorable to pre-stall warning or pre/post-stall control, other design factors can be considered. The factors often used are washout, aerodynamic washout, and stall strips.

A wing with washout is twisted forward so that the wing tips are at a lower angle of incidence than the wing roots. This means that in flight, the wing tips are at a lower angle of attack than the roots. The stall is therefore delayed at the tips, allowing for lateral control in the early stages of the stall.

Aerodynamic washout—also known as aerodynamic twist—has a similar function. Instead of physically twisting the wing, however, a different airfoil shape is used at the wing roots and tips. The airfoil used at the tips will have a higher critical angle than the airfoil used at the roots. This causes the wing roots to stall prior to the tips which, as discussed previously, is a favorable stalling pattern.

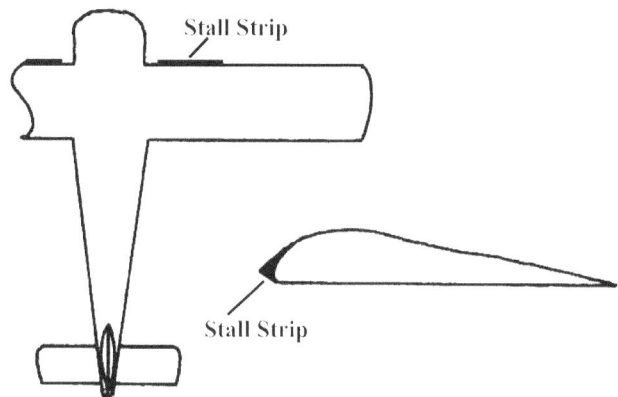

Fig 6-15 - *Stall Strips.*

Stall strips (Fig 6-15) also serve to stall the wing root before the tips. This time, however, the technique used is to induce premature airflow separation over the wing roots. Because the stall strips involve a pointed leading edge, sharp directional changes are required of the airflow. At high angles of attack (approaching the stall), these directional changes cause the airflow to separate. At the wing tips, however, the absence of stall strips allows the wing to continue flying to a higher angle of attack. This allows for a warning of the impending stall as the root stall causes buffeting, and it allows for lateral control in the initial stall.

Boundary Layer Control

Under certain conditions of flight, such as takeoff and landing, we want the aircraft to be able to fly as slowly as possible. Lower takeoff and landing speeds improve an aircrafts versatility since they allow the aircraft to operate in and out of shorter airstrips. Unfortunately, the lower end of the speed envelope for an aircraft can be higher than we want. A number of solutions have been developed to help solve this problem (including flaps, which we have already discussed). Boundary layer control is one of these solutions.

The idea behind boundary layer control is to prevent or delay separation of the boundary layer by re-energizing it. This can be done by a number of methods, including vortex generators, slots and slats, and the suction method.

Vortex Generators – Vortex generators (Fig 6-16) are miniature "wings" which are placed on the upper surface of the wing at an angle to the freestream airflow. These "wings" are just large enough to reach out of the boundary layer and into the freestream. Vortex generators act just like wings in that they produce wing tip vortices. These vortices cause a mixing action between the boundary layer and the freestream, and energy from the freestream gets transferred into the boundary layer—thus delaying separation.

Vortex generators will produce a small amount of parasite drag. However, they will also prevent drag by delaying separation of the boundary layer (recall from Chapter 3 that this will reduce form drag). As well, vortex generators will increase the lifting capacity of the wing by allowing it to fly to higher angles of attack.

Slots and Slats – Slots and slats both serve the same function. The difference is that slots are fixed in place, while slats are movable.

Slots, and slats when extended, produce a gap near the leading edge of the wing (Fig 6-17). At low angles of attack, air simply flows over the gap. At high angles of attack, however, the high pressure air below the wing is forced through the opening. This high energy air adds energy to the the boundary layer on top of the wing. The added energy allows the boundary layer to adhere to the wing surface for longer, thus separation is delayed.

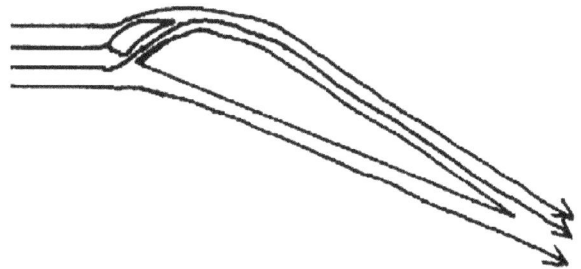

Fig 6-17. - Slots (or Extended Slats).

Often, control surfaces are slotted to improve their effectiveness. This can be done with any of the primary flight controls, but it is more common on flaps. Slotted flaps have increased effectiveness since high energy air is guided over the top of them. This prevents airflow separation and increases the lifting capacity of the flapped wing.

The Suction Method – With the suction method of boundary layer control, spanwise vents along the upper surface of the wing are used to suck the low energy air away from the boundary layer (Fig 6-18). This air is then replaced by the high energy air from the freestream or the upper boundary layer.

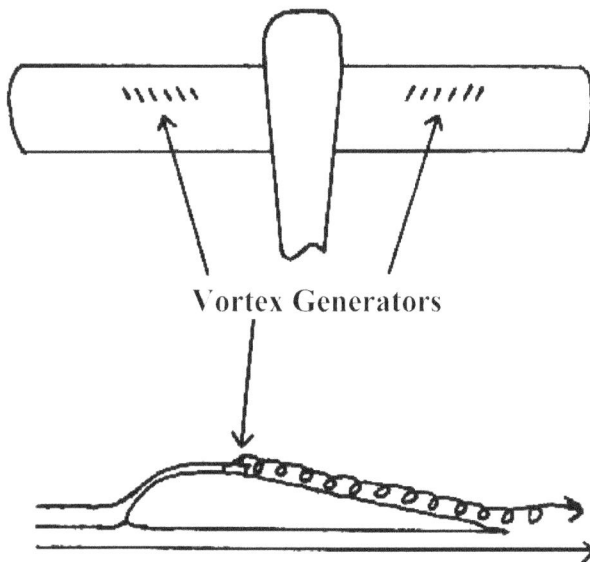

Vortex Generators

Fig 6-16. - Vortex Generators - The mixing action created by vortex blends the boundary layer air with the freestream air. This re-energizes the boundary layer and delays separation.

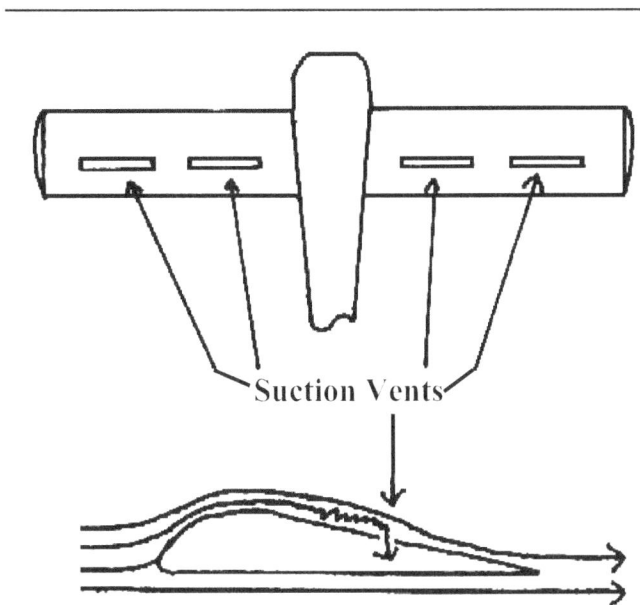

Fig 6-18. - The Suction Method of Boundary Layer Control.

Roll and Yaw

The presence of roll and/or yaw during a stall can be very unsettling. If unchecked, they will usually lead to a condition known as auto-rotation—or a spin.

If roll is introduced during a stall, the two wings will have relative airflows from different directions (Fig 6-19). The downgoing wing will have a relative airflow from below, while the upgoing wing will have a relative airflow from above. This results in the downgoing wing experiencing an

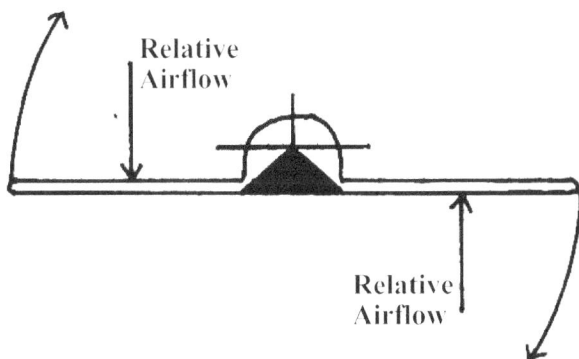

Fig 6-19. - The Effect of Roll on AOA - During a roll, the downgoing wing will experience an increase in AOA, while the upgoing wing will experience a decrease. In a stall, this aggravates the stall and may lead to a spin.

increase in angle of attack—deepening the stall. Meanwhile, the upgoing wing experiences a reduction in angle of attack—reducing or even fully recovering the wing from the stall.

If yaw is introduced during a stall, the yaw will result in a roll as we discussed in the previous chapter. The result of this roll will also be the deepening of one wings stall.

As the downgoing wing gets deeper into the stall, two thing happen. First of all, it experiences a greater loss of lift—thus it drops even more. Second of all, there will be an increase in drag, causing further yaw of the aircraft. Meanwhile the exact opposite is happening with the upgoing wing (Fig 6-20). The decreased angle of attack causes a reduction of drag and an increase in lift.

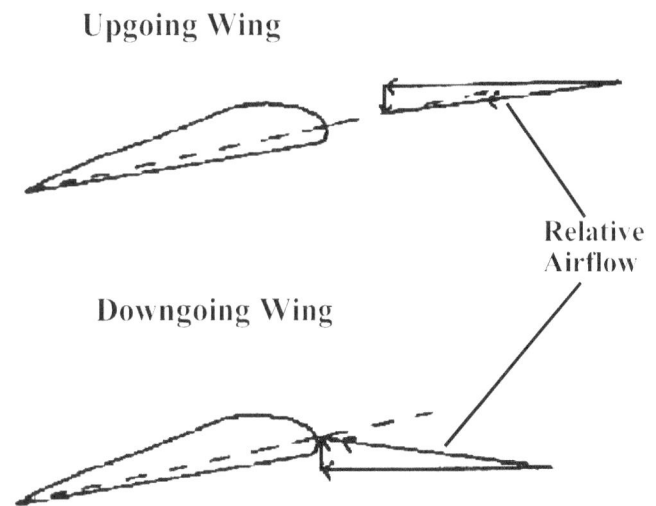

Fig 6-20. - The AOA Differences During Roll - When an aircraft is rolling, the upgoing wing has a lower AOA than the downgoing wing. During a stalled condition, this can lead to an unstable roll that results in a spin.

The net effect on the aircraft is for it to enter an initially unstable rotation which gets progressively worse until all of the aerodynamic and inertial moments balance. This is the beginning of a spin.

The Use of Ailerons

As mentioned earlier, if the wingtips stall first, ailerons will be ineffective. They may even be reversed due to the stall. This is due to the effect that aileron deflection has on angle of attack.

If a roll input is made at any time, the downgoing aileron will cause an increase in angle of attack while the upgoing aileron will cause a decrease in angle of attack (Fig 6-21). This means that in a stall, the wing with the downgoing aileron experiences a loss of lift due to the increased angle of attack. So instead of rising as it should, that wing will drop. Aggravating this situation is the fact that the more stalled wing will produce more drag. The imbalance in drag will cause a yaw towards the lowering wing. Remember that the secondary effect of yaw is roll, meaning that the roll becomes worse as yaw is introduced.

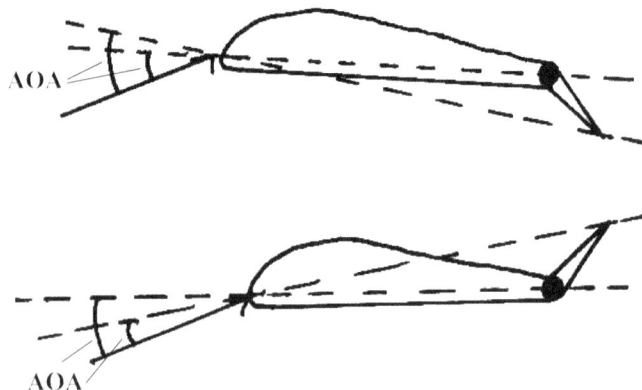

Fig 6-21. - *The Effect of Aileron on AOA* - If the stall is deep enough to affect the ailerons, an input to roll may reverse due to the stall. The downgoing aileron—which is supposed to increase lift—will deepen the stall and lead to a loss of lift on that wing. For the upgoing aileron, the opposite is true.

In the previous section, we saw the result of roll and yaw during a stall. In the event of an aileron input, some aircraft will spin in the direction opposite the direction of the input.

In the case of an aircraft that stalls at the wing roots first, however, this effect will not be as pronounced. In some cases it will be non-existent. This having been said, aileron drag is much more pronounced near or in the stall. This is because aileron drag is caused by induced drag on the downgoing aileron. At high angles of attack, induced drag is predominant, thus aileron drag is significant. So a roll input can cause a significant amount of yaw, which may lead to a spin, even in an aircraft that stalls at the wing roots first.

Climbing and Descending Turns

In a level turn, both wings have the same angle of attack. This means that they will stall at the same time and no roll or yaw will be introduced. For aircraft with dihedral or anhedral in climbing and descending turns, however, the wings no longer have equal angles of attack. So one wing will stall prior to the other, inducing a roll which can lead to a spin. Dihedral is when the wings are located with their wingtips above the wing roots (Fig 6-22a). Anhedral wings (Fig 6-22b) have the wingtips located below the wing roots. Reasons for dihedral and anhedral will be discussed in Chapter 7.

(a) Dihedral wings

(b) Anhedral wings
Fig 6-22. - *Anhedral and Dihedral Wings.*

For a dihedral aircraft in a climbing turn, the outside wing will stall first, leading to a spin to the outside of the turn. In a descending turn, on the other hand, the inside wing stalls first. This leads to a spin to the inside of the turn. For anhedral aircraft, this will be reversed. That is, in a climbing turn the aircraft will spin to the inside, while in a descending turn the aircraft will spin to the outside of the turn.

To understand why this is happening, it is important to remember that stalls occur as a result of exceeding the critical angle of attack. When the wings are experiencing different angles of attack, the one with the greater angle will stall first or will stall more. This results in a wing drop which eventually leads to a spin.

For a dihedral aircraft, the wing to the outside of the turn is at a higher angle of *bank* than the inside wing (Fig 6-23). If the aircraft is climbing, there is a component of the relative airflow that's coming from above. Because of this, the angle of attack is reduced. With straight wings or with the wings level, this reduction in angle of attack is easily corrected for by pitching the aircraft up. However, with a banked, dihedral aircraft, the downward component of the airflow is different on each wing. This results in a difference in angle of attack that cannot be prevented. The downward component

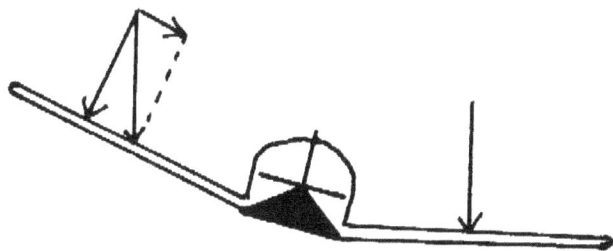

Fig 6-23. - *Dihedral in a Climbing Turn.*

of airfow is greater on the low wing, thus the low wing has a lower angle of attack.

For a descending turn, the relative airflow is now coming from below (Fig 6-24). This still results in different angles of attack for each wing for the same reason. This time, however, the angle of attack is being increased by the descent. The increase is more pronounced on the low wing, thus the low wing now has a higher angle of attack.

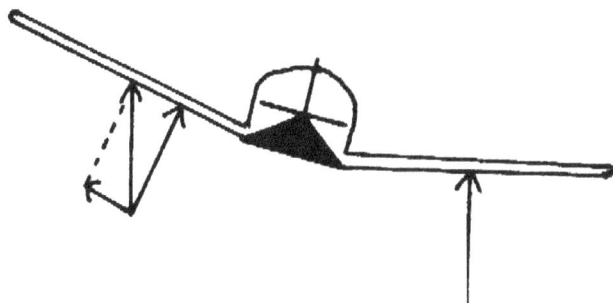

Fig 6-24. - *Dihedral in a Descending Turn.*

So, we can see that a change in angle of attack due to a climbing or descending turn will be more pronounced on the wing which is inclined less to the horizon. For a straight winged aircraft, this means there will be no difference. For a dihedral aircraft, however, this means that in a climbing turn the outside wing will stall first while in a descending turn, the inside wing will stall first. For an anhedral aircraft, on the other hand, the opposite will occur. In a climbing turn the inside wing will stall first and in a descending turn, the outside wing will stall first.

Stages of a Spin

A spin has three separate stages to consider. The first is the incipient stage, next is the fully developed stage, and next is the recovery stage.

During the incipient stage, the spin is still developing. Pitch, yaw, and roll rates are changing from zero in straight and level to the rate which will be maintained in the spin. As well, in the incipient stage, the aircraft will still have some forward motion—the spin axis will not be vertical yet.

As the pitch, yaw, and roll rates stabilize, the aircraft enters into the fully developed stage of the spin. At this point, the aerodynamic moments and the inertial moments will balance out. During the fully developed stage, the rotation of the aircraft becomes repetitive and cyclic. The spin axis in the fully developed stage is vertical—the aircraft no longer has any forward motion over the ground.

The fully developed stage of the spin is not what is commonly referred to as a fully developed spin in training circles. In flight training, a spin with a full rotation is usually referred to as fully developed. However, it usually requires two to three rotations before the spin is actually fully developed. Few students will see a full spin unless they do aerobatic training.

A recovery can be initiated from the incipient stage or from the fully developed stage. The pilot must introduce control inputs which will stop the rotation of the spin and then unstall the aircraft. This means upsetting the balance of aerodynamic and inertial moments that are maintaining the spin.

Spin Recognition and Recovery

Recognition of the spin is fairly simple. Rotation rates will be high and the aircrafts rate of descent will be high. In most aircraft, the nose will be low—although pitch attitude does vary from one aircraft type to the next. The key identifying feature of the spin is the low airspeed. Even though the aircraft is descending rapidly and is in a very nose low attitude (usually), the airspeed remains at or about the stall speed. The low airspeed is due to the high amounts of drag being produced by the stall.

Spin recovery procedures will vary from one aircraft to the next. The most common procedure for general aviation aircraft is to apply rudder opposite the direction of rotation. The use of ailerons is avoided because of their unreliability during a stall and the possibility that they can make the spin worse and/or delay the recovery. The rudder stops the rotation of the spin, while forward pressure on the control column is then used to lower the AOA and break the stall.

For some aircraft, the behavior of ailerons in a stall is used to recover from the spin. If an aileron input is made *into* the spin, the upgoing wing will then experience an increase in drag and a decrease in lift. This will tend to yaw and roll the aircraft to the outside of the spin—thus slowing or stopping the rotation. Forward pressure will then be needed on the control column to recover from the stall.

These recovery procedures will vary from one aircraft to the next. The best source of accurate information for the aircraft type is the Pilots Operating Handbook or the Flight Manual.

Stall and Spin Recovery Factors

There are several factors which will have an effect on the stall and spin recovery characteristics of an aircraft. They include, but are not limited to, center of gravity position, rudder effectiveness due to size and airflow, inertial moments due to weight distribution (not the same thing as C of G), weight, and power.

Center of Gravity – Any aircraft that are certified for intentional spins will have very specific center of gravity limits published. There are several reasons for this, most of which will be discussed at various points in this book. For the purpose of spins, it is an issue of the moment arm of the rudder.

Recall from Chapter 5 that the rudder rotates the aircraft around the vertical axis. This axis (as well as the other two) passes through the C of G. We can see from Fig 6-25 that this makes the rudder a sort of lever. The force available to be produced by the rudder is a constant for a given speed, so the effectiveness of the rudder will depend on the moment arm available.

With a longer arm, the yawing moment created by the rudder will be stronger, even though the force produced by the rudder is the same. This stronger moment will be more likely to overcome the aerodynamic and inertial moments which are maintaining the spin. A shorter arm will reduce the moment provided by the rudder, thus reducing the chances of recovery.

A longer rudder arm will be provided by a more forward C of G, while a shorter rudder arm would be the result of a more aft C of G. This is

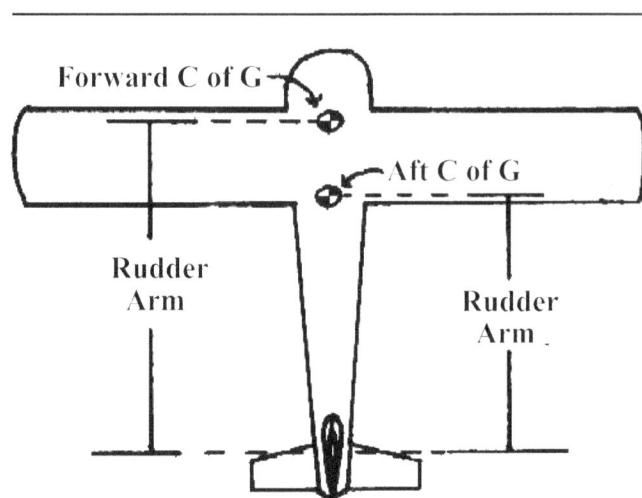

Fig 6-25. - *The Effect of C of G on the Rudder* - With an aft C of G, the rudder arm is shortened and thus the rudder is less effective. With a forward C of G, the rudder arm will be longer and the rudder will be more effective.

why aircraft with multiple weight and balance envelopes will be certified for intentional spins in the forward portion of the envelope, but not the aft.

The C of G position also affects the moment arm of the elevator in the same way as the rudder. This means that a forward C of G will make it easier to recover from the stall as well.

Rudder Effectiveness – The effectiveness of the rudder is a question of how much of a moment the rudder can create to yaw the aircraft. The force produced by the rudder is determined by a number of factors, including airspeed, the size of the rudder, and the deflection angle of the rudder.

This means that, all other things being equal, an aircraft that has a relatively high stall speed (thus a relatively high airspeed in the spin), a large rudder, and a large amount of rudder deflection will *generally* be more likely to recover from a spin. Remember that even with a large force on the rudder, we still need an arm between the rudder and the C of G for that force to be effective.

When flaps are extended—particularly on high wing aircraft—the turbulent air coming off of the flaps can "blanket" the rudder and reduce it's effectiveness (Fig 6-26). As well, flaps produce more drag, especially on the wing that is more stalled. This may prevent the rudder from recovering the aircraft from the spin while flaps are extended.

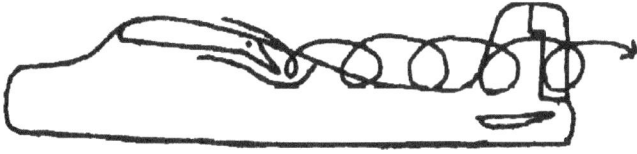

Fig 6-26. - Flaps Can Have an Adverse Effect on the Rudder.

Weight Distribution – The inertial moment of the aircraft determines how badly the aircraft—once established—wants to remain in the spin. Inertial moments are determined by the weight distribution. Be careful to note that weight distribution is not the same thing as C of G. To illustrate the difference, consider two rods with weights attached to the ends (Fig 6-27).

The total weight of the two rods is the same, and the C of G is at the center point of the rods. Both rods are rotating at the same rotational rate (RPM). The difference between the two is that rod *a* is only half the length of rod *b*. So even though the two rods have the same weight and C of G, the weight is distributed differently. Rod *b* will have a much higher amount of rotational inertia due to the higher speed of the weights.

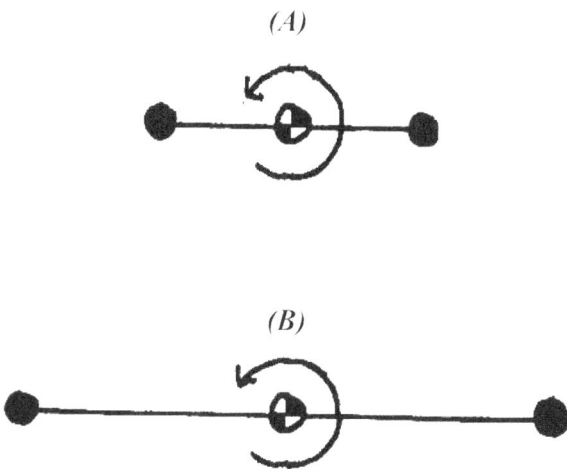

(A)

(B)

Fig 6-27. - The Effect of Weight Distribution - With the weight of an aircraft spread out, the rotational inertia of an aircraft will be increased at a given rotation rate.

Because of rotational inertia, an aircraft with wingtip fuel tanks will be more difficult to recover from a spin than the same aircraft with fuel tanks in the wing roots. With tip tanks, the weight is spread out more. So even though weight and C of G are the same for the two aircraft, the tip tanks increase the rotational inertia.

Moments of rotational inertia are one of the reasons that twin engine aircraft are not usually certified for spins. The mass of the engines out on the wings increases the inertial moments of a twin so much that spins are often unrecoverable.

Weight – Just as the weight *distribution* affects the moments of inertia, so does the actual *amount* of weight. Comparing two aircraft, if they both have the same C of G and weight distribution (proportionally), the heavier aircraft will have a stronger tendency to remain in the spin. For this reason (as well as the reason of load factor limits – see Chapter 9), aircraft certified for intentional spinning will often have a maximum allowable weight for spins.

Power – The use of power in a spin can have varying effects. The effects that occur will depend on how much power is used, and various design characteristics of the aircraft.

Generally speaking, power will increase the rotation rate of the spin. As well, it may "flatten" the attitude of the aircraft in the spin—that is, the pitch attitude will be tending more toward nose high.

Power can also lead to problems with the recovery from the spin. After the rotation stops and the angle of attack is reduced, the aircraft is usually in an extremely nose down attitude. This nose down attitude leads to a rapid build up of airspeed and loss of altitude. If power is carried or added during this phase of the recovery, the airspeed will build more quickly and may approach or exceed the Never Exceed Speed (V_{ne}). As well, with power applied and such a high airspeed, the propeller RPM could possibly exceed "redline".

Chapter Summary

With variations in the airspeed of an aircraft and the amount of lift being produced, the AOA will vary. A decrease in speed or an increase in the lift will dictate an increase in AOA. This increase in AOA increases the wings lift potential (or lift coefficient). Unfortunately, the AOA cannot be increased indefinitely. We eventually reach a limit where lift will *decrease* with an increase in airspeed. This is known as a stall and occurs at the critical

AOA.

On the aft portion of the top of a wing, there is an adverse pressure gradient which opposes the rearward movement of air. This adverse pressure gradient becomes worse at higher angles of attack and eventually causes the boundary layer to separate from the surface of the wing. The region of the wing in the separated flow only contributes minutely to the production of lift. As the AOA is increased further, the separation point moves forward until most or all of the wing is engulfed in separated flow. When the critical AOA is reached, the lift coefficient will drop due to the effect of the separated flow. This is the stall.

As the stall is approached, most aircraft will give warning signs to the pilot. The most significant pre-stall warning is the buffeting of the airframe due to the separated airflow striking the fuselage and tail. In the stall, the buffeting will intensify and the aircraft will lose altitude and pitch forward.

The airspeed at which the stall occurs will vary according the flight conditions since it will occur at a constant AOA. Changing the weight, C of G, load factor, or power setting will have a direct effect on the stall speed. For this reason, published speeds should be interpreted with due consideration to these variables. As well, a change in configuration will have a direct impact on the stall speed due to the change in the maximum lift coefficient available from the wing. Altitude will affect the *true* stalling speed due to changes in air density, but the *equivalent* airspeed for a given set of conditions will not change.

The planform shape of the wing will determine where the stall begins and how it spreads over the wing. It is desirable for the stall to begin at the wing roots and spread outwards from there. When the chosen planform shape is not conductive to this stalling pattern, other techniques (like washout, aerodynamic twist, and stall strips) can be employed to achieve the desired effect.

Boundary layer control can delay airflow separation over the wing and therefore delay the stall. Boundary layer control techniques include vortex generators, slots and slats, and the suction method.

The presence of roll and/or yaw during a stall can aggravate the situation and lead to a spin. A spin results when the two wings are at different AOA's and therefore are stalled by different amounts. The wing at the higher AOA (deeper in the stall) will drop while the wing at the lower AOA (less stalled—possibly unstalled) will rise. This rolling is made worse by the fact that the extra drag produced by the downgoing wing will induce a yaw.

Using the ailerons to correct an unwanted roll during a stall can be detrimental due to the effect the ailerons have on the wings chord line. The aileron which is deflected downwards will increase the AOA of that wing—thus deepening the stall. When one would expect the wing to rise, it may simply drop away more sharply. This potential for the ailerons to reverse in a stalled condition makes them unreliable and essentially useless during stalls on many aircraft.

Stalls during climbing and descending turns can lead to a difference in AOA between the two wings. The reason is that most aircraft are either dihedral or anhedral. The angle between the two wings will cause them to have different AOA's when the aircraft is banked and moving vertically (climbing or descending). In the case of a dihedral aircraft, a climbing turn will lead to the outside wing stalling first, and a descending turn will lead to the inside wing stalling first. The opposite is true for anhedral aircraft.

If roll and/or yaw are introduced in a stall and a spin is the result, that spin will have three stages. The incipient stage is the beginning of the spin and in most aircraft lasts for one to two rotations. During the incipient stage, the aircraft will still have forward motion over the ground and the rotation will not yet be cyclic. The fully developed stage follows, and involves a vertical spin axis and a repetitive rotation. The recovery is where the pilot used appropriate control inputs to stop the rotation and recover from the stall. The recovery can be initiated from the incipient or the fully developed stages.

During a spin, the aircraft will be descending rapidly, but at a low and steady airspeed. Rotation rates will generally be high and in most aircraft a nose low attitude will be prevalent. Recovery on most aircraft involves opposite rudder to slow or stop the rotation and forward elevator to recover from the stall. From there a normal dive recovery can be initiated.

An aircrafts ability to recover from a stall or spin is affected by a number of factors. The main ones are C of G location, rudder effectiveness, weight and weight distribution, power, and configuration.

List of Formulae

<u>Stall "Speed" Variables</u>

$$\text{Stall Speed} = V_s \sqrt{(GW/MGW)} \qquad (6.1)$$

$$\text{Stall Speed} = V_s(\sqrt{n}) \qquad (6.2)$$

$$n = L/W \qquad (6.3)$$

Questions and Problems

1) What is a stall? What causes a stall?

2) If an aircraft loaded to a gross weight of 3000 lbs will stall at an angle of attack of 17°, what angle of attack will the aircraft stall at when it is loaded to a gross weight of 2000 lbs?

3) What are some symptoms of the stall? Why do they occur?

4) What effect do flaps have on the stall characteristics of an aircraft?

5) What will happen if an aircraft stalls in a climbing turn? Descending turn? Why?

6) Is spin recovery more favorable with a forward or an aft C of G? Why?

7) Why does a spin develop after a wing drop during a stall?

8) What happens to the center of pressure approaching and during the stall?

9) A 2500 lb aircraft has a published stall speed of 55 knots. What speed will this aircraft stall at when loaded to 2000 lbs.?

10) The aircraft in Q10 is certified for a transatlantic flight to fly at 50% over MGW. What is the new stall speed?

11) What speed would the aircraft in Q10 stall at when pulling out of a dive at 3 g's?

12) What are the differences between the incipient and the fully developed spin?

Chapter 7
Aircraft Stability

The stability of an aircraft is it's tendency to return to it's original position after a displacement. Without stability, it would be difficult to maintain control of the aircraft since a small displacement of the controls would lead to an extremely large displacement of the aircraft. The displacement of an aircraft without stability would lead to an even larger displacement—even without a control input. This would make the aircraft difficult or even impossible to control.

On the other hand, an aircraft with too much stability would also be very difficult to control since it would have a strong tendency to remain stationary. Control inputs in an aircraft like this would have little or no effect.

Stability

Aircraft stability consists of static and dynamic characteristics. Static stability is the *initial* tendency for the aircraft to return to it's original position. Dynamic stability is behavior of the aircraft *over time* as it oscillates to and/or from it's original position.

Stability can be positive, negative, or neutral. Positive stability means that the aircraft will return to it's original position, while negative stability means that a displacement will cause a further displacement. With neutral stability, any displacement that the aircraft experiences will simply remain as is with no tendency to stray further or to return to it's original position.

Examples of positive, negative, and neutral stability can be seen with the classic "ball in the bowl" example. If we were to place a ball inside a bowl (Fig 7-1a) and then move it up the side and let it go, the ball would return to it's original position at the bottom of the bowl—this is positive stability. On the other hand, if we were to invert the bowl (Fig 7-1b) and place the ball on top, the slightest

displacement would cause the ball to roll even further—this is negative stability. If we placed this same ball on a flat surface (Fig 7-1c), moving the ball would have no effect. The ball would not have any tendency to move closer to or further away from it's original position—this is neutral stability.

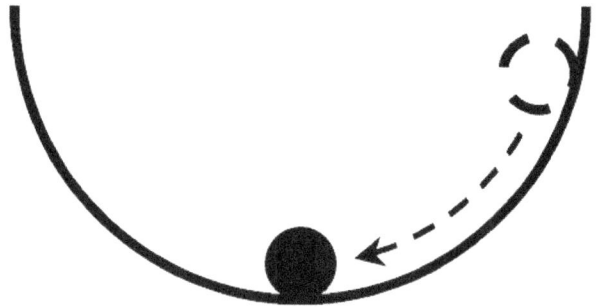

(a) A ball inside a bowl has positive stability.

(b) A ball on an inverted bowl has negative stability.

(c) A ball on a flat surface has neutral stability.

Fig 7-1. - *The Principle of Stability* - Placing a ball in different locations illustrates the concept of stability.

To understand the difference between static and dynamic stability, remember that static stability is only the *initial* tendency of the aircraft. If we look again at Fig 7-1a, the ball which is returning to it's position at the bottom of the bowl will initially roll right on past the bottom and up the other side. At the high point, the ball will once again begin to travel towards the bottom of the bowl. This oscillation back and forth is a result of dynamic stability, while the tendency for the ball to return to center in the first place is a result of static stability.

Each type of stability (static and dynamic) can be positive, negative, or neutral. Positive static stability does not necessarily mean positive dynamic stability. If the ball and bowl in Fig 7-1a were completely frictionless, the ball would keep rolling to the same height on the bowl again and again. This would be an example of positive static stability with neutral dynamic stability. On the other hand, in real life we always encounter friction. So the oscillations of the ball will eventually dampen out and the ball will come to rest at the bottom of the bowl. These dampened oscillations are an example of positive static stability combined with positive dynamic stability.

Trim

A discussion on stability is centered around the concept of trim. Trim in aerodynamics doesn't quite have the same meaning as it does to pilots—although the two meanings are closely related. The trimmed position for an aircraft is the position in which no moments exist—thus the aircraft can maintain equilibrium. Any displacement from the trimmed position should create moments that bring the aircraft back to or further from the trimmed position.

The trimmed position is measured and defined as a set of angles. The first angle of concern is the angle of attack, which is the angle between the wings chord line and the relative airflow (sometimes referred to as α – alpha). Second, we have the angle of sideslip (Fig 7-2), which is the angle between the longitudinal axis and the freestream airflow as seen from above (also referred to as β – beta)

With positive stability, the moments created by a displacement will bring the aircraft back to it's trimmed position. For negative stability, the moments will move the aircraft further from it's trimmed position. If the aircraft has neutral stability,

a displacement will not result in any returning or aggravating moments.

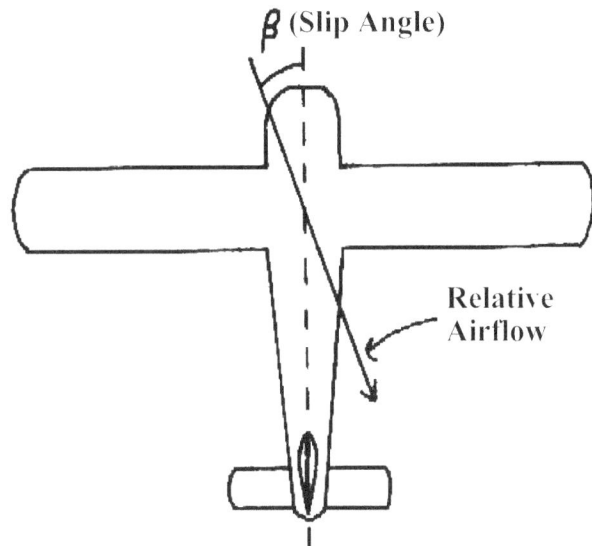

Fig 7-2. - *The Sideslip Angle* - The angle between the relative airflow and the longitudinal axis as seen from above.

Aircraft Axis'

We know from Chapter 5 that an aircraft has three axis' of rotation. These are the lateral axis, the longitudinal axis, and the normal axis. Movements about each of these axis' are termed pitch, roll, and yaw, respectively. Stability is also defined in relation to these three axis'.

Stability of the motion about the lateral axis is termed longitudinal stability, or pitch stability. Stability about the longitudinal axis is referred to as lateral stability, or roll stability. The stability of an aircraft around the vertical/normal axis is called vertical stability, or yaw stability.

Longitudinal Stability

The longitudinal stability of an aircraft is the stability about the lateral axis (Fig 7-3). Also referred to as pitch stability due to it's relationship to the pitching movement, longitudinal stability is often considered by pilots to be the most important type of stability. This is due in part to the fact that pitch control is so critical in the overall control of an aircraft. As well, pitch stability is the type of stability that pilots can have a direct influence over since, as we will see shortly, C of G position has such a large influence on it.

Negative longitudinal stability will lead to very "pitchy" flight conditions. Large deviations from the desired attitude, airspeed, and altitude will be experienced. These deviations could in some cases become dangerous or even uncontrollable (which of course would also be dangerous). Because of this, positive longitudinal stability is very important, and aircraft designers go to great lengths to insure that it is achieved. As pilots, however, we must be sure that the aircraft is flown in such a manner that the manufacturers designs are effective. In particular, we will see that the aircraft must be kept within it's certified weight and balance envelope (more about *how* this is done in Chapter 10).

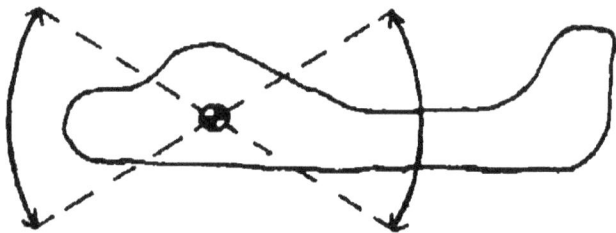

Fig 7-3. - *Longitudinal Stability* - Stability about the lateral axis, also referred to as pitch stability.

Pitching Moments

As we've seen from Chapter 2, any object that is producing lift will also produce a nose down moment which will tend to reduce the object's AOA. Because of this, a wing is an unstable device and must therefore have stabilizing devices associated with it.

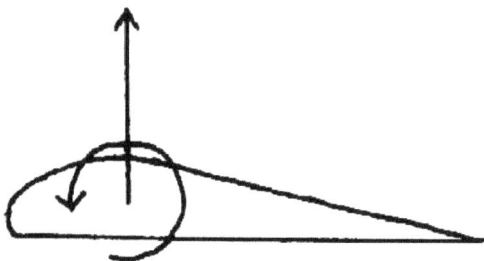

Fig 7-4. - *A Wing with no Attached Structure* - On it's own, a wing is unstable and will eventually assume a flight path directly toward the ground.

If we consider simply a wing with no attached structure (Fig 7-4), it can be seen that the production of lift will cause the wing to pitch forward causing a reduction in AOA. This reduction in AOA will cause a corresponding reduction in lift, thus the wing will accelerate downwards. A downwards acceleration will also cause an increase in airspeed thus increasing the nose down moment. This means that the wing will now pitch down even more, causing a further reduction in lift and more downwards acceleration. With more downwards acceleration there will be more of an increase in airspeed and thus more of a nose down moment. This unstable cycle will continue until eventually the wing reaches equilibrium— travelling straight down.

Center of Gravity

Obviously, a wing can't fly without some sort of stabilizing apparatus. Left to it's own devices, a wing will eventually assume an accelerating flight path directly towards the ground. What we need is some way to produce a nose up moment to counteract the nose down moment about the AC. It would *initially* appear that the lift/weight couple produced by placing the C of G behind the AC would do the trick (Fig 7-5).

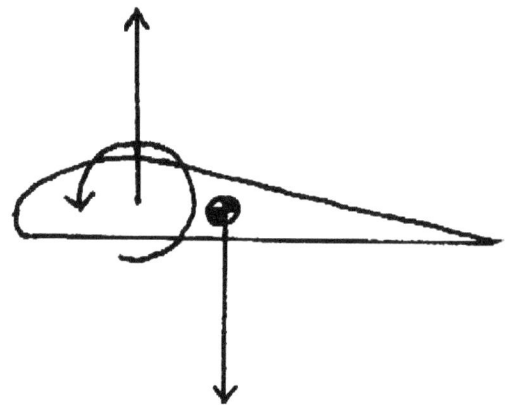

Fig 7-5. - *Placing the C of G Behind the AC* - This configuration will provide equilibrium at one particular airspeed/AOA combination. However, the arrangement is unstable.

Unfortunately, this arrangement will only maintain equilibrium at one particular airspeed/AOA combination. An increase in airspeed, will cause the nose down moment to increase. This brings us back to the "free wing" situation, where the wing eventually reaches equilibrium traveling straight down. On the other hand, an increase in the AOA will cause an increase in lift, thus causing the nose up moment to increase. When the nose pitches up, the AOA is increased even further, causing a further increase in lift which also increases the nose up moment. In this case, the wing will continue to pitch up until the stall is reached.

So in placing the C of G behind the AC we have created an equilibrium at one particular airspeed/AOA combination, but this equilibrium is unstable. If we instead placed the C of G ahead of the AC (Fig 7-6), stability becomes a step closer. With this arrangement, an increase in AOA will increase lift thus increasing the nose down moment. The nose down moment will tend to decrease the AOA back toward it's original value. Likewise, a decrease in AOA will decrease lift, thus decreasing the nose down moment.

So placing the C of G forward of the AC provides us with a configuration which approaches stability. However, stability has still not been achieved since all of the moments present are nose down moments. We have come up with a method of reducing or increasing the nose down moments as necessary, but we haven't balanced them with a nose up moment.

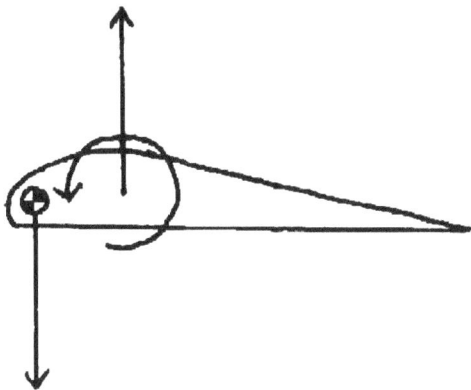

Fig 7-6. - *Placing the C of G Ahead of the AC* - This brings longitudinal stability a step closer. Unfortunately, though, all of the moments are nose down. This means that some sort of apparatus must be added in order to establish equilibrium.

Horizontal Stabilizer

The horizontal stabilizer (Fig 7-7) is the device which we use to provide the necessary nose up moment. The stabilizer is an inverted miniature wing placed on the tail of the aircraft. This wing produces lift in the downward direction. The downward lift at the tail's AC and the upward lift at the wing's AC create a nose up moment. This nose up moment is how we balance the nose down moment produced by the lift/weight couple and the wings inherent pitching moment.

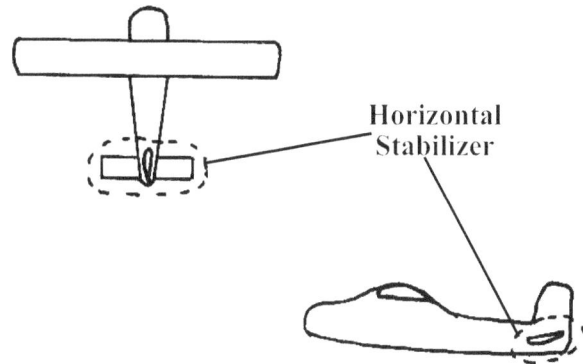

Fig 7-7. - *The Horizontal Stabilizer* - This device produces the nose up moment to establish equilibrium and contribute to pitch stability.

So, the horizontal stabilizer produces a nose up moment which allows us to place the C of G in a position that is conducive to positive pitch stability. As it would happen, the horizontal stabilizer also adds it's own measure of stability to the aircraft. If an upgust causes an increase in AOA from the trimmed value (Fig 7-8), the lift of the wing will be increased. As well, the down force on the tail will be decreased due to a decrease in AOA. The increase in lift will cause an increased nose down moment and the decreased down force on the tail will cause a decrease in the nose up moment. The net effect is for the nose to pitch down thus reducing the AOA.

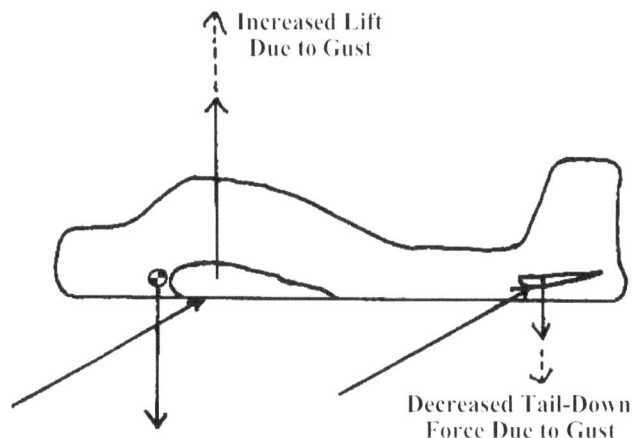

Fig 7-8. - *The Tails Contribution to Pitch Stability.*

So initially, the gust will cause an increase in the AOA, but the inherent stability of the aircraft will cause the AOA to be reduced back to the trim angle.

Combined Effect

As we can see from the above discussion, we have three primary items to consider in regard to longitudinal stability. They are the pitching moment of the wing, the lift/weight couple, and the horizontal stabilizer. Together, these three components will provide us with positive longitudinal stability. This, of course, is assuming that they are arranged properly. An improper arrangement could leave us with neutral or even negative longitudinal stability.

For example, if the C of G was placed aft far enough, the aircraft would reach an unstable configuration. This is the main factor that is used to determine the aft C of G limit in an aircraft's weight and balance envelope.

Stick Fixed and Stick Free

The mobility of the elevator has a significant effect on the dynamic longitudinal stability of an aircraft. This mobility is dependant on whether the control column (or stick) is released and allowed to move freely or held steady. The dynamic pitch oscillations of the aircraft will be heavily damped if the elevator is stationary. With the elevator "floating" freely, this dampening will be less pronounced and may even be non-existent.

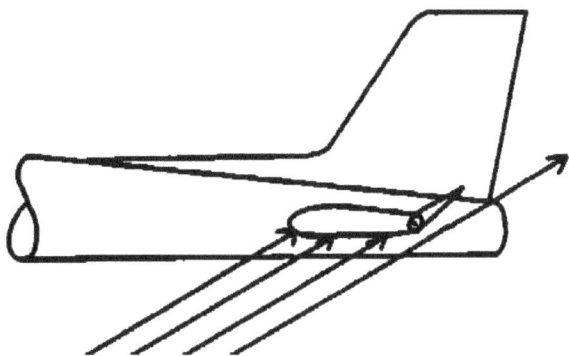

Fig 7-9. - *Relative Airflow's Influence on a "Stick Free" Condition* - With the elevator free to pivot, pitch stability will be reduced.

If the angle of attack is changed from the trimmed angle, the airflow on the elevator will tend to move it in the direction which will increase the deviation (Fig 7-9). This will reduce the restoring moment acting on the aircraft and as a result will reduce the stability of the aircraft. If instead the elevator is fixed in place (Fig 7-10), the AOA change on the stabilizer will have more of an effect will increase stability.

While the elevator is fixed in place, the pitch oscillations resulting from a disturbance will dampen out much more quickly. This is why dynamic oscillations are often not noticeable in the aircraft. Control inputs supplied by the pilot will correct for the deviation. This will often be the case even if no correction is made and the controls are simply held stationary.

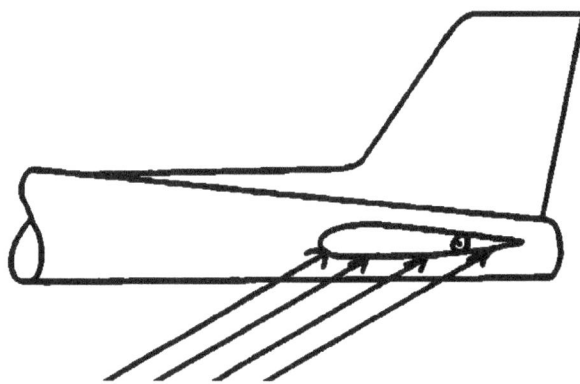

Fig 7-10. - *Relative Airflow's Influence on a "Stick Fixed" Condition* - With the elevator fixed in position, stability is improved.

Lateral Stability

Lateral stability is the stability of motion around the longitudinal axis (Fig 7-11). It is also referred to as roll stability, since motion about the

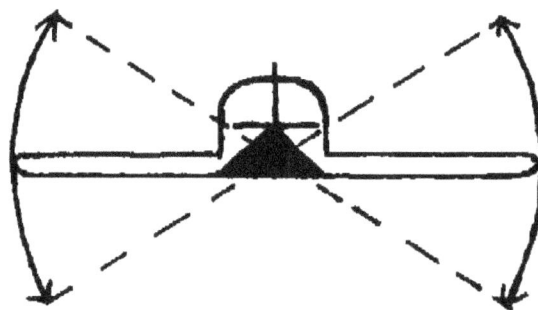

Fig 7-11. - *Lateral Stability.* - Stability about the longitudinal axis, also referred to as roll stability.

longitudinal axis is roll. An aircraft with positive lateral stability will maintain a sideslip angle of zero by rolling away from the slip.

Notice that lateral stability does not require the aircraft to maintain a wings level attitude. An aircraft can have very strong positive lateral stability, and it will simply roll away from a sideslip. This is because if no slip is present, no rolling moments will be produced. So the aircraft can have positive lateral stability and still remain in a variety of banked attitudes.

In this regard, lateral and directional stability are very closely related since directional stability is dependant on a sideslip as well. In fact, the two are not usually separated by aircraft designers since one affects the other. We will shortly see that there is a great deal of overlap between the factors that affect each.

Dihedral/Anhedral

Dihedral and anhedral are the terms we use for wings that are not straight along the span. Dihedral (Fig 7-12a) means that the wings are inclined upwards. That is, the wing tips are higher than the wing roots. Anhedral (Fig 7-12b) means that the wings are inclined downwards. That is, the wing tips are lower than the wing roots.

(a) Dihedral

(b) Anhedral

Fig 7-12. - *Dihedral and Anhedral Wings* - The incline of the wingspan relative to the lateral axis will influence the lateral stability of an aircraft. This is due to the effect that a sideslip will have on the AOA.

Dihedral wings have an effect on lateral stability because of the difference in the AOA's of the two wings during a sideslip. If the aircraft is rolled inadvertently, the weight or inertia of the aircraft will cause a slip towards the low wing (Fig 7-13). This slip, as we can see from the diagram will

cause a change in the relative airflow (the angles have been exaggerated here for clarity), but the change will be different on each wing. The low wing will experience an increased AOA, while the high wing will have a reduced AOA. This means that the low wing will produce more lift, inducing a rolling moment which rolls the aircraft back to it's original position. This is an example of positive lateral stability.

Fig 7-13. - *Sideslip Effect in a Dihedral Aircraft* - The slip causes a difference in AOA between the two wings and rolls the aircraft.

If a sideslip is induced on an aircraft with anhedral wings, the result will be a rolling moment which rolls the aircraft further from it's initial position. This will result in reduced (and possibly negative) lateral stability.

Note here that the dihedral/anhedral effect is only one factor that contributes to the overall lateral stability of the aircraft. Even though anhedral will reduce lateral stability it can still be useful, especially in aircraft that are too stable. Too much stability reduces the aircrafts maneuverability and is thus undesirable. As well, anhedral will reduce dutch roll tendencies, which will be discussed later in this chapter.

"Keel Effect"

In a high wing (AKA – low fuselage) aircraft, something known as "keel effect" will contribute to positive roll stability as. Keel effect is due to the effect that the fuselage has on the relative airflow during a sideslip.

The high wing configuration results in the change in relative airflow near the wing roots (Fig 7-14). In a sideslip, the airflow must separate to flow around the fuselage. This causes an upflow on the low wing and a downflow on the high wing. This change in relative airflow will increase the lift on the

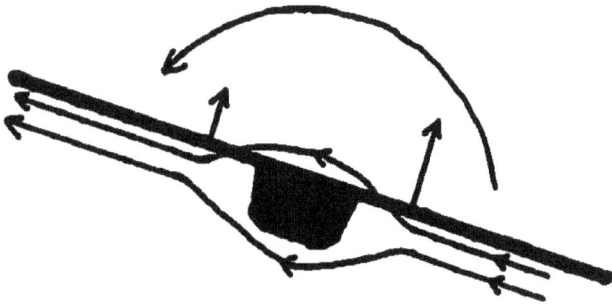

Fig 7-14. - *High wing Aircraft in a Slip* - The airflow separating around the fuselage leads to an increased AOA on the low wing and a decreased AOA on the high wing. This encourages a roll away from the slip.

low wing, rolling the aircraft back towards the high wing.

As for a low wing aircraft (Fig 7-15), the airflow change caused by the fuselage will increase the roll rate. These effects would result in lateral instability if nothing was done to counteract them. Because of this, low wing aircraft will almost always have more dihedral than high wing aircraft.

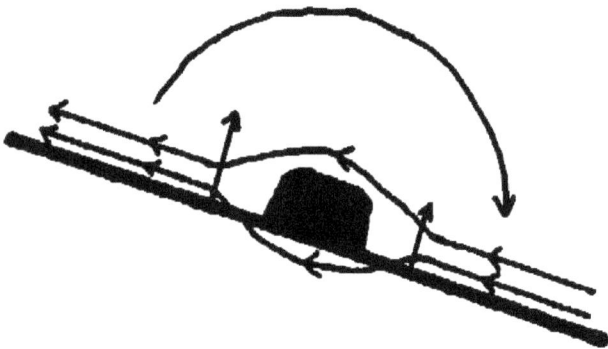

Fig 7-15. - *Low Wing Aircraft* - This configuration leads to less lateral stability due to the inverted "keel effect" and the airflow separation around the fuselage. The roll induced tends to increase the slip.

Sweepback

An aircraft with sweepback will have the wing tips located further aft than the wing roots (Fig 7-16). This means that the leading edges of the wings will be at an angle to one another as seen from above.

If an aircraft with swept wings is placed into a sideslip, the forward wing meets the relative

airflow more directly. As a result of this, more lift is produced by the forward (low) wing, inducing a roll toward the high wing.

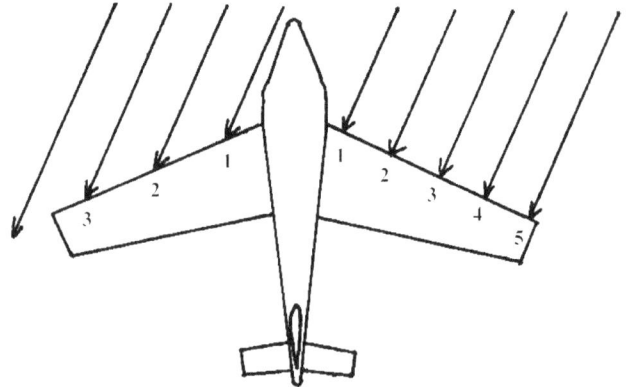

Fig 7-16. - *Sweepback.*

Vertical Stabilizer

The vertical stabilizer is intended to contribute to directional stability. However, due to it's typical location above the C of G a sideslip will produce a rolling moment.

Fig 7-17. - *Vertical Stabilizer* - The force produced by the vertical stabilizer in a slip will create a rolling moment that tends to eliminate the slip.

As before, inertia can be considered to act at the C of G. In a sideslip, the vertical fin will produce "lift" in a sideways direction due to it's AOA. The "lift" from the fin and the inertia at the C of G will produce a rolling moment (Fig 7-17).

Directional Stability

Directional stability—also known as yaw stability—is the stability of the aircraft around the vertical or normal axis (Fig 7-18). Just like lateral stability, directional stability depends on the production of a sideslip.

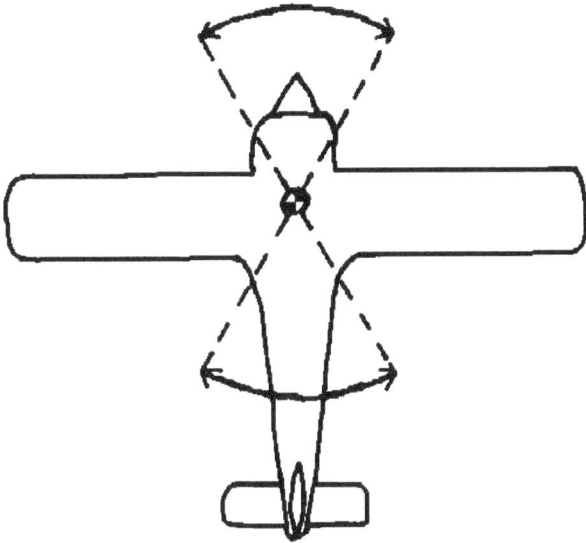

Fig 7-18. - *Directional Stability.* - Stability about the vertical/normal axis, also referred to as yaw stability.

Center of Gravity

The C of G will have a direct effect on directional stability because of the moment produced by inertial forces and the aerodynamic force of the fuselage. This aerodynamic force is dependant on the presence of a sideslip.

If the C of G is located at the central point for the aerodynamic force (Fig 7-19a), no moments will be produced in a slip, and thus directional stability will be neutral. However, if the C of G is positioned forward or aft of the center of pressure for the fuselage, the aerodynamic force will create a moment about the C of G.

With the C of G forward of the center of pressure (Fig 7-19b), a sideslip would create a restoring moment and positive stability will be the result. With the C of G behind the center of pressure (Fig 7-19c), a sideslip would cause a yawing moment which induces a further sideslip. This situation would be directionally unstable.

Aircraft are normally designed so that as the C of G moves aft, pitch stability is lost before directional stability. If this is not the case with an original design, the designers will normally increase the size of the vertical stabilizer (discussed in the next section).

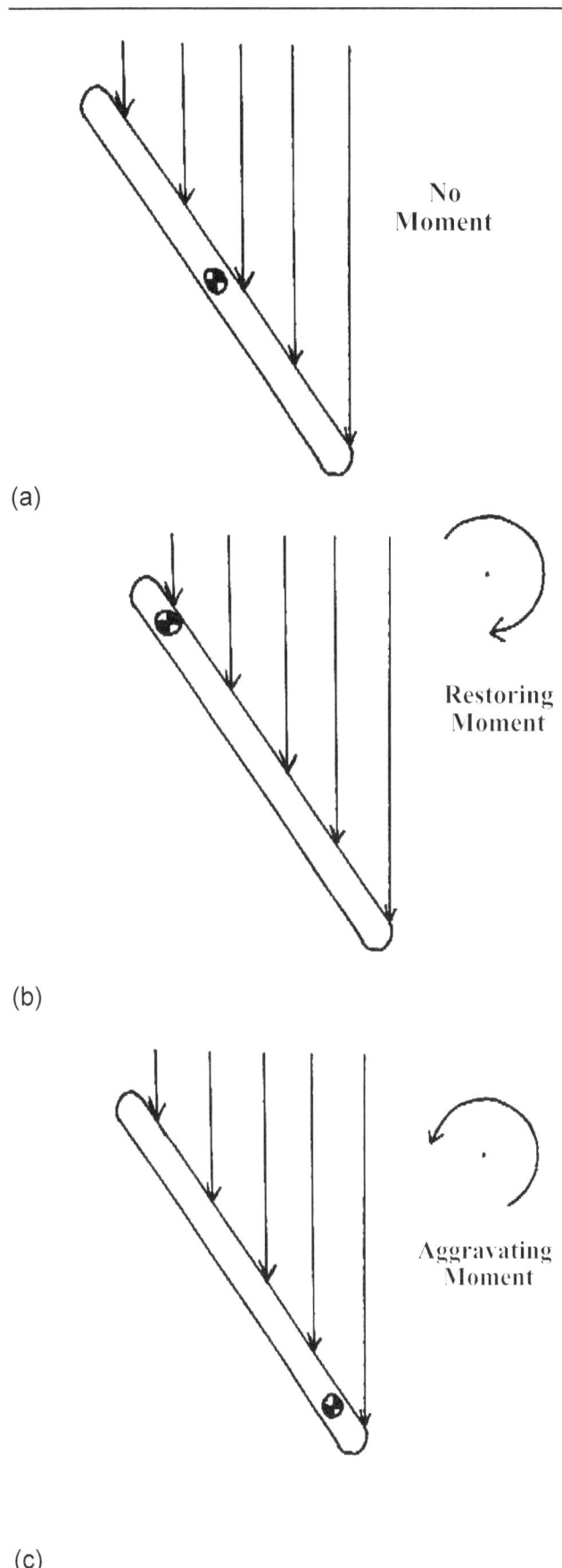

(a)

No Moment

(b)

Restoring Moment

(c)

Aggravating Moment

Fig 7-19. - *C of G Effect on Directional Stability.*

Vertical Stabilizer

The *vertical stabilizer* (*vertical fin*, or just *fin* – Fig 7-20a) is intended to move the fuselage's center of pressure aft by presenting more surface at the aft portion of the airframe. During a sideslip, this   fin experiences an angle of attack and produces "lift"—an aerodynamic force to the side. This force combined with the arm from the stabilizer to the C of G will produce a restoring moment (Fig 7-20b).

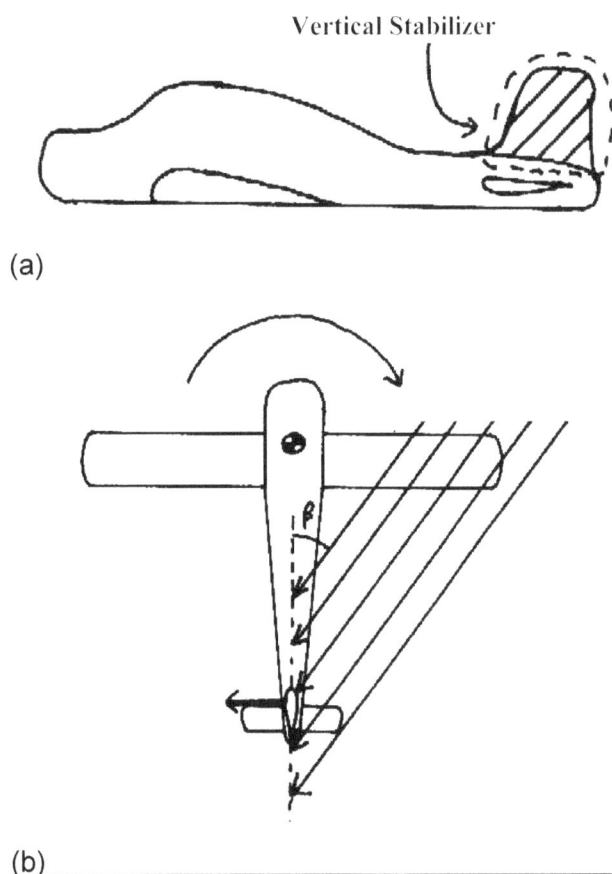

Vertical Stabilizer

(a)

(b)

Fig 7-20. - *Th Vertical Stabilizer* - In a sideslip, the "lift" produced by the stabilizer produces a yawing moment that tends to eliminate the slip.

A larger stabilizer, a longer arm, or a larger sideslip angle will each result in a larger restoring moment. So it can be seen that a larger fin which is placed further aft on the fuselage with a forward C of G will provide better directional stability.

Dihedral/Anhedral

We saw that the dihedral or anhedral of the wings will affect the lateral stability of an aircraft.

This happens because of the change in AOA that results from a sideslip, and therefore the change in the lift of the two wings. We also know from Chapter 3 that an increase in lift will lead to an increase in induced drag. So a slip will not only lead to a roll, but will cause an imbalance in the drag created by the two wings.

For a dihedral aircraft, the wing that is into the slip will experience an increase in lift and an increase in drag. This increase in drag will cause a yaw into the slip, thus creating positive directional stability.

Sweepback

We have already seen that an aircraft with swept wings will roll if it is placed in a sideslip. Another effect of a sideslip is that more drag will be produced by the forward wing due to the more direct relative airflow. This excess drag creates a yawing moment which tends to eliminate the sideslip.

Rudder Fixed and Rudder Free

The effect of the rudder on directional stability is very similar to the effect of the elevator on pitch stability. If the rudder is fixed in place, directional oscillations will dampen out much more quickly. On the other hand, if the rudder is allowed to float freely and move with aerodynamic loads, this dampening will occur much more slowly. Again, like pitch oscillations with a free elevator, directional oscillations may not dampen at all with a free rudder.

Lateral/Directional Coupling

It was stated earlier that lateral and directional stability are not normally considered to be separate from one another. This is because of the direct link that one has to the other. We can see from the factors affecting lateral and directional stability that there is significant overlap between the two.

Both lateral and directional stability depend on the presence of a sideslip. Laterally and directionally, the desired trimmed position is a sideslip angle (β) of zero. So when the aircraft is in a sideslip, yawing and rolling moments are produced. If the aircraft is stable, these moments will tend to reduce the sideslip angle, but if the aircraft is unstable, these moments will tend to increase the slip angle.

Even with the close relationship between lateral and directional stability, it is possible to have one which is positive while the other is negative. As well, the results of static and dynamic stability can lead to some undesirable combinations. These combinations are a result of lateral/directional coupling.

Directional Divergence

Directional divergence is a result of negative yaw stability combined with very low, neutral, or even slightly negative roll stability. If a disturbance causes an aircraft with directional divergence to yaw, the moments resulting will cause a further yaw. The flight path of this aircraft will begin to change, but the heading will change even faster, leading to a flight path similar to that in (Fig 7-21). In an extreme circumstance, this aircraft will in fact end up flying side-on or backwards through the air—obviously an undesirable situation.

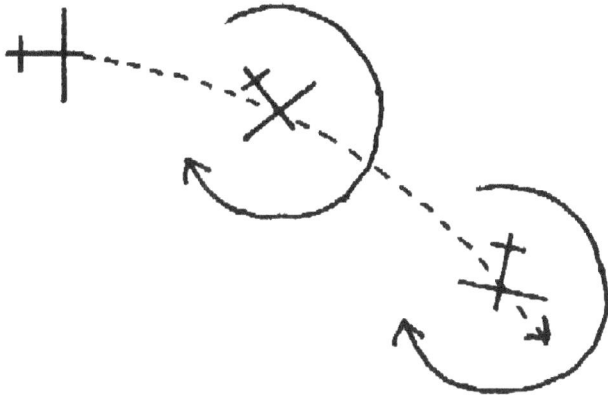

Fig 7-21. - *Directional Divergence* - When an aircraft has neutral or negative directional stability combined with very low, neutral or even negative lateral stability, directional divergence is normally the result. Notice that the orientation of the aircraft is changing faster than the flight path. This means that the sideslip angle is increasing.

Spiral Divergence

Spiral divergence is a result of yaw stability that is greater than roll stability. If a sideslip is introduced to an aircraft with spiral divergence, yaw stability will be predominant. The aircraft will tend to change direction into the slip readily, but because the slip doesn't get to develop, the roll response will be slight. The yaw will lead to a turn that is ever tightening—a spiral (Fig 7-22). The presence of the yaw without a significant restoring moment in roll can also cause the secondary effect of yaw to be dominant. This results in the aircraft rolling into the slip along with the yaw response.

Unfortunately, the spiraling path produced will not be level. Yawing while in a banked attitude will cause pitch changes. In this case, yawing in the direction of the bank will cause the nose to drop—leading to a spiral dive. As well, in the banked attitude, unless the angle of attack is adjusted with the elevator, the nose will be forced to drop by the pitch stability of the aircraft (more about this in Chapter 12).

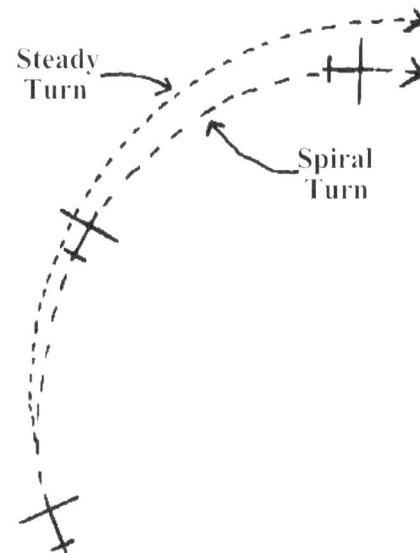

Fig 7-22. - *Spiral Divergence* - Strong directional stability combined with weak (or weak*er*) roll stability can lead to spiral divergence. In response to a sideslip, the aircraft yaws into the slip. When allowed to progress, this yaw can lead to a spiral dive. Unlike directional divergence, the aircraft's orientation will change at approximately the same rate as it's flight path—meaning that the sideslip angle remains low.

Notice also that the spiral turn is gradually diverging from the steady turn. This is because the angle of bank is gradually increasing.

The progression from a slight yaw to a spiral dive is very slow and can easily be prevented or corrected by a wary pilot. However, when attention is diverted to something other than flying the plane (reading a map, talking on the radio, etc.), a spiral dive can result.

Dutch Roll

Dutch roll is the result strong positive roll stability. If the aircraft enters a slip, the response is a roll away from it. This roll is combined with the directional response of yawing into the slip. So as the aircraft rolls in one direction, it is simultaneously yawing in the other direction.

The with the opposite roll and yaw, the dynamic response of the aircraft is an oscillating uncoordinated turn called Dutch Roll. The flight path followed will be similar to that in (Fig 7-23).

If the dynamic stability is positive, these oscillations will eventually dampen out. However, if the dynamic stability is neutral or negative, the oscillations will continue until corrective action is taken. In the case of negative dynamic stability, the oscillations will continue to get larger.

Fig 7-23. - *Dutch Roll* - With strong positive stability in roll, dutch roll will normally be the result. Dutch roll is an oscillating uncoordinated turn that results from the aircraft yawing into a slip and simultaneously rolling away from the slip. Notice that the aircraft orientation and flight path are both oscillating, but they are out of phase with one another.

Notes on Coupling

All aircraft will have coupled effects between the lateral and directional stability. Which of the above three tendencies is predominant will depend on the aircraft, but there are some general statements that can be made.

First of all, directional divergence is usually avoided at all costs—it is simply too dangerous to allow. It's also easy to fix—divergence can be eliminated simply by enlarging the vertical stabilizer and/or by increasing the distance between the stabilizer and the C of G.

Unfortunately, the solution for directional divergence is also the cause of spiral divergence. If we increase the yaw stability too much without a corresponding increase in roll stability, spiral divergence will be the result. So with an increase in tail size or tail arm, we have gone from the dangerous problem of directional divergence to the annoying problem of spiral divergence.

The solution to this problem is to increase the roll stability so that it matches or exceeds the yaw stability. This, however, leads to dutch roll since the aircraft will roll to compensate for a sideslip. Once again, we have solved a problem by creating another one. One of these last two (spiral divergence or dutch roll) is generally accepted, though, since it is impossible to eliminate all three coupling effects. The lesser of the three evils is chosen by the designer and dealt with by the pilot.

A neutrally or positively stable dutch roll could be selected by the aircraft designer, but this would be annoying to the pilot and the passengers (dutch roll encourages airsickness). Spiral divergence, on the other hand, often goes unnoticed even by the pilot (and certainly by the passengers) unless the controls are completely ignored for several seconds (sometimes minutes).

One thing to consider is that none of these effects can occur unless a sideslip is introduced. An aircraft with spiral divergence will not enter a spiral dive unless a slip is present to initiate the action. It could therefore be argued that an aircraft which is trimmed out perfectly can be released to fly on it's own with the pilot simply sitting back and watching. In theory this is true, but remember that spiral divergence is an unstable motion—once started it gets progressively worse—and only an infinitesimal sideslip angle is necessary to begin the spiral. To argue that a perfectly trimmed aircraft will maintain trimmed straight and level flight (or a turn, climb, or descent) is similar to arguing that a billiard ball can be balanced perfectly on top of another billiard ball—in theory it is quite possible, but the instability causes it to be impossible in practice.

Chapter Summary

An aircraft's stability is it's tendency to return (or not return) to it's trimmed position after a displacement of some sort. The trimmed position is measured as a set of angles—the angle of attack for longitudinal stability and the sideslip angle for lateral and directional stability. At the trimmed position, no net moments exist and the aircraft can maintain equilibrium.

Stability is considered around all three axis'—the longitudinal, vertical/normal, and lateral axis'. Stability about the lateral axis is called longitudinal stability or pitch stability. Stability around the longitudinal axis is referred to as lateral or roll stability. Stability around the vertical axis is directional or yaw stability.

In order to have positive longitudinal stability, the aircraft must produce pitching moments which tend to return it to the trimmed AOA. These moments are produced by keeping the AC behind the C of G and using a horizontal stabilizer to provide a balancing moment. Releasing the elevator to pivot according to aerodynamic forces (called a "stick free" condition) can reduce the returning moments.

Lateral and directional stability are established when a sideslip creates moments that will eliminate the slip. This is accomplished directionally by yawing into the slip. Laterally, there must be a roll away from the slip. Lateral and directional stability are provided for by a number of design features, including dihedral/anhedral, keel effect, sweepback, and the vertical stabilizer. Directional stability is also influenced by the C of G location.

Coupling effects between lateral and directional stability can lead to directional divergence, spiral divergence, and/or dutch roll. Directional divergence is considered unsafe and is eliminated in the design of the aircraft. However spiral divergence and dutch roll are both allowed to a limited degree since all three effects cannot be eliminated.

Questions and Problems

1) What does it mean for an aircraft to be in it's trimmed position? How does this relate to stability?

2) What is the difference between dynamic and static stability?

3) How is longitudinal stability achieved? Why is it important?

4) How is lateral stability achieved? Why is it important?

5) How is directional stability achieved? Why is it important?

6) What types of effects would one expect to encounter (in terms of stability) if flying an aircraft loaded with the C of G aft of the aft limit? Why?

7) Why are lateral and directional stability so closely related? What types of coupling effects could one expect to encounter?

8) Is longitudinal stability enhanced or degraded with the elevator floating freely as opposed to being fixed in place? Why?

Chapter 8
Aircraft Performance

The performance that can be anticipated from an aircraft is of great importance to a pilot. Being able to predict the takeoff distance required, or the fuel required for a flight, or the distance required to clear an obstacle can obviously prevent an embarrassing (and probably dangerous) mishap.

Most aircraft have performance figures for various conditions of altitude, weight, wind, etc. published in the POH or Flight Manual. This means that actual derivations and calculations are seldom required of the pilot. However, a thorough understanding of where these "book" numbers come from as well as the factors that affect them will allow a pilot to use them much more effectively.

The Power Curve

Most, if not all, performance characteristics of a prop driven aircraft are dependent on the

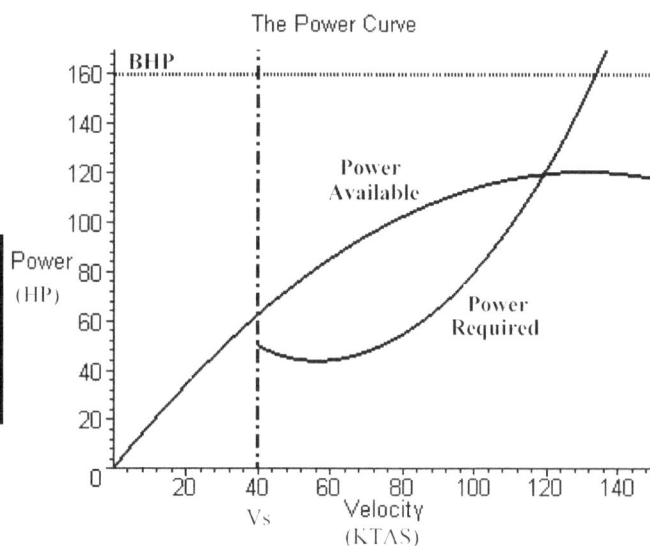

Fig 8-1. - *The Power Curve* - A representation of the power required to maintain level flight at various airspeeds and the power available at those airspeeds.

power curve. We have already seen the power required curve, as well as the power available curve. Putting these two curves together, we get the Power Curve (Fig 8-1).

The relationship between power required and power available determines many performance figures for an aircraft. For some other performance figures, just the power required is the determining factor. As for turning performance, power may or may not become a limiting factor—more about this in the next chapter.

Variables

Several variables will affect the shape of the power curve. They include anything that will affect the induced power required or the parasite power required. As well, anything that affects the thrust horsepower available will have an effect on the power curve.

To illustrate the effect of each variable, we will start with a power curve for an aircraft at MGW and at sea level (Fig 8-1). From there, we will look at the effect of each variable individually.

Weight – If we were to decrease the gross weight of the aircraft, the lift required would be decreased as a result. This reduction in lift will also mean a reduction in induced drag and thus induced power. The resulting power curve would look like (Fig 8-2). Notice here that the change in power required is much more noticeable at the lower end of the speed range than at the higher end. This is because at higher speeds, parasite power accounts for a much greater portion of the total power required than induced power. At lower speeds, induced power is predominant, so a change in induced power required will have a more significant impact on the total power required. Parasite power does not vary with weight.

The Effect of Weight on The Power Curve

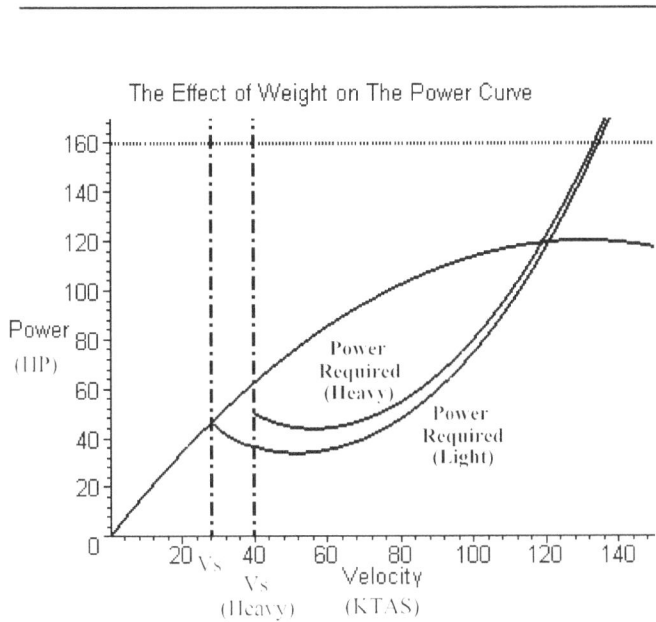

Fig 8-2. - The Power Curve at Varying Weights - A decrease in weight will lead to a decrease in the power required. This results from the reduced induced drag and is more pronounced at lower airspeeds where induced drag is predominant.

Center of Gravity – The position of the C of G will affect the power curve in a similar way to weight. As we've seen in Chapter 6, a forward shift of the C of G will increase the lift required due to the increased tail down force. This means that a forward C of G will increase the induced power required, while an aft C of G will decrease the induced power required.

Altitude – Altitude will have the effect of moving the power required curve up and to the right (Fig 8-3). To see why this is happening, we need to remember that the drag produced by the aircraft is a function of the dynamic pressure. However, as the altitude is increased, the air density decreases, thus a higher speed is required for the same dynamic pressure.

This is the reason that the curve shifts to the right. As for the up shift, this is caused by the fact that power is determined by multiplying force (drag) by the velocity. This velocity is the *true* airspeed. So we experience the same amount of drag at the same *equivalent* airspeed, but due to the higher *true* airspeed, the power required is higher.

Altitude will also affect the power available. Since as we climb the engine loses power due to the less dense air, the power available curve will be moved down.

Configuration – The effect of configuration changes will be to change the amount of drag being produced. This, in turn, will change the amount of power required.

So if we change the configuration in such a way that the drag is increased (e.g. – extend flaps or landing gear), the power required will also increase. If, on the other hand, we change the configuration in such a way that drag is reduced (e.g. – retract the flaps or landing gear), the power required will also be reduced.

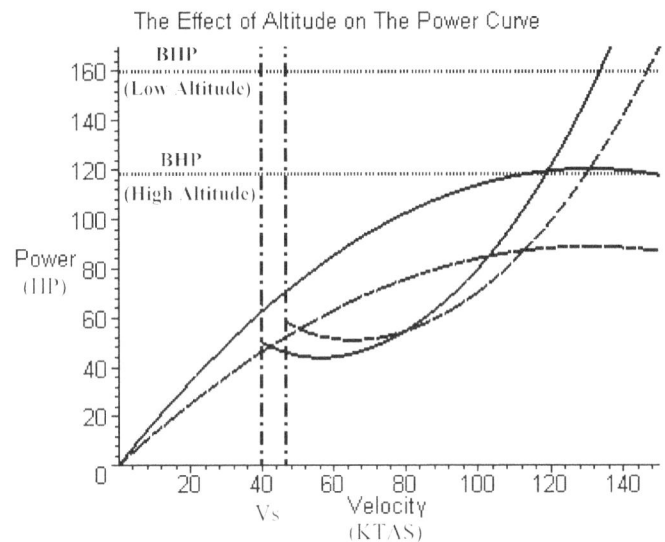

The Effect of Altitude on The Power Curve

Fig 8-3. - The Power Curve at Varying Altitudes - An increase in altitude will cause the power required curve to shift upwards and to the right. At the same time, the reduction in air density which leads to a reduction in BHP and therefore a reduction in THP as well.

Load Factor – The load factor is determined by the amount of lift being produced. Because of this, it will have a direct effect on the induced power required. Increasing the load factor has an effect similar to increasing weight. The power required increases across the entire range of airspeeds, but the change is most pronounced at lower speeds.

Takeoff

The takeoff is a critically important phase of flight. The ability to takeoff safely within the runway distance available and clear necessary obstacles is critical to a safe flight. In this section we will consider the factors which affect the takeoff roll and the initial lift-off. Obstacle clearance will be looked at in the next section, which covers climbing performance.

Thrust and Power

For most performance characteristics of a propeller driven aircraft, the power available is an important consideration. In the case of takeoff performance, however, it is easier to deal with thrust instead. As we know from Chapter 4, thrust and power are closely related. Power is a rate of transfer of energy (thrust multiplied by velocity), while thrust is a rate of transfer of momentum (thrust is a force, and thus causes an acceleration).

In the case of a takeoff, the aircraft is accelerating. Although power and thrust can both be used for takeoff calculations, it is much simpler for us to use thrust. Up to this point, we have been considering the propeller to be a power transmitting device, now we have to consider it's ability to transmit thrust.

Thrust Available and Thrust Required for the Take-Off Roll

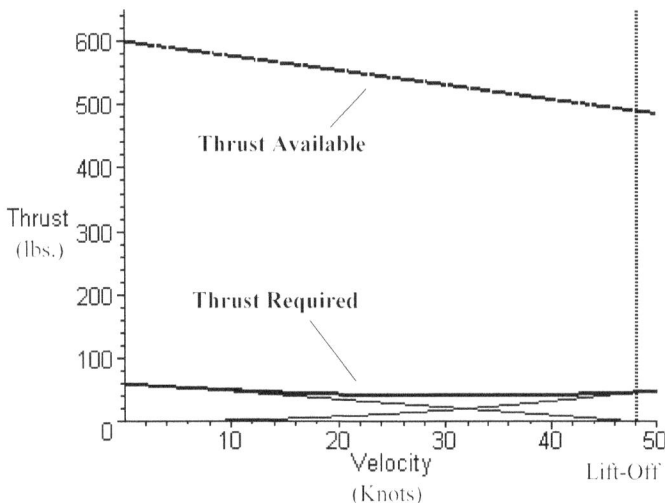

Fig 8-4. - *Thrust and Drag Available During the Takeoff Roll* - For prop aircraft, the thrust available is at a maximum when the aircraft is stationary. Thrust decreases gradually as the airspeed incenses during the takeoff roll. Drag, on the other hand, starts at a low value and increases at a rate that is somewhat lower than the increase in parasite drag.

The propeller produces thrust by accelerating air backwards. The force exerted to accelerate the air backwards results in an equal and opposite reaction which accelerates the aircraft forward. The maximum *thrust* available from a prop occurs at an airspeed of zero due to the high blade AOA. From there, the thrust decreases with an increase in airspeed (Fig 8-4). At low airspeeds (in

the takeoff range), this decrease is relatively slight, and for our purposes we will assume that the thrust available is constant throughout the takeoff roll.

This simplifying assumption will lead to inaccuracies for aircraft designers. However, to a pilot, the assumption is sufficient for practical purposes. This is especially the case since, as we will see shortly, the thrust available for the takeoff will be used to determine takeoff distance proportions —not absolute takeoff distances. The assumption also has the added benefit of simplifying things greatly for us.

Drag and Rolling Friction

The aerodynamic drag is constantly increasing as the aircraft accelerates down the runway. For the most part, we need only consider parasite drag during the takeoff roll. Induced drag is present, but due to the initially small amount of lift being produced and the presence of ground effect, it can be ignored for our purposes.

Another type of drag that we have to contend with during the takeoff roll is rolling friction. Rolling friction is caused by the contact between the wheels of the landing gear and the axles. This friction will retard the rotation of the wheels, and this loss in rotational speed of the wheels must be replaced by the contact between the wheels and the ground. This contact will cause a force in the direction opposite to the direction of travel.

As the aircraft gains airspeed, it's weight is gradually removed from the wheels. This causes a reduction in the rolling friction as there is less normal force acting on the wheel axles. So the net effect of gaining airspeed on the takeoff roll is to increase the total drag of the aircraft at a rate that is somewhat less than the increase in parasite drag alone (Fig 8-4).

Takeoff Performance

The overall takeoff performance of the aircraft will depend on a number of factors. Primarily, they are the total drag force (aerodynamic and rolling), the thrust force, the weight of the aircraft, and the lift-off speed. Ideally, we want low drag, high thrust, low weight, and a low lift-off speed.

Recall from Chapter 1 that the acceleration

of an object is dependant on the accelerating force and the mass of the object. In this case, the accelerating force is the thrust less the drag, and the mass is the weight divided by gravitational acceleration. So the acceleration of the aircraft can be determined from the following formula:

$$a = \frac{(T - D)g}{W} \tag{8.1}$$

We also know from equation (1.7) that the distance covered by an accelerating object starting from rest is equal to half of the acceleration multiplied by the time squared,

$$d = \tfrac{1}{2}at^2 \tag{1.7}$$

Since we know the final speed in this case (lift-off speed), the time can be determined by dividing this speed by the acceleration (8.1). Taking all of this into consideration, and after a few manipulations, we can come up with the following:

$$d = \frac{Wv_{LO}^2}{2g(T-D)} \tag{8.2}$$

Bear in mind that this equation is greatly simplified. Since thrust and drag both will vary with speed as the takeoff roll progresses, the actual takeoff distance equation would be far more complicated. However, This equation does illustrate some important trends for us to consider.

To look at each variable in equation (8.2), we can see that an increase in weight will result in a proportional increase in takeoff distance. If we were to alter our lift-off speed for any reason, our takeoff distance would change according to the *square* of the speed change. A change in net thrust available (thrust minus drag) would change the denominator, thus the takeoff distance would be affected *inversely*. Increasing the thrust available from the engine would increase the value of the denominator, thus decreasing the takeoff run. An increase in drag, on the other hand, would *decrease* the value of the denominator, thus *increasing* the takeoff run.

Configurations

The configuration used for takeoff will depend on the type of aircraft and the objective during the takeoff. Typically speaking for a general aviation aircraft, no flaps are used for a normal takeoff and varying amounts are used for specialty takeoffs (short fields and soft or rough fields).

There are pros and cons to using flaps and the amount to use will vary from one aircraft to another. Flaps will increase the lift being produced, and as such, they will reduce the weight resting on the wheels. This means that rolling friction will be reduced—reducing the retarding forces in the roll. Unfortunately, the flaps also create some drag of their own. On some aircraft the reduction in rolling friction is offset by the increase in drag from the flaps. Another advantage of the flaps, however, is that they reduce the stall speed of the aircraft. Since lift-off speed is based on some multiple of the stall speed, this means that we can reduce the lift-off speed safely, thereby reducing the takeoff roll.

The decision as to whether or not to use flaps during a takeoff also depends on the effect that they have on the aircraft's climb performance when clearing an obstacle. We will cover this consideration in more detail in the section on climbs.

Runway Condition

The condition of the runway is an important consideration when trying to determine the takeoff distance that's required. The surface condition of the runway, as well as the slope of the runway can have a serious impact on aircraft performance. Performance figures published in the POH or Flight Manual are normally based on a paved, level, and dry runway. Some manufacturers will publish correction factors for circumstances not meeting this criteria, while others will leave it up to the pilot.

Slope – If the runway being used is on an incline, the weight of the aircraft is no longer acting perpendicular to the aircraft's path. This means that a component of weight will be acting perpendicular to the aircraft's path, while another component will be acting parallel (Fig 8-5). This parallel component of weight will either help or hinder the takeoff roll depending on the direction of the slope.

In the case of an upsloped runway (Fig 8-5a), the weight will act in the direction opposite the takeoff roll. Thus the weight will contribute to the total retarding force that is acting on the aircraft. Referring back to equation (8.2), we can see that this will increase the takeoff run. If, however, we were to use a downsloped runway (Fig 8-5b), the parallel component of weight would contribute to thrust—decreasing the takeoff run.

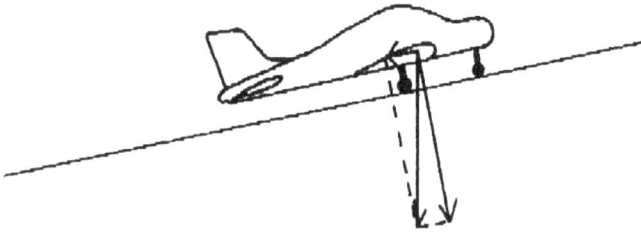

(a) An upsloped runway will increase the distance required to takeoff due to the component of weight acting against the takeoff path.

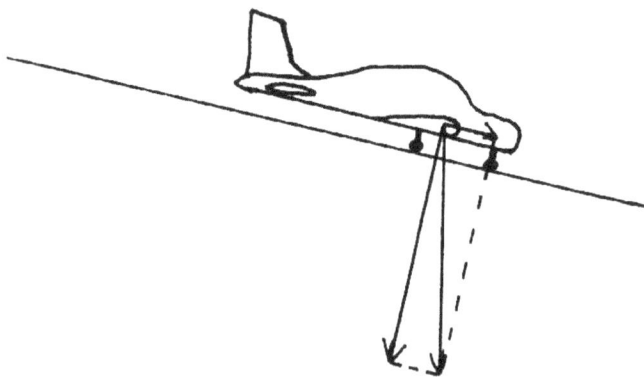

(b) A downsloped runway will decrease the distance required to takeoff due to the component of weight actg in the takeoff direction.

Fig 8-5. - Slope Effect on Takeoff.

Without the manufacturers raw performance data (or some correction provided by the manufacturer), the effect of slope cannot be accurately evaluated. However, there are some useful rules of thumb presented in Chapter 22.

Surface – The surface condition of the runway will also have direct impact on the distance required to takeoff. On a smooth surface, the rolling friction is a result of the friction between the wheel assembly and the wheel axle of the landing gear. On a rough surface, however, we have this rolling friction combined with some additional resistive force.

For example, on a grass strip, the tires will sink into the ground slightly. This means that there will be surface material (grass and/or gravel) in front of the tire as well as below it (Fig 8-6). As the aircraft rolls down the runway, this material must be removed from the tires path. The tire forces the material out of the way, and as a result, the reaction

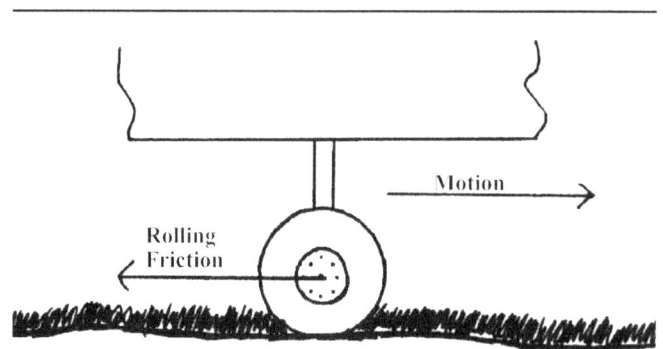

Fig 8-6. - Takeoff From a Soft (Grass) Surface - Surface material in front of the tire must be removed from the path. This creates a retarding force and lengthens the takeoff roll.

force (Newton's third law) acts against the aircraft's motion.

Once again, without the manufacturers raw performance data or some sort of correction, the effect of surface condition cannot be evaluated properly. Thumb rules are presented in Chapter 22.

Other Variables

There are several other factors which will affect the takeoff performance of an aircraft. They all relate back to equation (8.2) in some way.

Altitude – An increase in altitude from sea level will be associated with a decrease in air density. As we know from Chapter 4, this decrease in air density will cause a proportional decrease in the power produced by the engine. A reduction in power will unfortunately mean that the net thrust available to be used during the takeoff (or any other phase of flight) will be decreased as well. We can see from equation (8.2) that this reduction in thrust will increase our takeoff distance.

Altitude will also have another undesirable effect on our takeoff performance. Recall from Chapter 3 that the lifting capacity of the wings is dependant on dynamic pressure, and at higher altitudes (lower densities) we need a higher *true* airspeed to achieve this dynamic pressure. That is, we need a specific *equivalent* airspeed to lift off, but at higher altitudes, our actual lift-off speed (TAS) will be higher. From equation (8.2) we can see that this increase in speed will increase our takeoff run proportional to the *square* of the speed change—even though the speed we read off of the ASI has not changed.

Wind – Generally speaking, a takeoff run is performed into the wind. In equation (8.2), we have so far considered the lift-off speed to be airspeed, but the distance covered over the ground is in fact a function of *groundspeed*. This means that taking off with a headwind will decrease our takeoff roll since the lift-off groundspeed will be reduced.

A tailwind, on the other hand, will increase our lift-off groundspeed, thus increasing our takeoff roll. Note here that since our takeoff distance will change with the *square* of our lift-off speed, a tailwind will increase the takeoff run more than a headwind will decrease it. For example, let's take an aircraft that rotates at 60 knots and requires a distance of 500 ft in a no wind condition. If this aircraft takes off into a 10 knot headwind it will require 347 ft to lift off—a reduction of 153 ft. On the other hand, if this same aircraft were to take off with a 10 knot tailwind, the distance required to lift off would be 681 ft—an increase of 181 ft.

Weight – Weight will have a direct impact on the takeoff distance according to equation (8.2). In addition to this, weight will also have some indirect effects which will increase the takeoff roll even more.

An increase in weight will also increase the rolling friction—both at the axle and at the surface if the surface is soft. A heavier aircraft has a larger normal force which will cause more friction, plus the extra weight will cause the tires to sink some more in soft ground. This increase in drag force will slow the acceleration of the aircraft, and thus increase the takeoff roll.

Another effect of weight is that it affects the stall speed of the aircraft. Since lift-off speed is determined as some multiple of the stall speed, a change of stall speed will also change the takeoff distance. Stall speed changes with the square root of the weight change while the distance increases by the square of the of the speed change. The net effect is that the distance will increase proportionally to the weight change.

These two weight effects combined lead to a significant change in takeoff distance with weight. The distance will change proportionally to the *square* of the weight change.

Pilot Skill – When calculations are done by engineers to determine the performance of an aircraft, these calculations must be demonstrated in practice by test pilots. These pilots are highly trained and experienced, and are capable of flying the aircraft to very specific tolerances. The performance figures in the POH or Flight Manual for the aircraft are based on the performance observed in these test flights.

Most general aviation pilots are not capable or experienced enough to fly to the tight tolerances required in test flights. Rotating slightly early will create extra drag and will deteriorate the performance of the aircraft. Rotating slightly late will increase the lift-off speed and thus increase the takeoff distance. Any unnecessary steering or braking will also create drag and lengthen the roll. Any control inputs that are not *exactly* "by the book" will increase the distance required to get off the ground. As well, any circumstances that do not conform to the conditions which the performance numbers are based on will have an effect on the roll.

Clearly, some judgment is required of the pilot in gauging his/her skill level. It is important that a realistic figure be added to the "book" figure when considering ones experience and skill level.

Aircraft Condition – The published performance figures for an aircraft are also based on a brand new airplane. That is, the aircraft will have a fresh and smooth paint job with a fresh coat of wax, no dents in the aircraft surface, a brand new engine straight off the factory line, etc. These sound like minor items, but when the paint is chipped, the leading edge of the wing has minor dents in it, and the engine has several hundred hours on it, the effect can be quite significant.

Climbing

Another important performance consideration is climb performance. As we will see shortly, the climbing capability of an aircraft is based on either excess power, or excess thrust, depending on what type of performance is being considered. *Rate* of climb is determined by excess power, while *angle* of climb is determined by excess thrust.

The rate of climb of the aircraft is the amount of altitude gained per unit of *time*. The angle of climb is the amount of altitude gained per unit of *distance covered*. To see the difference more clearly, refer to Fig 8-7. Notice that an

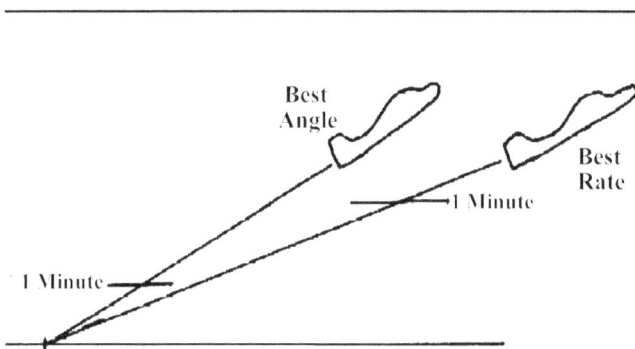

Fig 8-7. - Rate vs. Angle of Climb.

aircraft climbing at the best angle speed will in fact be climbing at a lower rate than an aircraft climbing at best rate.   Initially, this doesn't seem quite right—shouldn't an aircraft that is climbing steeper also be climbing faster?   The answer to this question is in the forward speed of the aircraft. Even though the aircraft at best angle is climbing slower, it is also covering ground more slowly. The net effect is that more altitude will be gained per unit of ground covered.

Best Rate

Power—or more correctly, *excess* power—is the primary consideration when determining rate of climb. The power required is the power that must be used to sustain level flight at a particular airspeed. Any power available that is above and beyond the power required will either accelerate the aircraft to a higher airspeed or cause the aircraft to climb.

To understand why this happens, remember from Chapter 1 that power is the rate of doing work. Work, in turn, is the transfer of energy. Remember also that the height of an object determines it's potential energy. So excess power can be used to add to the potential energy (the height) of the aircraft.

The standard unit used to measure rate of climb is feet per minute (FPM). If we were to multiply the weight of the aircraft by the rate of climb, we would come up with some number in units of foot pounds per minute. As we have seen in Chapter 1, this is one of the units of power. One horsepower is defined as 33,000 foot pounds per minute. From this, we can determine the following:

$$W \times ROC = 33{,}000 \times ETHP,$$

therefore,

$$ROC = \frac{33{,}000 \times ETHP}{W} \qquad (8.3),$$

where (ROC) is the rate of climb in FPM, (ETHP) is the excess thrust horsepower, and (W) is the weight of the aircraft. So the aircraft's rate of climb is determined by the weight of the aircraft and the excess thrust horsepower available. Since weight is a constant for a given point in time, the rate of climb will be determined by the excess power available.

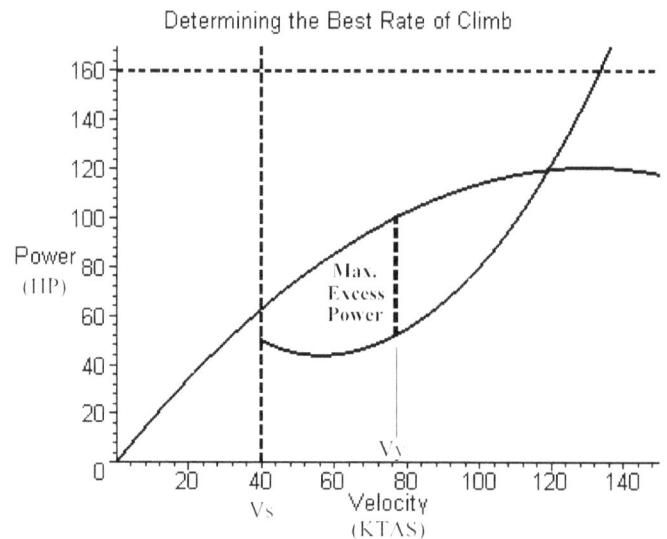

Fig 8-8. - Best Rate of Climb Speed (V_y) - Since rate of climb is determined by excess thrust horsepower, the maximum rate of climb will occur at the speed which provides the most ETHP. On the power curve, this speed is seen as the largest space between the power required and the power available. Note that this speed is not necessarily the speed with the lowest power required or the highest power available.

The best rate of climb speed has several practical applications, and some of these will be discussed in more detail in Chapter 14. Since rate of climb is determined by excess thrust horsepower, the best rate of climb speed (V_y) is determined as the speed which allows the *most* excess thrust horsepower. On the power curve, this speed can be seen as the speed where the two curves are the furthest apart (Fig 8-8).

Notice that V_y is not necessarily at the point of lowest power required or highest power available.Since prop efficiency changes with speed,an aircraft may need to be flown at a speed requiring more power to achieve a greater *difference* between power required and power available.

Best Angle

The angle of climb is determined by the excess thrust available. Just like power, the thrust required is the thrust needed to maintain level flight. Any thrust that is available and not required will either accelerate the aircraft to a higher airspeed or cause the aircraft to climb.

With an inclined flight path, the weight of the aircraft—which always acts straight down—can be divided into two perpendicular force vectors (Fig 8-9). One of these vectors will be perpendicular to the flight path, while the other will be opposite the flight path. The aircraft now maintains equilibrium with lift equal to the "downward component of weight and thrust equal to the drag plus the aft component of weight.

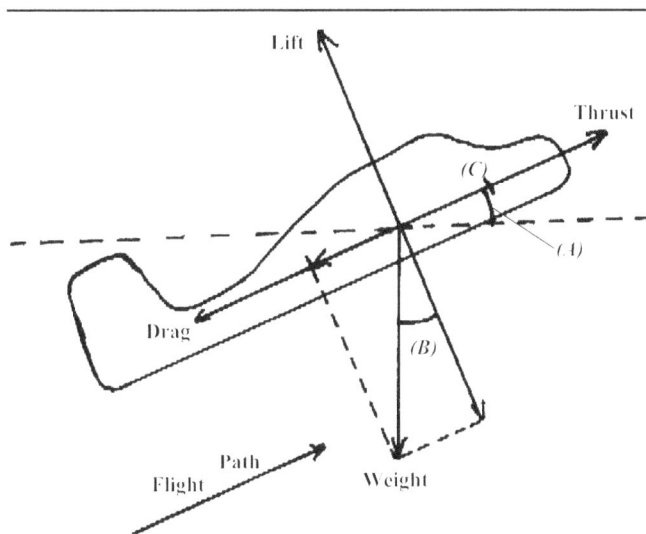

Fig 8-9. - *Forces in a Steady Climb.* - In an equilibrium climb, the thrust is equal to the drag plus the aft component of weight. The thrust beyond point 'c' is termed "excess thrust". This means that it is thrust that is not required to maintain level flight—so it will be used to accelerate or climb (in this case, climb). It can be shown that the more excess thrust is available, the steeper the climb angle will be.

Referring to Fig 8-9, angle *a* is the angle of climb and angle *b* is also equal to the angle of climb. It can be seen from the diagram that:

$$\text{Sin}b = \text{Sin}a = \frac{ET}{W} \qquad (8.4),$$

where *a* is the angle of climb, ET is the excess thrust, and W is the weight.

For power producing aircraft, the angle of climb is determined somewhat differently. The angle of climb is a function of the ratio between the true airspeed and the vertical speed (Fig 8-10). From the diagram, the following formula can be determined:

$$\text{Sin}a = \frac{ROC}{TAS} \qquad (8.5)$$

where *a*, once again, is the angle of climb, ROC is the rate of climb, and TAS is the true airspeed. It is important, of course, that the units of ROC and TAS be compatible. If one is in FPM, so must the other, and if one is in knots, so must the other. This equation is really no different from equation (8.4). If we were to substitute the rate of climb equation, (8.3), into equation (8.5), it would reduce back to the original angle of climb equation, (8.4).

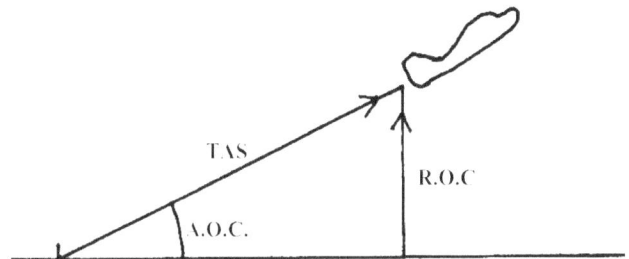

Fig 8-10. - *Rate of Climb and True Airspeed.*

When deciding on the speed for the *best* angle of climb, we need to find either the speed at which we have the highest amount of excess thrust available, or we need to find the speed which provides for the highest ratio between the ROC and the TAS. The two methods are in fact the same, but the ease of use—and thus which method gets applied—will depend on the type of aircraft.

For thrust producing aircraft (jets), it is easier to determine the speed for maximum excess thrust by looking at the drag/thrust curve (Fig 8-11). The best angle of climb for these aircraft will occur at the point of lowest drag—also the speed for maximum endurance on jet aircraft.

For power producing aircraft, a new graph is used. This is the ROC curve which plots ROC against airspeed (Fig 8-12). This new curve is derived from the power curve since ROC depends on excess thrust horsepower and will indicate the rate of climb that can be achieved at various airspeeds with full power. In using the ROC curve, we need to find the speed which allows for the highest ratio of ROC to TAS.

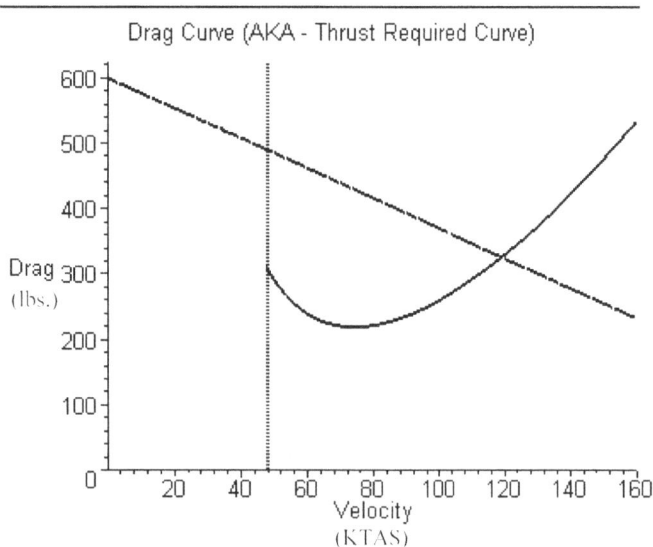

Fig 8-11. - *A Typical Drag Curve for a Prop Aircraft* - Also referred to as the thrust required curve, the drag curve can be used to determine the best angle of climb speed by finding the speed with the most excess thrust. This curve is for a prop aircraft (note the decreasing thrust available). For a jet, the thrust available will remain approximately constant as the airspeed varies.

Looking at lines that originate at the origin, it can be seen that the angle θ (theta) depends on this ratio:

$$Tan\theta = \frac{ROC}{TAS}.$$

The tangent of an angle increases as the angle increases. So we can see here that the larger the value of θ is, the larger the angle of climb will be. The limiting factor is that the line from the origin must cross the curve of ROC vs. TAS. So the largest value of θ that occurs within the aircrafts possible flight envelope is when the line just touches the top left of the curve. This line is called a tangent line since it just touches but does not cross the curve.

The speed determined by the contact point of the tangent line is the best angle of climb speed (V$_x$).

Obstacle Clearance

It is sometimes necessary during takeoff, or some other phase of flight, to climb over an obstacle. Several factors come into play when we want to avoid an obstruction. Clearly, the first will be speed—what speed should be used during an obstacle climb? On initial consideration, the speed selected will be V$_x$—the best angle of climb speed. This speed will allow us to gain the largest amount of altitude for a given amount of ground covered. If we consider some other factors, however, we may discover that V$_x$ is in fact not the most optimum speed to use.

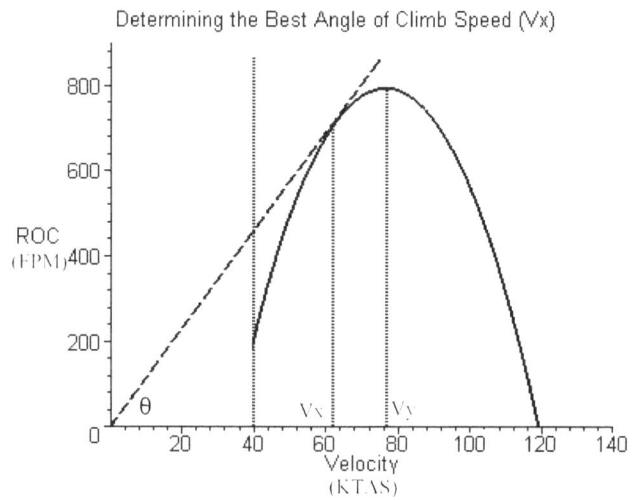

Fig 8-12. - *Rate of Climb vs. Airspeed* - A tangent line drawn from the origin will indicate the best angle of climb speed since it gives us the best ROC to speed ratio.

V$_x$ is the best angle of climb speed *once established*. During the takeoff roll, however, the aircraft is accelerating and is therefore not maintaining a constant airspeed. Since in this instance, part of the engines power is being used to accelerate the aircraft while the rest is being used for the climb, the climb angle will be more shallow. Once the aircraft is established in equilibrium at the best angle of climb speed, the climb angle will be at a maximum. However, during the initial climb, the lack of equilibrium may dictate the use of a lower climb speed.

Take Fig 8-13 as an example. Aircraft *a* rotated at a normal lift-off speed and climbed at a speed below V$_x$. This resulted in an earlier lift-off with a shallower climb. Aircraft *b* remained near the ground to accelerate to V$_x$, and a climb was then established and maintained at V$_x$. This resulted in a longer takeoff run and a steeper climb. As we can see from the diagram, there is a point where both techniques result in the same amount of clearance. Prior to this point, aircraft *a* will accomplish better obstacle clearance, and after this point, aircraft *b* will have better clearance.

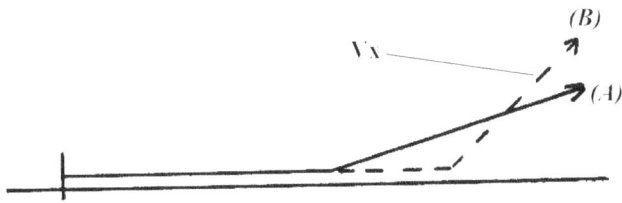

Fig 8-13. - *Obstacle Clearance Climb on Takeoff.* - Accelerating to Vx and then climbing results in a steeper climb, but a longer distance required near the ground. Whether or not this is advantageous depends on where the obstacle is.

Which of the two above techniques is more appropriate will depend on the aircraft and the height of the obstacle. In performance calculations done by aircraft designers, the standard obstacle height is 50 ft. So if at 50 ft. the aircraft using technique *a* requires a shorter distance, a climb speed lower than V_x will be provided in the POH for obstacle clearance takeoffs. On the other hand, if technique *b* requires less distance to clear 50 ft., the climb speed provided will be V_x.

Another consideration during an obstacle clearance takeoff is whether or not to use flaps. It was discussed during the section on takeoff performance that the use of partial flaps can decrease the aircrafts takeoff roll. Initially, it would appear that this would also decrease the obstacle clearance distance. Unfortunately, the drag produced by the flaps will decrease climb performance. As such, flaps may or may not assist obstacle clearance performance.

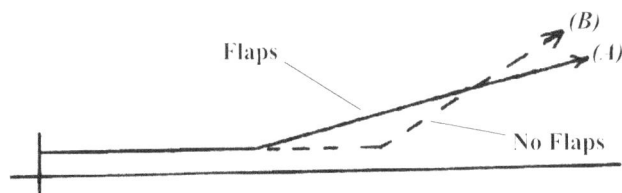

Fig 8-14. - *The Effect of Flaps on Takeoff and Climb.*

In Fig 8-14, aircraft *a* used flaps for the takeoff, and aircraft *b* didn't use flaps. As we can see, flaps in this case shortened the takeoff roll, but

they also caused a more shallow climb. Eventually we reach a point where the two aircraft are at the same height at the same distance from their starting point. Prior to this point aircraft *a* is higher, and after this point aircraft *b* is higher. So since a standard obstacle is considered to be at 50 ft., the choice of using flaps will depend on whether the crossing altitude for the two configurations is above or below 50 ft. This will vary from one aircraft to another, so the best source of information is the POH or Flight Manual.

As for an obstacle clearance procedure at some point other than takeoff, V_x will be the best speed to use. This statement doesn't take into account cooling properties of the engine. The aircraft manufacturer may suggest in the POH that climbing at V_x for prolonged periods of time could overheat the engine due to a lack of cooling airflow. From the perspective of aerodynamics, however, V_x will always be the best obstacle climb speed *once it is established.*

Variables

Anything that will alter the power curve of an aircraft will also affect the climb performance—both rate and angle. This means anything that will affect the power required (weight, C of G, density altitude, or configuration), or anything that will affect the power available (density altitude). As well, weight will have an extra effect on the climb performance since the rate of climb is inversely proportional to weight.

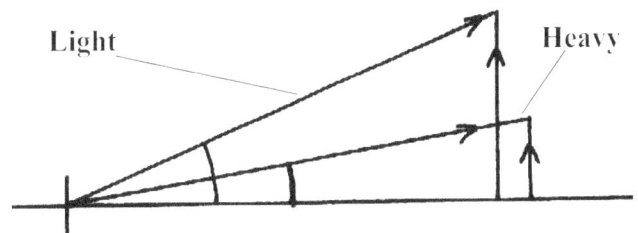

Fig 8-15. - *The Effect of Weight on Rate and Angle of Climb.*

Weight – For a moment, we will assume that the weight of the aircraft has no effect on the shape of the power curve. This means that V_y and V_x will be the same, and the amount of excess thrust horsepower will not change. An increase in weight will have the effect of decreasing the rate of climb, as well as the angle of climb.

Referring back to equation (8.3), we can see that doubling the weight will half the rate of climb. As a result of this, the angle of climb will be reduced as well. We can see this from equation (8.4) or from Fig 8-15.

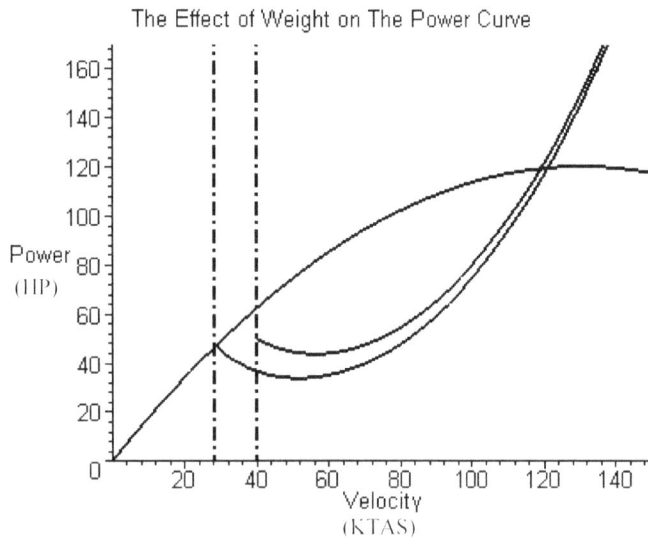

Fig 8-16. - *The Effect of Weight on the Power Curve* - The increase in power required will magnify the effect that weight has on rate of climb and therefore on angle of climb.

Now we can move away from our assumption that weight will not affect the power curve. As we know, a change in weight will result in a change in induced power required which is proportional to the square of the weight change. This change in induced power required will also result in an equal change in total power required (Fig 8-16). In reducing the weight from maximum gross weight, the total induced power required will be reduced.

As we can see from the diagram, this means that even more excess power will be available. Combined with the direct effect of reduced weight, this will mean that the rate of climb will be increased. Something else to notice from the new power curve is that V_y has been decreased from it's original value. Notice also that the angle of climb has been increased for a given airspeed due to the higher rate of climb. V_x will also be lower than the original value for V_x due to the change in the ROC curve (Fig 8-17).

C of G – The center of gravity location will also have a direct influence on the climb performance of the aircraft. A forward C of G will

increase the total power required by increasing the induced power required. This is caused by the increase in tail down force necessitating an increase in lift. As we know, any change in the power required will also change the amount of excess power available. This change in excess power will change the rate of climb of the aircraft.

So a forward C of G will decrease climb performance due to the increase in induced power required, and an aft C of G will increase climb performance due to the decrease in induced power required.

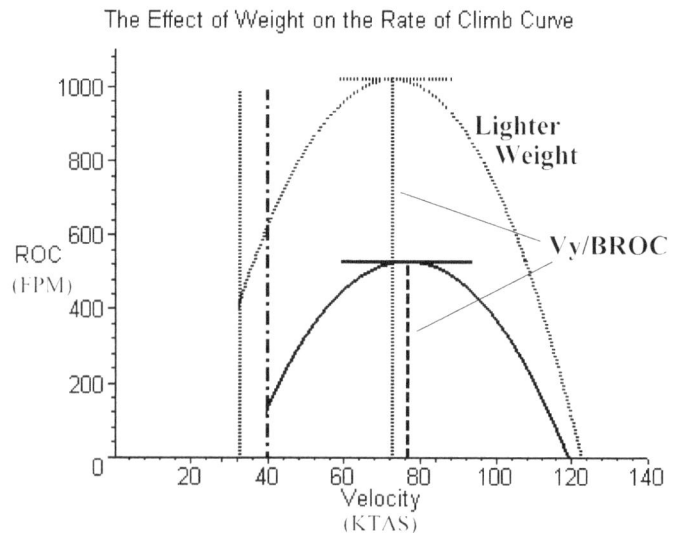

Fig 8-17. - *Weight Effect on ROC Curve.* - Obviously, climb performance improves at lighter weights. This graph quantifies this statement somewhat. Notice that both Vy and Vx are lower at the lighter weight.

Configuration – The configuration will affect the amount of drag being produced by the aircraft and thus the amount of power required by the aircraft. For example, if the landing gear is extended, the amount of parasite drag will be increased significantly. This will cause a corresponding increase in the parasite power required (Fig 8-18).

Flaps will have an effect on the climb performance for the same reason. Extending flaps will change both the parasite power required and the induced power required.

Wind – Since the aircraft moves through a mass of air with no consideration as to how the air is moving relative to the ground, wind will not affect the power curve in any way. As such, the wind will not affect the *rate* of climb in any way. However,

Configuration Effects on The Power Curve

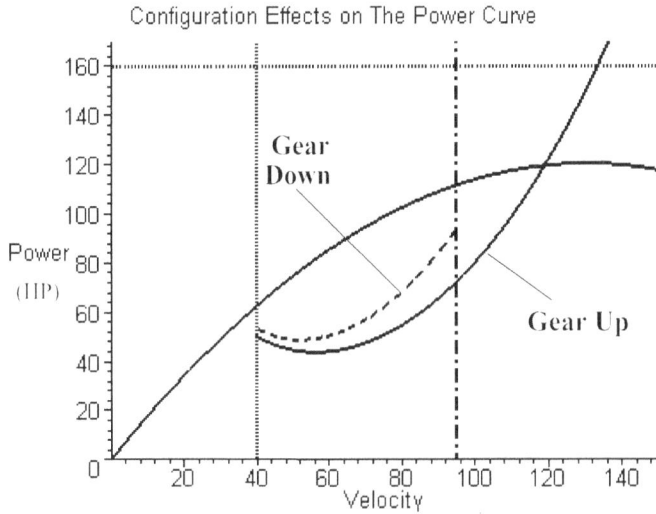

Fig 8-18. - *Configuration Effect on the Power Curve.* - This particular example illustrates the effect of landing gear on a retractable-gear aircraft. Notice that the increase in power required is more pronounced in the high speed range. Also notice the new limiting speed at the high speed end. This is the Vle—discussed in the next chapter.

due to the change in groundspeed, the *angle* of climb will be altered (Fig 8-19).

We have considered up to this point that the angle of climb is determined by the TAS and ROC. However, since the distance travelled over the *ground* is the important consideration it is in fact the *ground*speed combined with the ROC which is the determining factor.

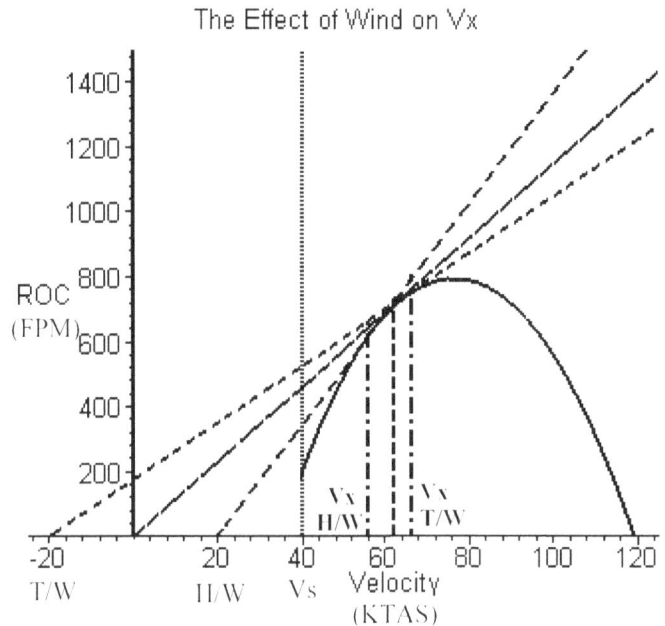

Fig 8-19. - *Wind Effect* - The wind will not affect the *rate* of climb, however, due to the change of groundspeed the *angle* of climb will be changed.

Wind will also change V_x. If we take another look at the ROC vs. TAS curve, we can determine the effect of wind. The groundspeed will change as a function of TAS plus a tailwind or minus a headwind. If we take this into consideration on the curve (Fig 8-20), it can be seen that a headwind will decrease V_x (provided that V_x wasn't at the stall speed to begin with), and a tailwind will increase V_x.

The Effect of Wind on Vx

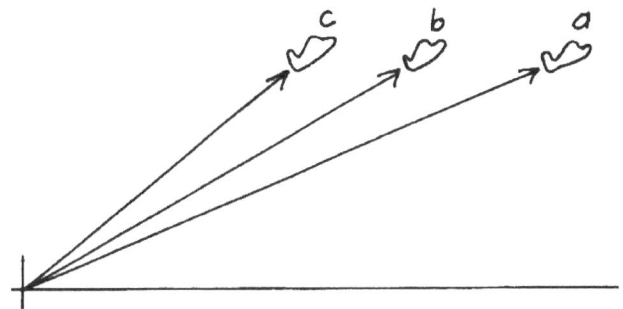

Fig 8-20. - *Wind Effect on V_x* - The change in groundspeed caused by wind will also change the Best Angle of Climb Speed (Vx).

As an extreme example to illustrate the point, consider an aircraft with a stall speed of 45 knots and a V_x of 65 knots. If this aircraft were to take off into a 50 knot headwind and climb at V_x, it would climb quite steeply. However, if instead the speed was reduced to 50 knots, the climb would be completely vertical due to having a groundspeed of zero. This is clearly an unrealistic example, but it illustrates the point. Fig 8-21 provides a more realistic illustration.

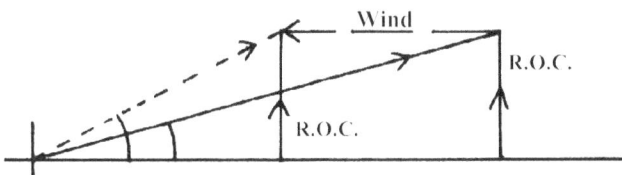

Fig 8-21. - *Illustration of Wind Effect on V_x* - (a) best angle without any wind, (b) best angle with a headwind, (c) best angle with a headwind and a speed adjustment.
This principle can be used to reduce the loss of climb angle when climbing with a tailwind. A slight increase in speed will reduce the effect of a tailwind. For example, this diagram would illustrate a tailwind if aircraft (c) was climbing without wind, aircraft (a) had a tailwind, and aircraft (b) had a tailwind and was correcting for it by increasing speed slightly.

Altitude – As we've discussed several times already, the air at higher altitudes is less dense. This means that less fuel can be burned and thus less power can be produced. Due to the lower brake horsepower available, the thrust horsepower will be decreased (Fig 8-22). This loss in thrust horsepower combined with an increase in power required will result in a loss of climb performance.

This loss of performance will be in both rate of climb and angle of climb. Rate of climb suffers because of the loss of excess thrust horsepower. Angle of climb is reduced for two reasons—the loss of rate of climb, and the increase in TAS for a given EAS. This means that the aircraft is moving forward more quickly and upward less quickly—resulting in a shallower angle of climb.

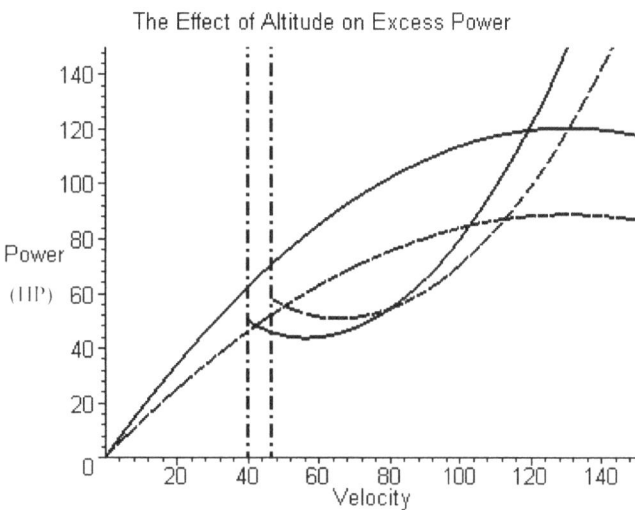

Fig 8-23. - *Using the Power Curve to Determine Ceilings* - The altitude at which the power curves just touch is the absolute ceiling. At this altitude, the aircraft cannot climb any further. The speed marked as 'Level Cruise' is the only speed that can be maintained without a descent.

an accurate airspeed would decrease the excess power and therefore the climb performance.

In determining the service and absolute ceilings, we can consider a series of maximum gross weight power curves at differing altitudes (Fig 8-23). As we can see, the amount of excess power available is decreasing as we climb, this means that the rate of climb will also decrease as we climb.

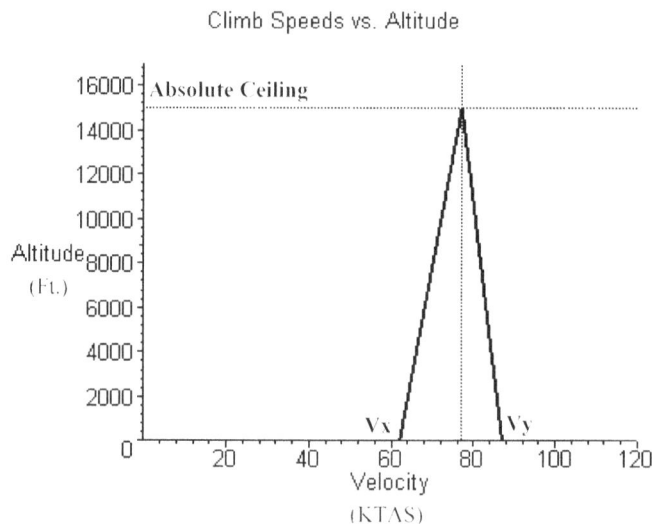

Fig 8-22. - *Altitude Effect on the Power Curve* - This change will lead to reduced climb performance at higher altitudes.

Altitude Limits

Because of the decrease in power available with an increase in altitude, we will eventually reach an altitude where the aircraft can no longer climb. This altitude is known as the absolute ceiling. Another limiting altitude is called the service ceiling. At the service ceiling, the aircraft can maintain a rate of climb of only 100 FPM. The service ceiling serves two purposes. First of all, it would be impractical to select a cruising altitude any higher, since the time to climb would be prohibitive. Second, the service ceiling is considered to be the absolute ceiling in turbulent air. This is because when flying in turbulence, the inability to maintain

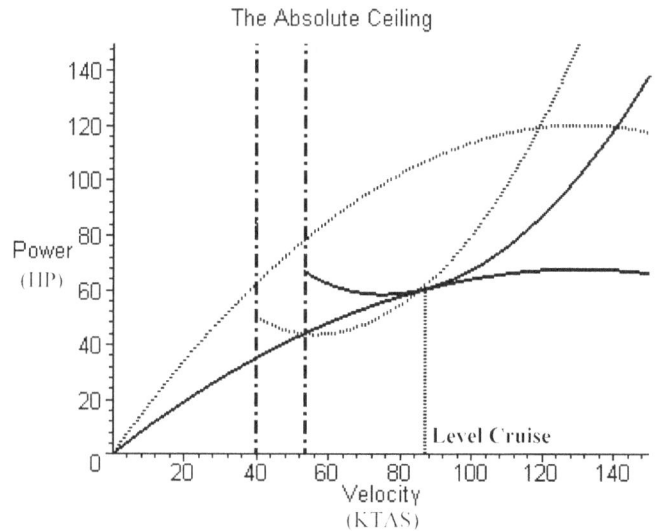

Fig 8-24. - *V_y and V_x vs. Altitude* - At higher altitudes, V_y and V_x become closer to one another. At the absolute ceiling, the two speeds are one and the same. This is the only speed that can be used to maintain level flight.

The altitude where the maximum rate of climb is 100 FPM is our service ceiling. Continuing to a higher altitude, we eventually reach a point where the two curves just touch each other. This is the absolute ceiling, and the speed at which the two curves touch is the only speed at which altitude can be maintained. To fly faster or slower would require a descent since a power deficiency would result.

If we consider altitude effects on V_y and V_x, we can see that they will change with altitude and become closer to one another (Fig 8-23 & 8-24). At the absolute ceiling, the two speeds meet and become one and the same. This speed is the speed that is required to maintain altitude at the absolute ceiling.

Cruise

One of the primary considerations when designing an aircraft is cruise performance. This is usually the case because most people use airplanes simply as a mode of transportation. Cruise performance entails a number of different aspects including the topics we will cover here.

We will look at the speed an aircraft can maintain in level flight, the endurance of the aircraft, and the range of the aircraft. Like other performance considerations, these items can be related to the power curve.

Speed

Speed is of primary concern in aircraft design. Most people who use aircraft outside the training industry and the military use them simply because they are a rapid method of transportation—much faster over long distances than cars for example. This means that the faster an aircraft can fly, the more likely that people will want to use it. The maximum speed of an aircraft, along with the more practical cruising speeds, are determined by the power curve.

On the high speed side of the power curve, the power required is increasing rapidly with speed, and the power available is decreasing due to the loss in prop efficiency (Fig 8-25). At the point where the two curves cross, the aircraft has reached it's maximum level cruise speed. Speeds above this cannot be realized without a loss of altitude.

The maximum level cruise (MLC) speed is

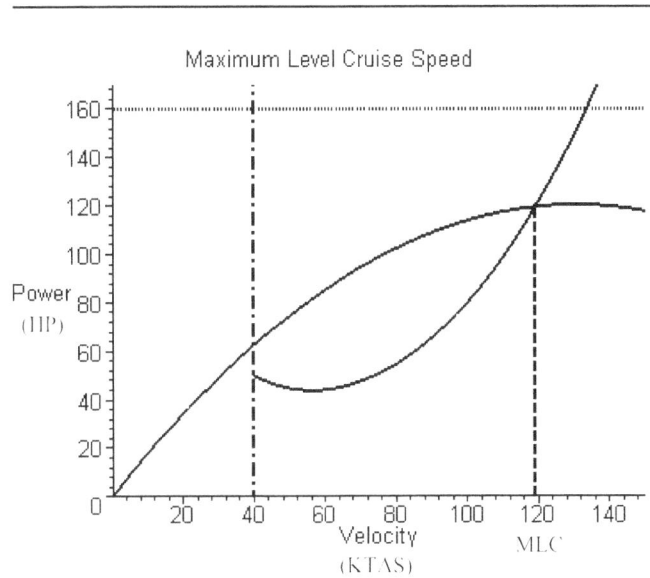

Fig 8-25. - *Maximum Level Cruise Speed* - At speeds higher than MLC, a power deficiency is unavoidable and a descent is required.

determined largely by the parasite drag of the aircraft characteristics of the aircraft along with propeller efficiency characteristics at high speeds. Induced drag has little bearing on high speed operations since it is minimal in this range.

Unfortunately, the maximum level cruise speed of an aircraft is not often a practical speed due to the increased inefficiency in fuel consumption. A power producing engine will burn fuel at a rate that is approximately proportional to the BHP being produced. At higher speeds, however, the speed increase will not be proportional to the power increase. In the area near the bottom

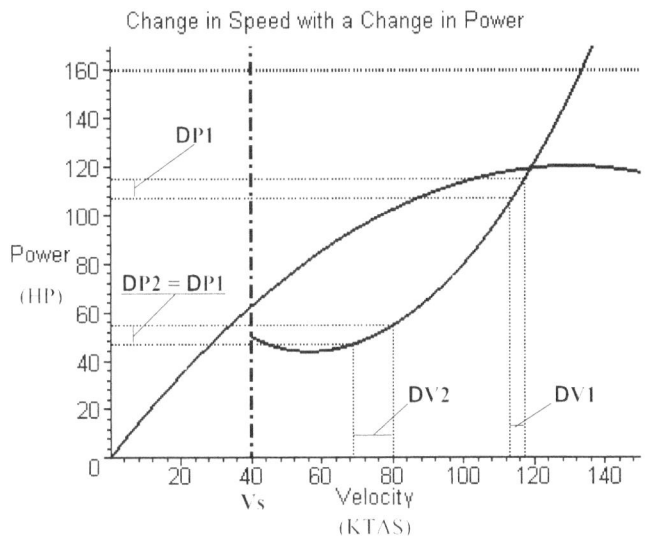

Fig 8-26. - *Power Changes vs. Speed Changes.*

of the power required curve (Fig 8-26), a small increase in power—and a correspondingly small increase in fuel consumption—will yield a relatively large increase in speed. At higher speeds near the maximum level cruise speed, a large increase in power—with a correspondingly large increase in fuel consumption—will result in a relatively small increase in speed. So even though the aircraft is capable of operating at higher airspeeds, the range of the aircraft will be sacrificed due to the disproportional increase in fuel flow.

In the next two sections we will consider two options available in the lower speed range. These options—flight for range and flight for endurance—both improve fuel consumption in their own way. Unfortunately, though, speed is sacrificed in order to improve fuel economy. This is a dilemma that designers and pilots alike face on a regular basis. In order to improve one aspect of performance, another must be given up—everything is a compromise. Under normal operations, where speed is not of essence and fuel economy is not critical, pilots often use a cruising speed that is at a happy medium between maximum range and maximum level cruise.

Another consideration when choosing a cruise speed is the toll that high power settings exact on the engine. Most piston engines are not meant to be operated at power settings above 75% for prolonged periods of time. This means that even though the aircraft may be aerodynamically capable of higher speeds, these higher speeds may not be available due to engine management considerations.

Low Speed Flight – Considerations of low speed flight include slow flight, stalls, and spins. Stalls and spins have already been covered in Chapter 6 and will be covered some more in Chapter 19. Slow Flight will be covered in detail in Chapter 18.

Endurance

Flight for maximum endurance is defined as obtaining the maximum amount of time aloft per unit of fuel consumed. Notice that with endurance, the distance travelled is not an issue. This means that the speed travelled is not an issue either. What is important for maximum endurance is the rate of fuel consumption and thus the power setting. A lower power setting will allow for a lower rate of fuel

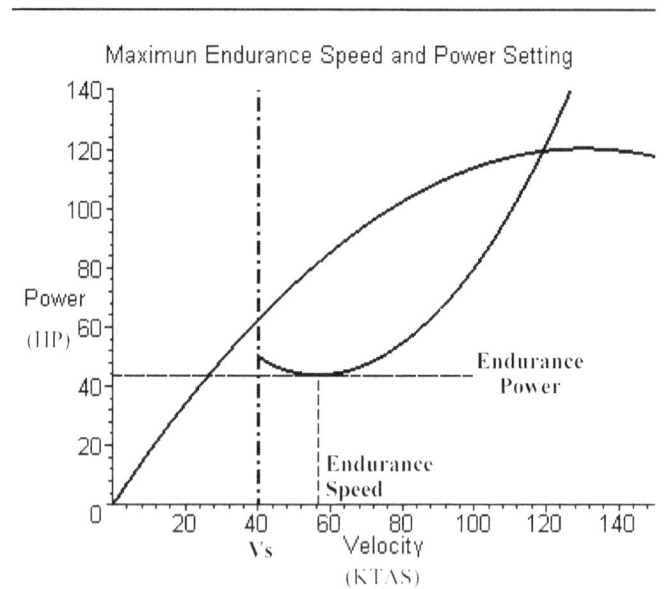

Fig 8-27. - Determining Maximum Endurance on the Power Curve - The lowest power required will provide the maximum endurance by reducing the fuel flow to a minimum.

consumption, and therefore greater endurance with a given amount of fuel.

On initial consideration, it would seem that a lower power setting simply means that we must fly at a lower speed. However, if we look again at the power curve (Fig 8-27), it can be seen that there is a lowest point on the curve. Speeds above this speed require more power, and speeds below this speed require more power. So simply flying as slow as the aircraft will go will not allow us the maximum possible endurance since these low airspeeds will still require a higher power setting.

Endurance will be achieved at the speed which requires the lowest power setting to maintain level flight. On the power curve, this is the speed corresponding to the lowest power required.

Range

Flight for maximum range is another form of fuel conservation. Unlike endurance, however, range deals with obtaining the maximum amount of distance from the aircraft per unit of fuel consumed. This means that the lowest power setting will no longer do the trick due to the low airspeed, while at a higher airspeed, fuel consumption can increase dramatically. What we need is a speed/power combination that provides us with the most distance covered for the least amount of fuel burned. Considering that the distance covered divided by the time taken will give us velocity, and the fuel burned divided by the time taken will give us the fuel flow, we get the following relationship:

$$\frac{\text{Distance Covered}}{\text{Fuel Burned}} = \frac{\text{Velocity}}{\text{Fuel Flow}}$$

Once again, the fuel flow is *approximately* (for our purposes, we can assume *exactly*) proportional to the power setting. So for maximum range we want the highest velocity to power ratio. The power curve relates for us the power that must be used to maintain level flight at different airspeeds. So from the power curve, we can determine the speed for maximum range by determining the speed for the maximum velocity/power ratio.

On the power curve, each speed has it's own velocity/power ratio. If we were to draw a line from the origin of the graph to any given point on the curve and then draw a line from that point to the power axis (Fig 8-28), a triangle would be formed. The tangent of angle θ will be equal to the velocity/power ratio. Since the tangent of an angle increases as the angle increases, we will want angle θ to be as large as possible. The line from the origin must cross the power required curve at some point, so the largest possible value for θ is when the line is a tangent to the curve. The speed determined by this tangent line is the speed for maximum range.

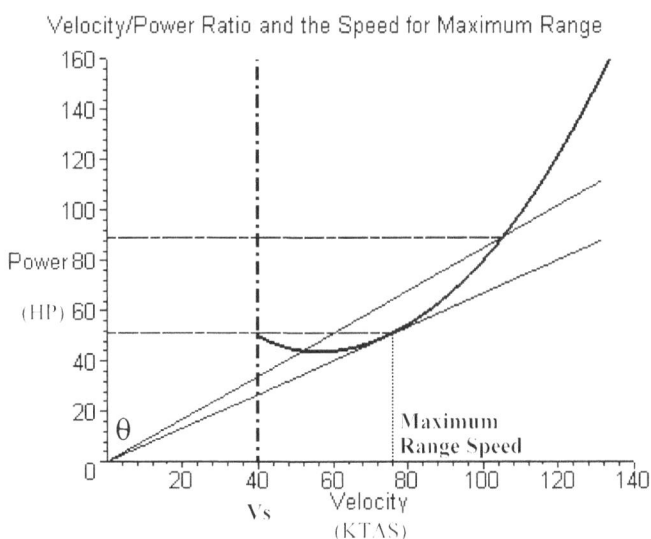

Velocity/Power Ratio and the Speed for Maximum Range

Fig 8-28. - *Velocity/Power Ratio* - The higher this ratio is, the better the range of the aircraft will be. The maximum V/P ration is achieved when a line from the origin is tangent to the power curve.

Recall that the power required is the drag multiplied by the velocity. So, starting with the velocity/power ratio, we get the following:

$$v/P = v/vD = 1/D.$$

This ratio becomes higher as the drag is reduced. So, for the optimum range performance we want the lowest amount of drag possible.

Recall as well, that the lift being produced at various speeds is always equal to the weight as long as the aircraft is in equilibrium. So as the drag decreases, another ratio increases—this is the lift/drag ratio. The lift/drag ratio is a measure of how efficiently the aircraft is operating. Flight for maximum range occurs at the AOA for the aircraft's best L/D ratio.

Variables

Just like the takeoff and climb performance of an aircraft, cruise performance will be affected by anything that affects the power curve. Each of these variables will have a combined effect, but we will look at them individually.

Weight and C of G – The weight of the aircraft will of course have an effect on the induced power required for the aircraft. Higher weight will mean a higher amount of lift is being produced, thus there will be more induced power required. This will have an effect on the total power required. A reduction in weight will have a similar effect, except the induced power required will be reduced (Fig 8-29). This change in the power curve will have an effect on all of the cruise performance aspects of the aircraft.

At the high end of the curve, the change will be much less noticeable than at the low end. This is due to the fact that at high speeds, parasite power required is predominant, and at low speeds, induced power required is predominant. None the less, an increase in weight will decrease the maximum level cruise speed slightly, and a decrease in weight will improve the maximum speed slightly.

The effect of weight on endurance is also apparent from Fig 8-29. The minimum power required will be reduced with a reduction in weight. This minimum will also occur at a lower airspeed. So now the aircraft will be able to maintain altitude

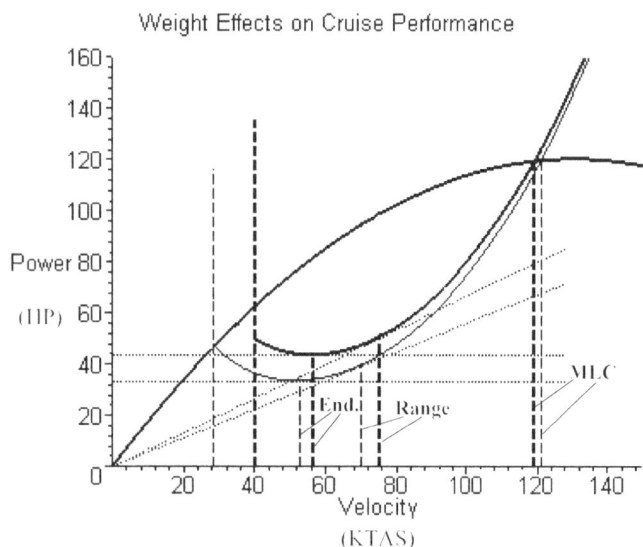

Fig 8-29. - *Weight Effects on Cruise Performance.* - The weight of the aircraft will affect all aspects of cruise performance, including, MLC, Maximum Range, Maximum Endurance, and the cruise speeds that result from any given power setting.

at a lower speed than before, and at a lower power setting than before. This means that the maximum endurance will improve with a decrease in weight due to the lower fuel consumption rate.

A change in weight will also have an effect on the range of the aircraft. Reducing the total power required will lower the tangent line drawn from the origin. This means that the velocity/power ratio will be increased, and thus the range will be increased.

As we can see from the diagram, the speed at which maximum range occurs will also be decreased. We know that maximum range will occur at the speed which allows the maximum lift/drag ratio. Since the best lift/drag ratio will always occur at the same angle of attack, we can determine what the new speed will be. Recall that equation (6.1) is based on the fact that stalls always occur at the same angle of attack. Since the best lift/drag ratio will always occur at the same angle of attack, the new range speed after a weight change can be determined with a similar equation.

$$\text{Stall Speed} = V_s \sqrt{(GW/MGW)} \quad (6.1)$$

becomes,

$$\text{New } V_{Range} = V_{Range} \sqrt{(GW/MGW)} \quad (8.6).$$

Altitude – As we discussed earlier, an increase in altitude will shift the power curve up and

to the right (Fig 8-30). The tangent line to the curve will still be at the same angle, meaning the velocity/power ratio remains the same. This means that the maximum range of the aircraft will not actually change with altitude. This statement does not, however, take into consideration the fuel burned in the climb. Fuel flow rates are relatively high in the climb, and groundspeed is relatively low. This will lead to an overall reduction in range with a climb to a higher altitude. As well, winds are not considered here. Wind can increase or decrease range and they can vary significantly from one altitude to the next.

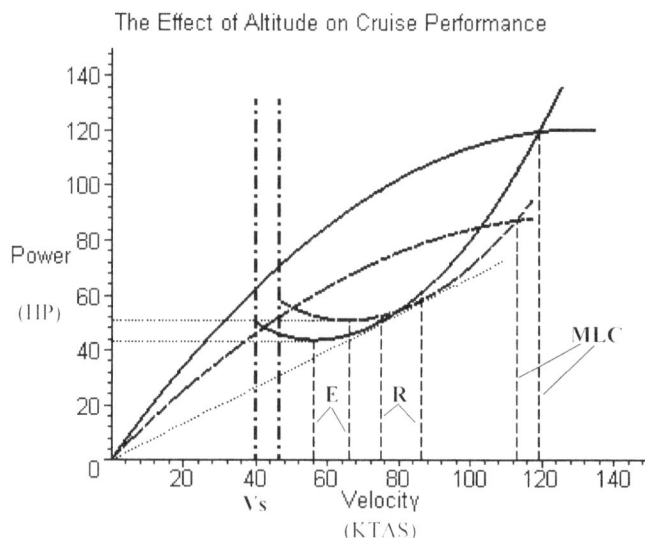

Fig 8-30. - *Altitude Effects on Cruise Performance.*

You may notice that in the POH or Flight Manual for most aircraft, the range performance graph indicates an increase in range with an increase in altitude (Fig 8-31). Notice, however, that these curves are for a specific power setting, not for the maximum range condition of the aircraft. As we can see from Fig 8-30, the speed of the aircraft will increase for a *given power setting* at higher altitudes. Since the speed increases but the power remains the same (therefore fuel flow remains the same), the range will increase. The *optimum* range condition will not change—the speed will be higher and the power setting will be higher, but the ratio will not change. Sometimes the graphs will be curved so that range is optimum at one particular altitude, while at higher and lower altitudes the range of the aircraft is not as good. This means that the graphs account for the fuel used in the climb, but they still refer to fixed power settings in cruise.

As for range, the wind plays an important role. Since the range is the distance travelled, it is reliant on groundspeed. Wind will alter the groundspeed and will thus affect the range of the aircraft. A headwind will slow the aircraft down and thereby reduce range. A tailwind will increase the groundspeed and thus improve range.

The wind will also affect the speed at which best range is achieved. Previously, we based the tangent line on the origin of the graph which was an airspeed of zero. If instead we based the origin on a *ground*speed of zero (Fig 8-32), the new tangent point would give us maximum range. A groundspeed of zero means that with a headwind, the airspeed must be equal to the wind, and with a tailwind the origin must be at a negative airspeed.

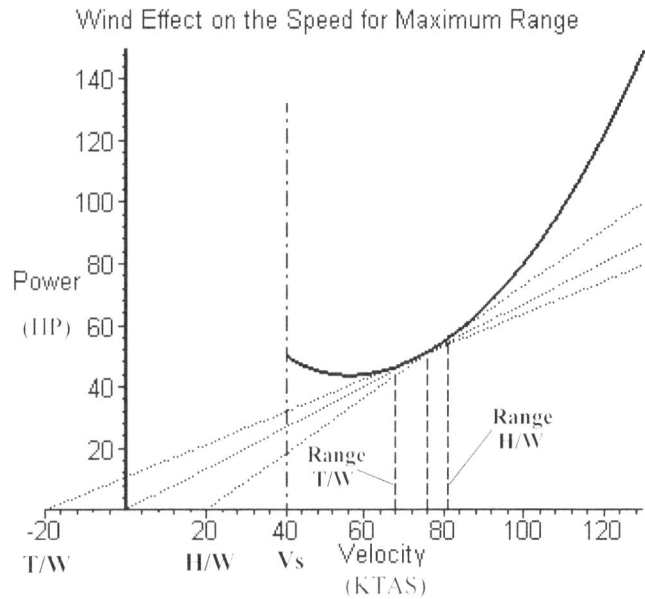

Typical P.O.H Graph of Range vs. Density Altitude

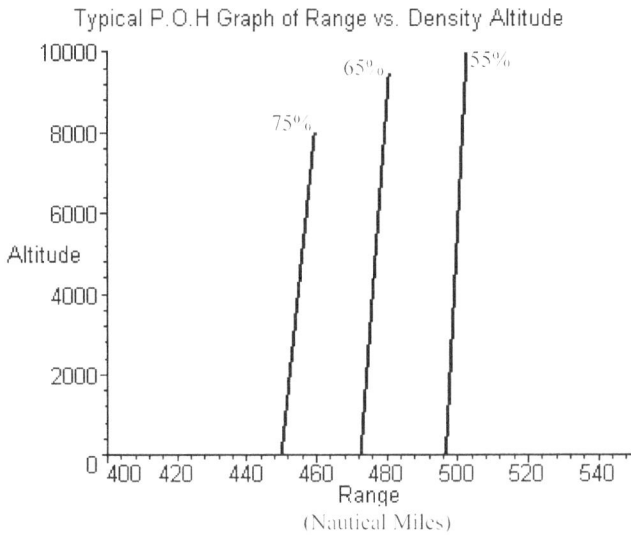

(a) A *typical* RANGE performance graph.

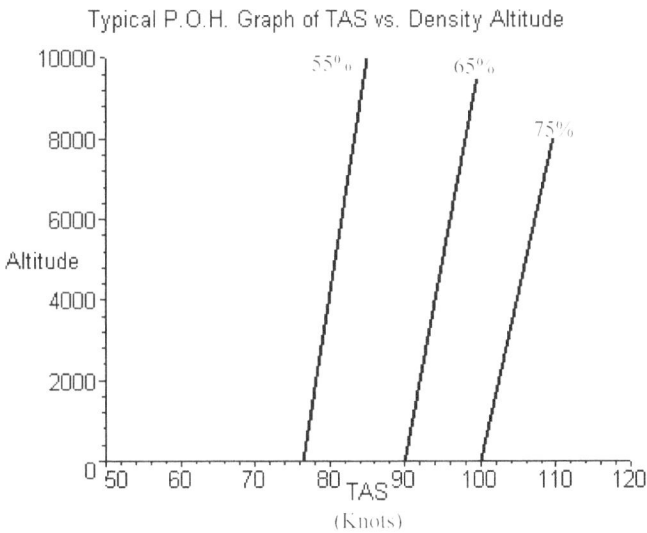

Typical P.O.H. Graph of TAS vs. Density Altitude

(b) A *typical* CRUISE SPEED performance graph
Fig 8-31. - Typical POH Cruise Data.

Wind Effect on the Speed for Maximum Range

Fig 8-32. - Correcting the Max Range Speed for Winds - Instead of using the origin as the tangent line start point, use a *groundspeed* of zero. This will lead to an increase in the best range speed with a headwind and a decrease with a tailwind.

The maximum endurance of the aircraft will decrease at higher altitudes. This is because the upward shift of the power curve increases the minimum power required to maintain altitude. This increase in the minimum power required will also increase the minimum fuel consumption and therefore reduce the maximum endurance.

Wind – The wind will have no effect on endurance. The power required to remain aloft is dependant on airspeed not groundspeed. So the shape and location of the power curve is not changed by wind, and thus the minimum power required will not change.

Configuration – In order to minimize the power required for any given speed, we need to keep the drag to a minimum as well. This means that it is preferred to keep the aircraft in the cleanest configuration possible. Flaps, landing gear, speed brakes, etc., will increase drag and as such should be retracted if possible.

Turbulence – Turbulence will cause fluctuations in speed and altitude. In the case of maximum range, this reduces the efficiency of the aircraft and will thus reduce the range of the aircraft

Turbulence is more critical while flying for maximum endurance. The speed fluctuations will cause the airspeed at some point to drop below the maximum endurance speed. In this condition, the aircraft requires more power than is being used (Fig 8-33), and will thus slow down or descend. If the aircraft is allowed to slow down, the power required will increase even more—leading to a greater deceleration. Eventually, this can lead the aircraft into a stall. This condition is know as slow flight and will be covered in more detail in Chapter 18.

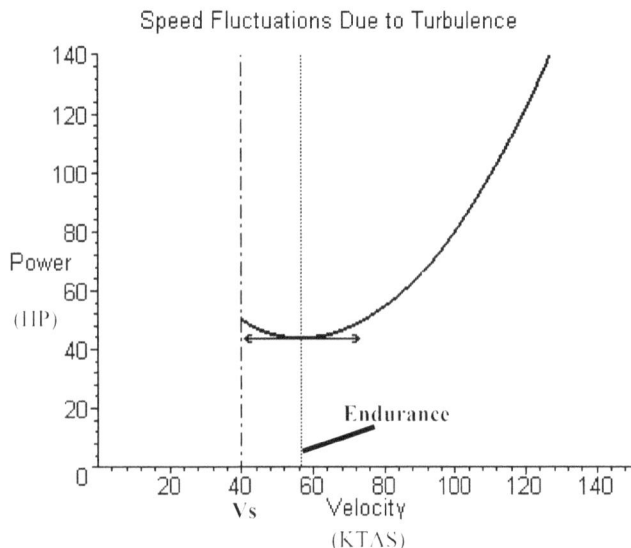

Fig 8-33. - *The Effect of Turbulence on Endurance.* - Once established in flight for maximum endurance, slowing down will cause a power deficiency. This change in airspeed can be caused by turbulence (among other things).

In order to avoid the slow flight speed range, it is advisable when flying for endurance in turbulence to increase the airspeed slightly. This will decrease the endurance of the aircraft somewhat, but will allow for much better control of the airspeed.

Descending

Every flight must end with a descent. In some types of flight operations (e.g. – flight training, sight seeing tours, etc.), several descents can be required on each flight. Understanding how and why an airplane descends is important, and sometimes critical, in completing a descent which suits the pilots current objective(s).

Descents are rather analogous to climbs. Two important measurements of a descent are the

rate of descent and the angle of descent. Where rate of climb is determined by excess power, rate of descent is determined by a power deficiency. As well, where angle of climb is determined by excess thrust, the angle of descent is determined by a thrust deficiency.

Rate of Descent

As stated above, the rate of descent is determined by a power deficiency. On the power curve (Fig 8-34), this means that for a given speed the power being produced will be less than that required. In order to maintain this given airspeed, the power required *must* be available. Since the power is not being produced by the engine, it must be provided by some other source. This source is the altitude of the aircraft. Remember that the height of the aircraft determines it's potential energy, and thus can be used to "produce" power—if altitude is given up.

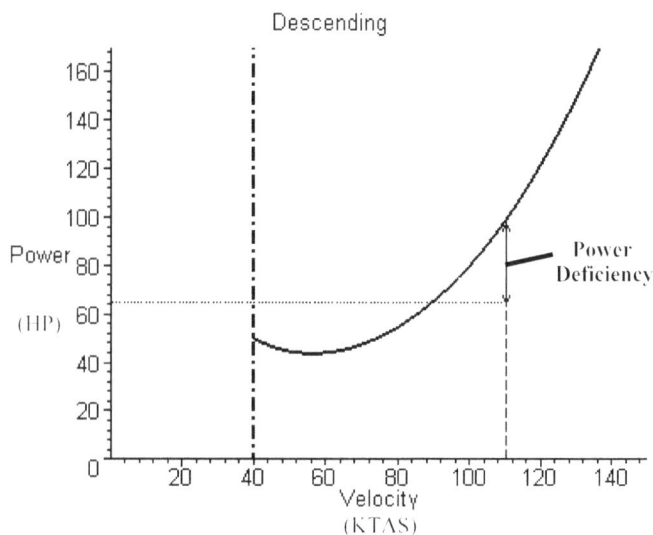

Fig 8-34. - *Flying With a Power Deficiency* - A deficiency in power will lead either to a deceleration or a descent. In maintaining equilibrium, a descent will result.

Similar to the climb, the equation for rate of descent is as follows:

$$ROD = \frac{33,000 \times THPd}{W} \qquad (8.7),$$

where THPd is the thrust horsepower deficiency.

For a given speed, the power required is a constant. This means that the rate of descent will be determined (if a constant airspeed is maintained) by the power setting. On the other hand, for a given power setting, the rate of descent will be determined by the airspeed.

Glide Range

The angle of descent is determined—like the angle of climb—by a combination of airspeed and rate of descent (Fig 8-35). Equation (8.5) can also be applied to descents:

$$\sin b = \frac{ROD}{TAS} \qquad (8.8).$$

As we can see, a low airspeed with a high rate of descent will result in a large angle of descent, while a high airspeed and a low rate of descent will result in a small angle of descent.

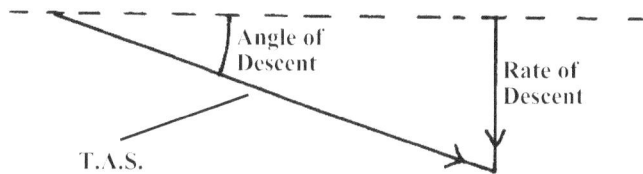

Fig 8-35. - *Airspeed and Vertical Speed Determine the Angle of Descent.*

Just as the angle of climb is determined by excess thrust, the angle of descent is determined by a thrust deficiency. The drag being produced by the aircraft must be countered by some force. In a descent, this force is provided by a forward component of the weight (Fig 8-36). The larger the thrust deficiency, the more weight is required to counteract the drag—this means that a larger angle of descent is required.

To obtain the most glide range out of an aircraft, a speed must be established which allows for the minimum thrust deficiency. This means that we want the speed which requires the least thrust. On the thrust curve, of course, this is the lowest point. This is also the speed which provides the best lift/drag ratio—just like the speed for maximum cruise range.

As it turns out, the lift/drag ratio is numerically equal to the glide ratio. In Fig 8-37, the weight vector is parallel to the height, and the forward component of weight acts along the same

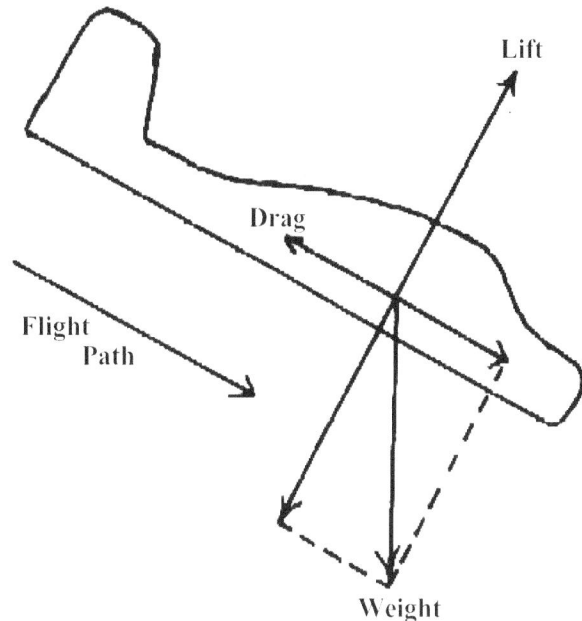

Fig 8-36. - *Forces in a Descent* - During a descent, the forward component of weight replaces the thrust (or some of the thrust) from the engine.

line as the flight path. From this, it can be seen that angles *a* and *b* are equal to one another. Therefore:

$$\tan a = \frac{height}{distance}$$

and,

$$\tan b = \frac{drag}{lift}.$$

Since the two angles are equal,

$$\frac{drag}{lift} = \frac{height}{distance}$$

Inverting this, we get:

$$L/D = Glide\ Ratio \qquad (8.9).$$

On the power curve, as we know, the speed for the best lift/drag ratio is determined by the tangent line from the origin. The tangent line also allows for the best TAS/ROD ratio. As we've seen from equation (8.9), this will allow for the minimum glide angle or best glide range.

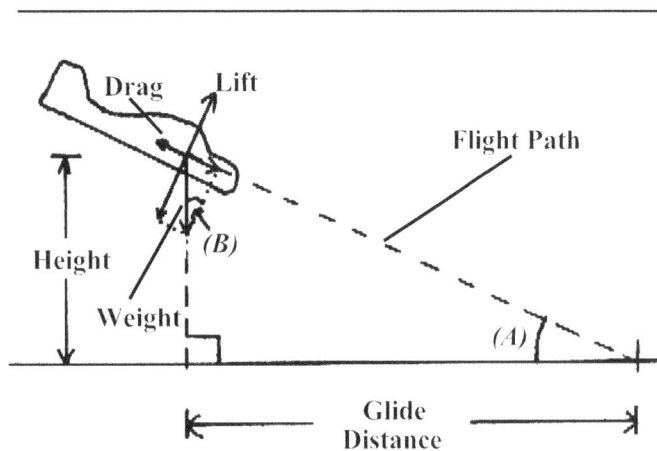

Fig 8-37. - *Lift/Drag Ratio is Equal to the Glide Ratio.*

Obstacle Clearance Descents

Obstacle clearance descents are normally used on landings over an obstruction. When landing, it is undesirable to waste runway once clear of the obstacle. So a descent speed which is slower than "best glide" is normally used. Higher descent speeds are avoided because, as we will see later in the chapter, low speeds are desirable on landing.

Variables

Weight – We can see from equation (8.7) that an increase in weight will in fact decrease the rate of descent for a given power deficiency. That having been said, the total power required will be increased due to the increase in induced power required. Which effect is predominant will vary with speed so that at some speeds the resulting rate of descent is higher while at other speeds the resulting rate of descent is lower.

The effect that weight has on angle of descent is a little bit different. As an example, let's reduce the weight of a fictitious aircraft by half. The induced drag being produced by this aircraft will now be quartered for a given airspeed. Referring to Fig 8-38, we can see that the minimum total drag point now occurs at a lower airspeed, and that the minimum drag is half of it's previous value. So we have halved the weight and at the same time halved the drag. This means that the L/D ratio has not changed, and the optimum glide angle will not be changed—however, the speed at which it occurs will be lower than before.

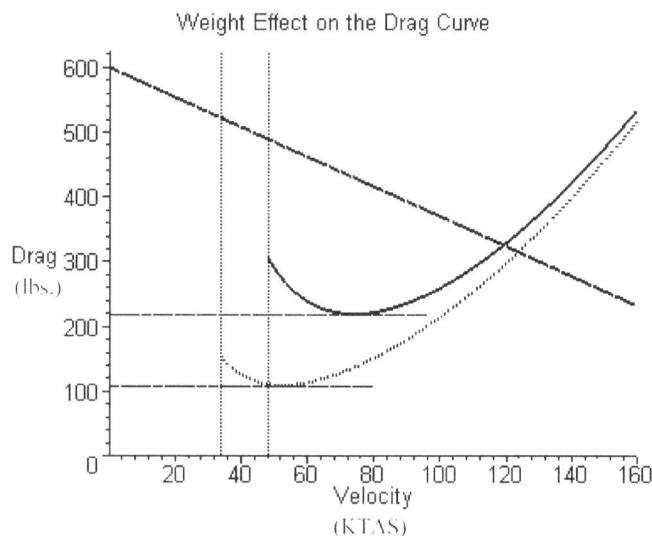

Fig 8-38. - *Weight Effect on the Drag Curve* - Halving the weight will also halve the total drag if the speed is adjusted to maintain the best L/D AOA. This means that the L/D ratio does not change and therefore the maximum glide ratio will not change.

The best L/D ratio always occurs at the same angle of attack. The speed for best L/D can therefore be determined using equation (8.6).

Configuration – The configuration has a direct influence on how much drag is being produced, and hence how much power is required. In a high drag configuration, the power deficiency will be greater. As a result, the rate of descent will increase. With an increase in rate of descent, the angle of descent will be increased as well.

Wind – Wind has no effect on the power curve, and as a result cannot influence the rate of descent (just like climbs). However, the wind will affect the aircraft's groundspeed. Because of this, the *angle* of descent will change. A headwind will decrease the groundspeed, and thus increase the angle of descent (decrease the glide range). A tailwind, on the other hand, will increase the groundspeed, and thus decrease the descent angle (increase the glide range).

Turning

Being able to turn an aircraft is obviously a useful skill. Unfortunately, most aircraft don't carry enough fuel to fully circle the globe and then land back at their point of origin. So unless an airport presents itself directly in front of our aircraft, a turn will be necessary at some point.

Turns in an aircraft are a little more complicated than they are in a car or even a boat. It is important to understand the forces that act on the aircraft in a turn since they determine the rate and radius of the turn, as well as the acceleration loads imposed on the airframe and the pilot's and passenger's bodies.

Why Does an Aircraft Turn?

During straight and level, unaccelerated flight, the lift produced by the wings is equal to and opposite to the weight. This is because the lift is produced perpendicular to the relative airflow and perpendicular to the wingspan. In a turn, however, the aircraft is banked. Since the lift remains perpendicular to the wingspan, the force produced by the wings (i.e. – lift) is now inclined to the side. This lift can be resolved into two perpendicular components, one of which still opposes the weight, and one of which is now perpendicular to the weight (Fig 8-39).

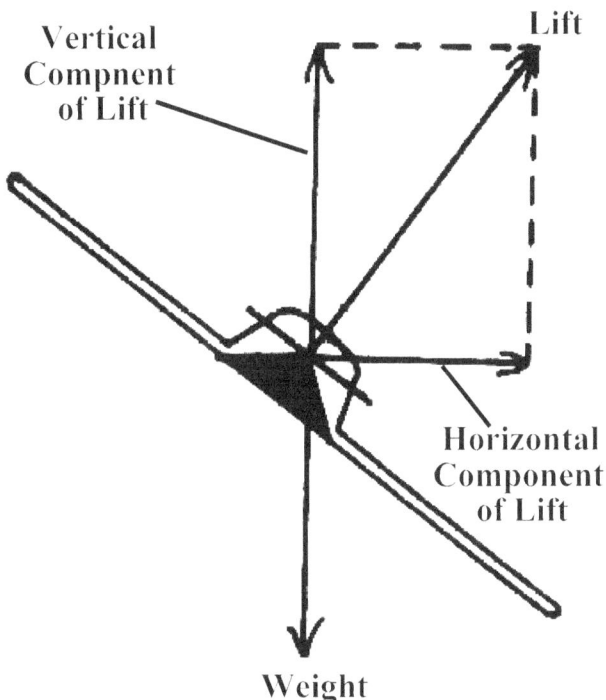

Fig 8-39. - Lift in a Banked Attitude - Since lift acts perpendicular to the wingspan, it will be inclined to the side. Dividing it into components, we get a vertical component of lift (which supports the weight of the aircraft) and a horizontal component of lift (which is unbalanced and accelerates the aircraft into the turn).

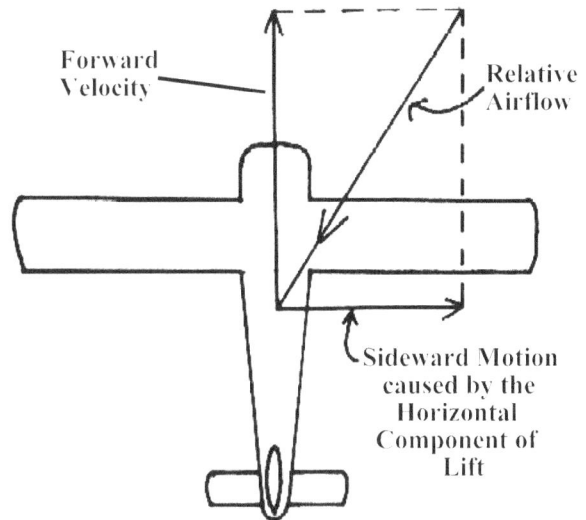

Fig 8-40. - Top View of Forces in a Turn - The horizontal component of lift causes an acceleration to the side. The sideslip created causes yaw via directional stability and the aircraft turns.

The vertical component of the lift will support the weight of the aircraft so that vertical equilibrium can be maintained. The horizontal component of lift, on the other hand, is unopposed. This creates an imbalance of forces, and as we know from Chapter 1, an imbalance of forces will cause an acceleration in the direction of the net force. So now the aircraft will accelerate in the direction of the bank (Fig 8-40).

As the aircraft accelerates to the side, a sideslip is produced. Due to directional stability, the sideslip causes a yaw. The total effect is for the aircraft to change it's direction of travel as well as it's heading.

The horizontal component of lift is the centripetal force which accelerates the aircraft into a circular flight path (Fig 8-41).

During a turn, the aircraft will tend to pitch down. This is a result of pitch stability—the aircraft tends to maintain a constant angle of attack. Once established in the turn, if the AOA and the airspeed are not adjusted, the vertical component of lift will not equal weight. This will cause the aircraft to accelerate downwards—changing it's flight path. With a downwards shift in flight path, the AOA will be increased and the aircrafts pitch stability will cause a downwards pitch. The aircraft will now be in a descending turn and will gain some airspeed due to the loss of altitude. This gain in airspeed will

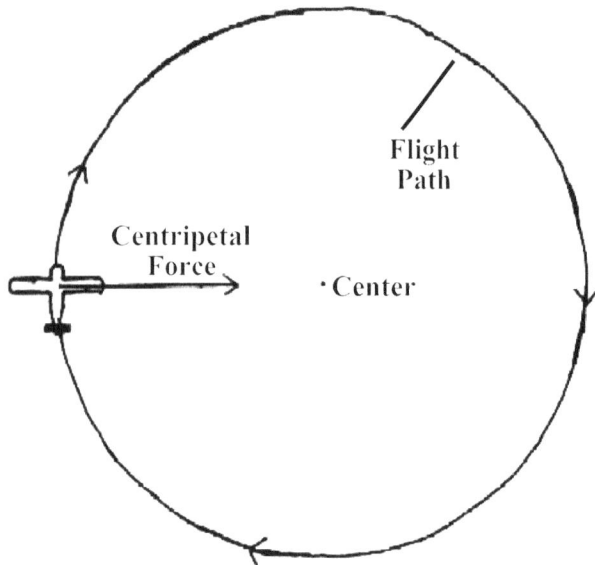

Fig 8-41. - *Centripetal Force* - In the case of a turning aircraft, the horizontal component of lift causes the centripetal acceleration.

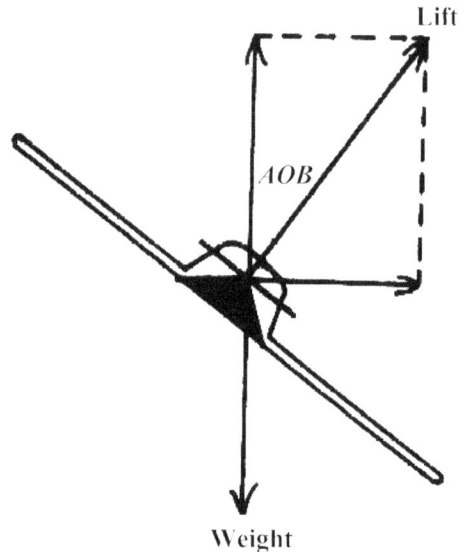

Fig 8-42.

allow the aircraft to establish enough lift to support it's weight at the trimmed AOA.

This only occurs, of course, if the pilot allows it to. An increase in back pressure during a turn will allow the AOA to be increased, thus producing enough lift to support the aircraft.

Load Factor

As we can see from Fig 8-42, the lift being produced by the aircraft in a turn is greater than the weight, which is equal to the vertical component of lift. The acceleration this leads to results in a sensation of being "pulled" down into ones seat (the sensation is in fact the seat *pushing* against *you*). This sensation is known as load factor and is usually measured in "g's", or multiples of the acceleration due to gravity.

The load factor (n) can be determined by dividing the lift that is being produced by the weight of the aircraft:

$$n = L/W \qquad (6.3).$$

So if a 2000 lb airplane is producing 4000 lbs of lift, 2 g's will be experienced. This equation for load factor can be applied to any symmetrical (non rolling) condition of accelerated flight.

In a turn, the load factor can be determined

by comparing the lift and weight in a different (but equivalent) way. As we can see from Fig 8-42, the lift vector and the vertical component of lift can be used to create a triangle with the angle of bank as one corner. We know that the vertical component of lift is equal to the weight of the aircraft. So we can see that the cosine of the angle of bank is equal to the weight divided by the lift, which is the inverse of the load factor. From this, we get the following:

$$n = 1/CosAOB \qquad (8.10),$$

where AOB is the angle of bank.

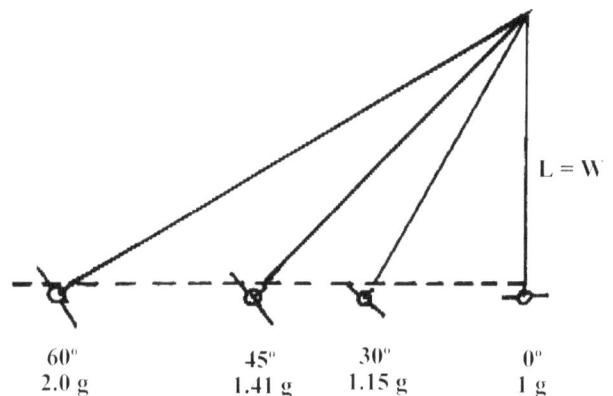

Fig 8-43. - *Load Factors at Various AOB's.*

So we can see from equation (8.10), and from Fig 8-43, that as the angle of bank increases, so does the load factor. This is indeed the case for level coordinated turns. In theory, the load factor should increase to infinity as the angle of bank approaches 90°. Obviously, however, there must be some practical limits to the loads on the aircraft. These limits, which include the power required in a turn, the stall speed in a turn (which will be discussed in the next two sections) and the load limits acceptable to the airframe (which will be discussed in the next chapter), must be considered.

Power Required

We know from Chapter 3 that as the lift is increased, the induced power required will be increased by an amount proportional to the square of the lift increase. In other words, if the aircraft is experiencing a 2g acceleration in a turn, the induced power required will be increased by a factor of 4. If the induced power required is increased, the total power required must increase by the same amount. The result will be a change in the power curve looking something like Fig 8-44. Notice that the increase in power required is most pronounced at lower airspeeds. This is due to the fact that the induced power required is more predominant at lower airspeeds than at higher airspeeds, while at high airspeeds, the parasite power required is predominant.

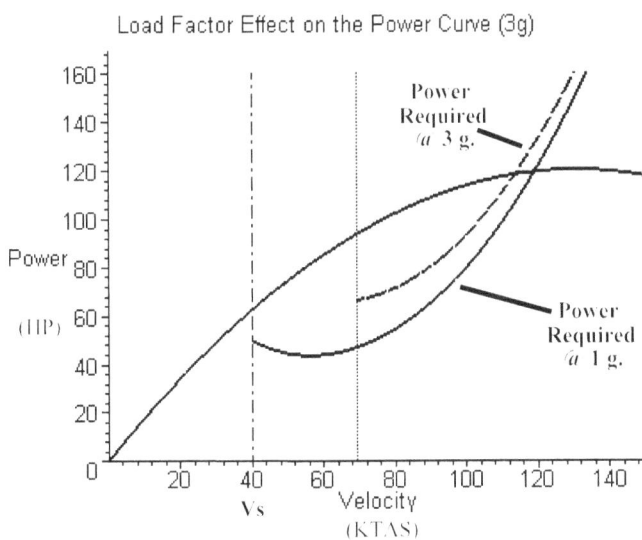

Fig 8-44. - *Load Factor Effect on the Power Curve* - Induced power required increases due to the increase in lift. The effect on the total power required is more pronounced at the low speed end since this is where induced power is predominant.

Since the power required has increased, but the power available hasn't changed, the possible range of sustainable airspeeds has been reduced. In steep turns where power available becomes a limiting factor, *higher* airspeeds are more likely to be sustainable than lower airspeeds. Once again, this is due to the large increase in power required at lower airspeeds with a less noticeable change at higher speeds.

Stall Speed

As we already know from Chapter 6, an increase in load factor will increase the stall speed of the aircraft proportional to the square root of the load factor. This means that regardless of whether or not power available is restricting us, the minimum speed in a turn will be higher than that in level flight. Recall, of course, that the term "stall speed" is a misnomer—stalling angle is more appropriate. This, of course, is the reason that the stall "speed" increases with load factor.

Fig 8-45 is a graph of load factor and stall speed multiple vs angle of bank for a level coordinated turn.

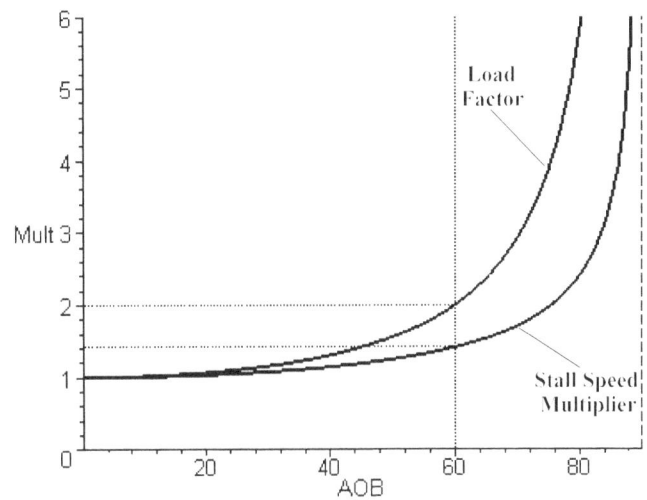

Fig 8-45. - *Load Factor and Stall Speed Multiple in a Turn.*

Radius of Turn

Way back in Chapter 1, we discussed the phenomenon of uniform circular motion. An aircraft which is maintaining a constant angle of bank and a constant airspeed will be experiencing just such motion. The equations which we applied to a body undergoing a centripetal acceleration can also be applied to an aircraft in a steady turn.

In the case of an aircraft, we can refer to Fig 8-46 and determine the centripetal force from the angle of bank:

$$CF = WtanAOB \qquad (8.11).$$

Substituting this into equation (1.27) and rearranging slightly, we get the following:

$$r = \frac{v^2}{gTanAOB} \qquad (8.12),$$

where (v) is the true airspeed, and (g) is the acceleration due to gravity.

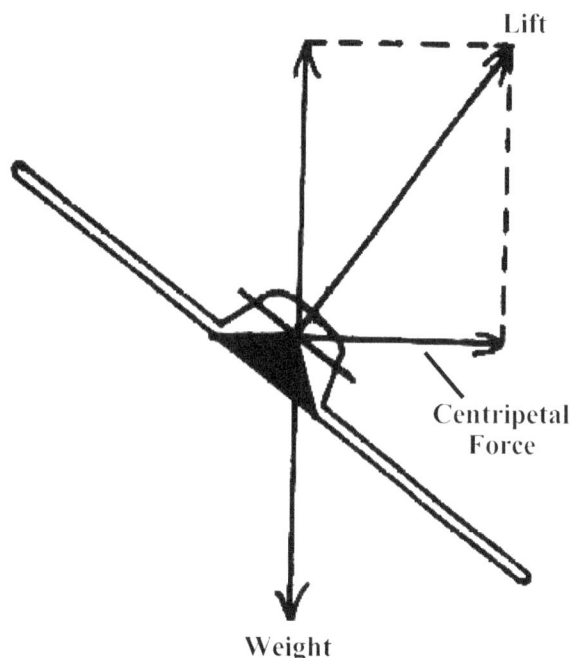

Fig 8-46. - *Determination of Centripetal Force in a Turn* - The force accelerating an aircraft into the turn is the horizontal component of lift.

So we can see here that the turn radius of the aircraft will increase with the square of the true airspeed, and will decrease with an increase in the angle of bank. These factors become an important consideration when determining the conditions necessary for a minimum radius turn. We will be discussing minimum radius turns some more in the next chapter. It's also important to understand the relationship between airspeed, AOB, and turn radius when pilots upgrade to faster aircraft.

Rate of Turn

In flying, particularly in instrument flying, the radius of turn is not the only important consideration in a turn. The rate of turn is also an item of concern to pilots. The rate of turn is measured in degrees per second. The "standard" turn (rate one) is 3 degrees per second or 360 degrees in 2 minutes.

The rate of turn can be determined from the radius of turn equation and some simple geometry. Knowing the radius of turn, the circumference of the turn can also be established:

$$C = \frac{2\pi v^2}{gtanAOB}.$$

Since the circumference is the distance to be travelled during the turn, if we know the velocity of the aircraft, we can determine the amount of time required to travel the full circle (360°):

$$\frac{C}{v} = \text{time to turn } 360° = \frac{2\pi v}{gTanAOB}.$$

From this, we can determine the rate of turn in degrees per second, and if we convert for an airspeed in knots, we get the following:

$$ROT = \frac{1091TanAOB}{v} \qquad (8.13).$$

So as we can see from equation (8.13), the rate of turn will increase with an increase in angle of bank and it will decrease with an increase in airspeed.

Landing

The landing performance of an aircraft is of significant importance. Generally, landing performance is measured in distance required—the shorter the distance, the better the performance. Sometimes, however, the ability to land on a soft or rough surface, or the ability to land in a crosswind are more important considerations. Soft surfaces and crosswinds will be covered in Chapter 23. For now, we will focus on the landing distance required.

During the landing maneuver, the aircraft is being decelerated to a stop. Because a deceleration is occurring, the equation used for takeoff distance (8.2), can be modified to use for the rolling portion of the landing:

$$d = \frac{Wv_{td}^2}{2g(D-T)}$$

For our purposes, we can assume that the thrust is zero, since in most cases the engine will be brought to idle for the landing. This yields the following:

$$d = \frac{Wv_{td}^2}{2gD} \qquad (8.14).$$

In this equation, the drag (D) is the combined effect of aerodynamic drag and rolling friction.

We can see from equation (8.14), that two fundamental variables will affect the landing distance for a given aircraft. They are the touchdown speed and the effectiveness of the braking (aerodynamic and frictional). These are the variables which we will consider in the following sections.

Approach Speed

The touchdown speed of the aircraft will be approximately the same no matter what the approach speed is (assuming that the flare is completed properly—more about this in Chapter 23). However, any excess speed that is carried in the approach must be bled off during the flare. Although this does not affect the actual landing "roll", the total landing distance will be increased due to the distance used up in the flare (Fig 8-47).

Landing distances published in operating Manuals are based on the rolling portion of the landing, so the distance covered in the flare is something that the pilot must take into consideration. In the case of published obstacle clearance distances, the flare is included, but the required speed must be adhered to.

As for the approach speed which should be used, there is some debate. If a speed is not provided by the manufacturer, you must determine one on your own. When deciding this speed, remember that we are limited by the stall speed of the aircraft—we can't possibly approach slower than this. Even speeds immediately above the stall speed are usually unacceptable since some variation must be allowed for—especially in turbulent weather. This having been said, a high

approach speed is something to be avoided as well. With a high approach speed, the aircraft will use up much more runway during the flare. As well, the high speed may lead to ballooning and /or bouncing on the landing.

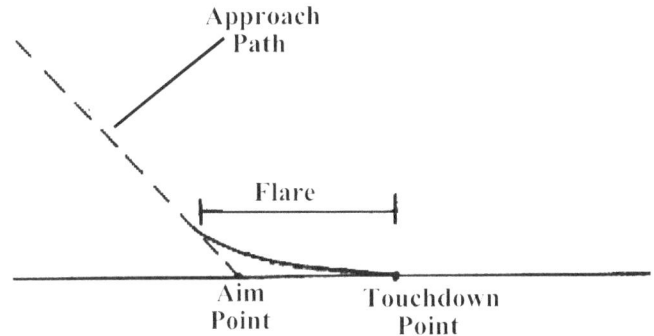

Fig 8-47. - *Flare Effect on Landing Distance. -* The actual touchdown point will be beyond the aiming point due to the distance used up in the flare.

So we don't want an excessively low approach speed, and we don't want an excessively high approach speed. A generally accepted rule of thumb to use is to multiply the stall speed by 1.2 to 1.4 and to add a factor for turbulence or gusting winds—usually half of the gust factor is acceptable. The choice of multiplying the stall speed by 1.2, 1.3, or 1.4 will depend on the type of landing being performed, as well as the pilots skill and experience level.

Drag

When trying to complete a maximum performance landing, the more drag that is available, the better. This is one of the only maneuvers in which more drag is actually desirable. From equation (8.15), we can see that an increase in drag will decrease the rollout. As well, increased drag will hasten the loss of airspeed during the flare, thus decreasing the total landing distance even further.

In the air, this drag is provided aerodynamically. On the ground, this drag is maximized aerodynamically as well as with maximum braking.

Rolling Friction

In order for maximum braking to be effective, it's necessary to apply as much weight as possible to the main wheels. This is a consequence of the friction equations which we discussed in Chapter 1. An increase in the normal force (weight in this case) will increase the frictional force available between the pavement (or other runway surface) and the rubber of the tires.

As long as the tires are rotating, *static* friction is slowing the aircraft down because the tires and the pavement are not actually "rubbing" against each other. However, if the static friction is too low—and/or the brakes are applied too enthusiastically—and the wheels stop rotating, the pavement and the tires will be in motion relative to one another. This means that the braking is being produced by *kinetic* friction and will be much lower, as we discussed in Chapter 1. This is known as "locking" the brakes and will reduce the total amount of drag slowing the aircraft. The more weight that is acting on the tires, the more effective the braking action will be, and the less likely it will be that the brakes will lock.

On initial consideration, it seems that the weight of the aircraft is a constant, and thus the weight acting on the landing gear is out of the pilots hands. It is true, of course, that the weight is a constant for a given landing. However, the amount of weight that acts on the landing gear can be controlled—or at least influenced—by the pilot via control inputs.

Lift, as we know, counteracts weight. So any lift that is present during the rollout will reduce the amount of weight resting on the landing gear and transfer it to the wings. Because of this, a reduction of lift is necessary during the landing roll to maximize braking.

This reduction in lift, unfortunately, is not the only consideration in increasing the normal force on the landing gear. It would seem that reducing the lift simply means lowering the nose of the aircraft and applying brakes. If instead, however, we were to maintain back pressure and try to hold the nose off the ground, the weight supported by the nose gear would then be transferred to the main gear or the wings. Along with this, back pressure means that the tail is producing lift in the downwards direction. This negative lift contributes to the normal force acting on the main landing gear. One more benefit of back pressure during the roll is the reduced stresses on the nose gear, which tends to be less durable than it's main gear counterparts on most aircraft.

So all in all, it seems that maintaining back pressure during the landing roll has more benefits than consequences (at least in terms of rolling friction). For most light aircraft this is the case. For information on a particular aircraft, consult the POH or Flight Manual.

Configurations

Due to the need to slow the aircraft with drag, the use of drag producing control surfaces is necessary during a performance landing. Items such as flaps, speed brakes, spoilers, etc. should be extended as per the POH.

As stated earlier, the touchdown speed of the aircraft will be approximately the same no matter what the approach speed is—provided the flare is executed properly. This touchdown speed will usually be slightly less than the stall speed out of ground effect. Since we want to touch down at the slowest speed possible, reducing the stall speed of the aircraft would be favorable to a shorter landing. This is another good reason to approach in a configuration which will reduce our stall speed (e.g. – flaps extended).

Spoilers, if available, will reduce the roll as well because of their effect of reducing (or "spoiling") lift. This allows more weight to rest on the landing gear, thus increasing braking effectiveness as we discussed previously. Spoilers also create some drag, although their primary function is to spoil lift.

Yet another consideration during the landing roll is the use of flaps. Flaps will increase the aerodynamic drag, but they also have the unfortunate effect of increasing lift, thus *decreasing* brake effectiveness. At high speeds (early in the roll), the increase in aerodynamic drag will generally be greater than the decrease in braking. At lower speeds (later in the roll), however, the drag is less likely to outweigh the loss in braking. So generally, the flaps should be retracted at some point late in the first half of the rollout. In the case of electrically actuated flaps, this means throwing the switch at the beginning of the roll due to the time it takes for the flaps to retract (NOTE: This is ill advised in an aircraft with retractable landing gear due to the possibility of confusing the flap switch and the gear switch during the rush of the landing roll).

Runway Condition

The state of the runway being used will also have a direct impact on the landing distance of the aircraft. Just like on take off, the slope of the runway will contribute to the thrust or to the drag of the aircraft, depending on whether the slope is up or down. If the runway is upsloped, a component of the weight will be acting to the rear of the aircraft upon touchdown (Fig 8-48a). This will contribute to the drag and will stop the aircraft in a shorter distance. If instead the runway is downsloped (Fig 8-48b), the weight will be contributing to thrust and will increase the landing roll. A downsloped runway also has the added disadvantage of "dropping away" from the aircraft during the flare. This will tend to increase the total landing distance required even further.

Combined with the effects of runway slope,

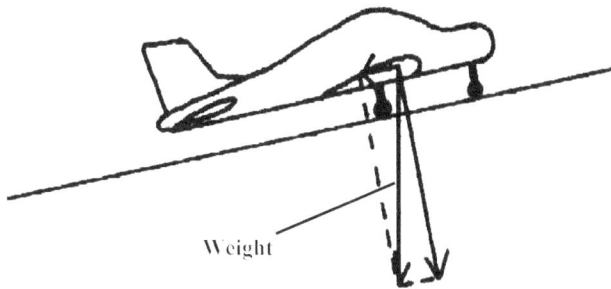

(a) On an upsloped runway, the weight acts to the rear and helps to slow the aircraft. This shortens the landing roll.

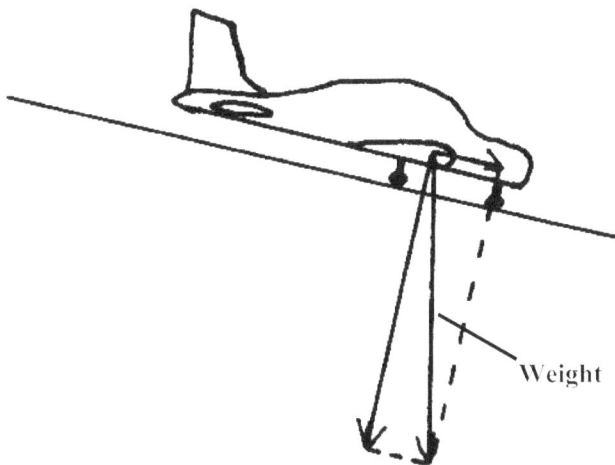

(b) On a downsloped runway, the weight acts to accelerate the aircraft. As such, the landing roll will be increased.

Fig 8-48. - *Runway Gradients* - Due to the weight acting along the landing path, a gradient will change the landing distance required.

the surface condition of the runway will affect the braking effectiveness of the tires. During the take off, a rough or soft surface increased the take off roll due to surface resistance. It would therefore *seem* that a rough or soft surface would decrease the landing roll due to that same increased resistance. Unfortunately, this is not the case.

A surface that is soft and therefore allows the tires to sink slightly will also have a reduced friction coefficient with the tires. The increased resistance of the "sinking" is more than offset by the loss of braking. All in all, a soft or rough surface will increase both the take off roll *and* the landing roll. This having been said, if we were to consider a landing roll *without* braking, the rollout would be shortened on a soft surface.

Other Variables

There are several other variables which will affect the landing performance of an aircraft. These variables are important to consider when executing a performance approach and landing. They include the weight of the aircraft, the wind, and the pilot's skill level.

Weight – The effect of weight on the landing distance is often misunderstood by pilots. It is generally believed that a heavier aircraft requires a greater stopping distance due to the increased momentum. A heavier aircraft contains more mass than a lighter aircraft, and will thus possess a greater amount of inertia at a given speed. Assuming that the force available to decelerate the aircraft remains constant, the deceleration will be slower due to the greater mass, and thus the stopping distance will be greater. We must, however, consider the effect of increased weight on the effectiveness of the brakes.

Recall from Chapter 1 that a force is equal to the mass of an object multiplied by the acceleration experienced:

$$F = ma \qquad (1.1).$$

The force we are discussing here is the decelerating force acting on the aircraft. Assuming for the moment that the decelerating force is entirely a result of braking friction, this decelerating force can be determined from our discussion on friction in Chapter 1:

$$F_{fr} = \mu_s N \qquad (1.18).$$

The normal force (N) in this case is the weight of the aircraft, and the coefficient of friction is that between the runway surface and the rubber of the tires. We also know from Chapter 1 that the weight is equal to the mass of the aircraft multiplied by gravitational acceleration (g). Since the force decelerating the aircraft is a result of friction, we get:

$$ma = \mu_s mg.$$

The mass cancels out on both sides of the equation. So we get the following:

$$a = \mu_s g \qquad (8.15).$$

From equation (8.15), we can see that the mass of the aircraft has no effect on the deceleration of the aircraft. The deceleration is dependant only on the friction coefficient between the tires and the runway surface (this coefficient is a constant) and gravitational acceleration (also a constant).

Now we must qualify the above discussion somewhat. First of all, some of the braking action available during the landing roll is a result of aerodynamic drag. This drag will not vary with weight, and as such, the landing roll will in fact be lengthened slightly at heavier weights. For light aircraft, this effect can be ignored for practical purposes, especially since other factors will often cancel this effect out.

One of these factors is the "sinking" in soft ground. This sinking will increase the drag acting against a heavier aircraft because the heavier aircraft will tend to sink slightly further. On solid ground, the compression of the tires as they roll will also absorb some of the energy of the aircraft. For practical purposes, these effects can be considered to cancel out the increase in the landing roll due to the aerodynamic drag.

So we can see that the weight of the aircraft has no *direct* effect on the landing distance—*all other things being equal*. We know, however, that the weight of the aircraft changes the stall speed of the aircraft. Since both the approach and the touchdown speeds are some multiple of the stall speed, *this* change will affect the distance required for landing.

We know that the stall speed is a function of

the *square root* of the weight. We also know that the landing distance changes with the *square* of the speed. The square root and the square will cancel one another out, and the landing distance will be directly proportional to the weight:

$$LD = D_{MGW} \times \frac{GW}{MGW} \qquad (8.16).$$

So to sum up, the weight of the aircraft has no *direct* effect on the landing distance. However, if the approach and touchdown speeds are adjusted appropriately for the new weight, the landing distance will be changed as a result.

Wind – The effect that wind has on a landing will be similar to the effect that wind has on the takeoff. A headwind during landing will reduce the touchdown groundspeed, and will therefore reduce the landing roll. A tailwind during landing will increase the touchdown groundspeed, and will therefore increase the landing distance required. Just like with the takeoff, a tailwind will increase the distance required by a greater amount than a headwind will decrease it.

Thrust Reversal – Some aircraft have the capacity to reverse their thrust. This can be done with propellers by rotating them to a negative pitch (often referred to as the "beta range"). Jet aircraft usually have doors around the exhaust nozzle that can redirect the exhaust to produce thrust forward (often referred to as "clamshells").

Thrust reversal contributes to the drag during a landing. So from equation (8.14), the landing distance will be reduced. Thrust reversal has the added advantage of being independent of runway surface condition. This makes it invaluable on large aircraft operating on contaminated runways.

Unfortunately, due to cost and complexity, few if any small aircraft have reverse thrust.

Pilot Technique – Just as with the takeoff, landing distances published by the manufacturer are calculated by engineers and demonstrated by highly experienced test pilots. It is unlikely that the average general aviation pilot can exactly duplicate the required techniques for a perfect landing. As a result, it is necessary to add a reasonable safety factor to the "book" numbers to account for the pilot's skill and experience level.

Chapter Summary

Aircraft performance includes a number of factors, including takeoff, climb, cruise, descent, landing, and turning performance. Each of these types of performance can be subdivided further (e.g. – cruise performance can refer to speed, range, or endurance). Most performance characteristics of an aircraft can be related back to the power curve in some way. Variables that affect the power curve will also affect the performance of the aircraft. These variables include aircraft weight and C of G, altitude, and configuration.

The takeoff involves an acceleration from rest to the lift-off speed and then the initial climb. Anything that will change the rate of acceleration or the lift-off speed will have an influence on the takeoff distance required. Factors that can contribute to or degrade acceleration include aircraft weight, density altitude, aircraft condition, and runway gradient and surface condition. Factors that can increase or decrease the lift-off speed include weight and C of G, density altitude, and wind.

Climb performance is based on the availability of excess power or excess thrust. *Rate* of climb is determined by excess power, and the best rate of climb speed is the speed which provides the most excess power. *Angle* of climb is determined by excess thrust, and the best angle of climb speed is the speed which provides the most excess thrust. Best angle of climb is normally used to clear obstacles. However, during takeoff the distance used up in accelerating to the best angle of climb speed may make it necessary to climb at a lower speed. Factors that can affect climb performance include aircraft weight and C of G, configuration, density altitude, and in the case of angle of climb, wind.

Cruise performance is often expressed in terms of speed, but it can also include range and endurance. The cruising speeds available are determined largely by the parasite drag characteristics of the aircraft. Cruising range is determined by the ratio of speed to fuel flow. Since fuel flow is proportional to the power being produced lower power settings and higher speeds are favourable during flight for range. For this reason, maximum range occurs at the speed which allows the highest velocity/power ratio. This speed is determined on the power curve by drawing a tangent line from the origin. Endurance is the amount of time an aircraft can remain aloft and is determined by the fuel flow which is in turn determined by the power setting. In order to fly for maximum endurance, the power setting must be minimized. The speed for maximum endurance is determined by the lowest point on the power curve. Factors that will affect the cruise performance include weight and C of G, density altitude, configuration, and in the case of range, wind.

When an aircraft is in a descent, two measurements are of importance. The first is the rate of descent, which is determined by the power deficiency. To increase the rate of descent, a higher airspeed and/or a lower power setting can be selected—this will increase the power deficiency. As well as the rate of descent, the angle of descent is often important. This angle is usually expressed as a glide ratio or just as a glide range. The glide angle is determined by the TAS and the rate of descent. So for a fixed TAS, power will determine the glide angle. In a power off glide for maximum range, the airspeed which provides for the most optimum glide ratio must be used. This airspeed is the airspeed for the best L/D ratio since the glide ratio is equal to the L/D ratio. Variables that affect the descent performance of an aircraft include weight and C of G, configuration, and in the case of range, wind.

During turning flight, an aircraft is banked so that a component of the wings lift is acting to the side. This component of lift accelerates the aircraft to the side and produces a sideslip. Directional and pitch stability cause the aircraft to yaw and pitch simultaneously in the direction of the bank. As the aircraft turns, it will experience a load factor due to the acceleration. This load factor increases with the angle of bank. As a result of the increased load factor, the stall speed increases and the power required increases. Aside from the load factor, the rate and radius of the turn can be of importance to a pilot. The turn radius increases at higher airspeeds and lower bank angles. The turn rate increases at lower airspeeds and higher bank angles.

The landing distance required is another important consideration in the performance of an aircraft. The landing distance will be determined by the touchdown speed and the amount of retarding force which can be applied to the deceleration. High touchdown speeds and a gradual deceleration will lead to a long landing roll. The factors which will affect the landing distance required include the aircraft's weight and C of G, density altitude, runway gradient and surface condition, wind, and the approach configuration.

List of Formulae

Takeoff
Takeoff Performance

$$a = \frac{(T - D)g}{W} \tag{8.1}$$

$$d = \frac{Wv_{LO}^2}{2g(T-D)} \tag{8.2}$$

Climbing
Best Rate

$$ROC = \frac{33,000 \times ETHP}{Weight} \tag{8.3}$$

Best Angle

$$Sina = \frac{ET}{W} \tag{8.4}$$

$$Sina = \frac{ROC}{TAS} \tag{8.5}$$

Cruise
Variables

$$New\ V_{Range} = V_{Range}\sqrt{(GW/MGW)} \tag{8.6}$$

Descending
Rate of Descent

$$ROD = \frac{33,000 \times THPd}{W} \tag{8.7}$$

Glide Range

$$Sinb = \frac{ROD}{TAS} \tag{8.8}$$

$$L/D = Glide\ Ratio \tag{8.9}$$

Turning
Load Factor

$$n = 1/CosAOB \tag{8.10}$$

Radius of Turn

$$CF = WtanAOB \tag{8.11}$$

$$r = \frac{v^2}{gTanAOB} \tag{8.12}$$

Rate of turn

$$ROT = \frac{1091TanAOB}{v} \tag{8.13}$$

Landing

$$d = \frac{Wv_{td}^2}{2gD} \tag{8.14}$$

Other Variables

$$a = \mu_s g \tag{8.15}$$

$$LD = D_{MGW} \times \frac{GW}{MGW} \tag{8.16}$$

Questions and Problems

1) How do the following factors affect the power curve?

   a) weight   b) C of G
   b) altitude   c) configuration
   c) load factor   d) wind

2) Why do the factors in Q1 have the effects that they do?

3) How does rolling friction develop? How does it affect takeoff performance?

4) How do runway gradients affect the takeoff distance?

5) If an aircraft weighs 2400 lbs, and is producing 260 lbs of drag at V_Y = 87 kts, what is the aircraft's ROC if the THP available is 120 HP?

6) Where does best rate of climb occur on the power curve? Why?

7) Why do some Flight Manuals require a climb speed different from V_x for an obstacle clearance climb after takeoff? Will this speed be lower or higher than V_x?

8) What variables can affect the climb performance of an aircraft? How and why?

9) How is the maximum level cruise speed determined from the power curve? Why is it determined in this way?

10) Where does maximum endurance occur on the power curve? Why?

11) Where does maximum range occur on the power curve? Why?

12) How does wind affect the endurance of an aircraft? Range?

13) When flaps or landing gear are extended, what happens to the angle of descent? Why?

14) Why does load factor increase at higher angles of bank?

15) How do rate and radius of turn vary with respect to airspeed and angle of bank?

16) How does weight affect landing distance? Why?

17) How does the runway gradient affect landing distance? Why?

18) A soft surface increases the takeoff roll by increasing the rolling friction. Why then, does a soft surface also increase the landing roll?

Chapter 9
Aircraft Limitations

In designing and building an aircraft, it is important that a manufacturer achieves a certain level of structural strength in the machine. In flying an aircraft, it's important that the pilot is aware of these structural limitations, and that he or she abides by them. Failure to do so could possibly lead to a in-flight failure, or if the pilot is more fortunate, to damage that is expensive to repair. In some instances, it is possible to overstress the aircraft in such a way that the damage done is not apparent—this could lead to a structural failure in the future.

Clearly, it is important that some definitive limitations be known and adhered to. These limits are published by the manufacturer in the form of weight, load factor, and speed limitations. Any non-compliance to these speed and load restrictions can lead to structural damage or failure.

Load Factor Limits

The load factor, as we know, is the ratio of lift to weight. If the aircraft is producing more lift than the weight it contains, the load factor will be greater than one. On the other hand, if the aircraft is producing less lift than it contains in weight, the load factor will be less than one. In the event that the aircraft is producing lift in the downwards direction (if the angle of attack is sufficiently negative) the load factor will in fact be negative. When this is the case, the aircraft will be accelerating downwards at a rate faster than gravitational acceleration (assuming the aircraft is right side up and not inverted).

When the load factor is not equal to one, the aircraft will accelerate in the direction of the net force (Fig 9-1). The effect of this acceleration is to apply stress to various aircraft components. These stresses must be sustainable for the acceleration to be safe. The airflow forces the wings up (assuming positive loading), and the wings support the load of the aircraft. The wing therefore must be able to support the weight of the aircraft multiplied by the limit load factor:

$$LWL = W \times LLF \qquad (9.1).$$

In this equation, (LWL) is the limit wing load, (W) is the weight of the aircraft, and (LLF) is the limit load factor of the aircraft.

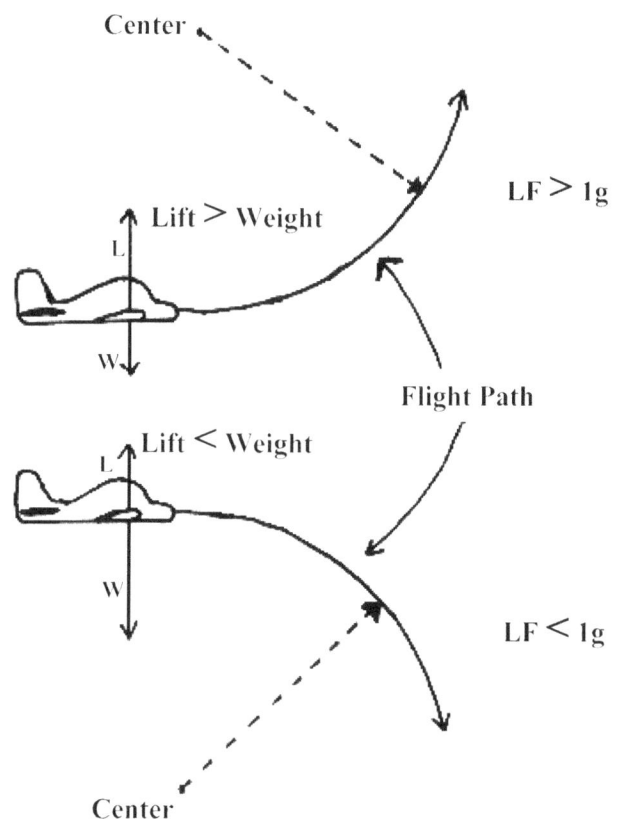

Fig 9-1. - *Load Factor* - When the load factor is something other than 1, the aircraft will accelerate in the direction of the net force. This leads to a curving flight path—either vertically (i.e. – pulling out of a dive or into a climb) or horizontally (i.e. – a turn).

When the airframe is under a load, it will flex to absorb that load. When the load is removed, the airframe will return to it's original shape. However, if the load is too large the deformation of the airframe will be permanent. The materials science behind what is going on is beyond the scope of this book, but it can be compared to a toy spring. If we stretch the spring only so far and let it go, it will return to it's original shape. However, if we were to stretch that same spring beyond a certain point, it would simply remain in it's new shape.

The amount of load that the wing can sustain is a constant for a given aircraft. So at first glance, it appears that a decrease in weight should also increase the limit load factor. This does not, however, take into consideration the constant weight components of the aircraft. For example, if an aircraft is certified to sustain 3.8 g's and has a maximum gross weight of 3000 lbs, it would appear that a weight reduction to 2000 lbs would allow the aircraft to accept 5.7 g's. However, even though the gross weight of the aircraft has changed, there are some constant weight components like the engine that can still only accept 3.8 g's. These components would sustain structural damage and possibly failure at 5.7 g's. So we can see that equation (9.1) only applies to an aircraft at maximum gross weight.

This brings us to the difference between the limit load factor and the ultimate load factor. The published limit load factor of an aircraft is the highest load factor that can be safely achieved. This is the load factor that engineers and test pilots have determined can be absorbed safely by the aircraft. Loads beyond the limit load factor will cause permanent structural damage. This damage will probably manifest itself as wrinkled skin on the wings, bent or twisted engine mounts, or possibly some sort of undetectable damage. The ultimate load factor is at least 1.5 times the limit load factor. Once the ultimate load factor is reached, structural failure is a possibility. At this point, the wings or other aircraft components may begin to depart your company.

As was just mentioned, it is possible for the damage due to overloading to be undetectable to a cursory inspection. If an aircraft is overstressed in flight, it is critical that it be inspected properly and thoroughly by an AME. If the aircraft is brought beyond the limit load factor, the damage done will weaken the airframe structure. On some future flight(s) this weakening can cause further damage

or even a failure at loads well below those that the aircraft is certified for.

One thing to note about these load limits is that they apply to the *structural integrity* of the aircraft. In the case of negative load limitations, the airframe may fare well, but the engine may not. Without an inverted fuel system, the engine will fail due to fuel starvation very shortly after a negative load is imposed on the airframe. Without an inverted oil system, prolonged inverted flight may lead to the engine seizing due to lack of proper lubrication. So when considering what maneuvers are safe, keep in mind not only the structural strength of the airframe, but also the ancillary systems to the engine. As well, if the aircraft is so equipped, hydraulic systems may be plagued with similar problems during inverted flight.

Load Factor Limitations

Different aircraft will be certified to different load limits depending on their mission objective and operating environment. There are several different load factor categories, and some aircraft can fit into two or more depending on how they are loaded. For example, many training aircraft are in the normal category, but with the weight and balance brought to within certain tolerances, the aircraft can be operated in the utility category.

Knowing the load limits of an aircraft is an important part of safe operation. A point to keep in mind is that with a change in configuration the limit load factor may change. For information on this, refer to the POH or Flight Manual.

Normal – Normal category aircraft are certified to limit load factors of between +2.5 and +3.8 g's and between -1.0 and -1.52 g's. Most general aviation aircraft fall into the normal category.

Utility – Utility category aircraft are certified to limit load factors of at least +4.4 g's and -1.76 g's. Several training maneuvers require that the aircraft be in the utility category, so most trainers are either utility or normal/utility. Utility aircraft are often certified for intentional spins. So not only is the weight restricted to allow for higher loads, but the C of G is often restricted to create a longer rudder arm and facilitate recovery. This is why on aircraft that are normal/utility, the weight and balance envelope looks similar to Fig 9-2.

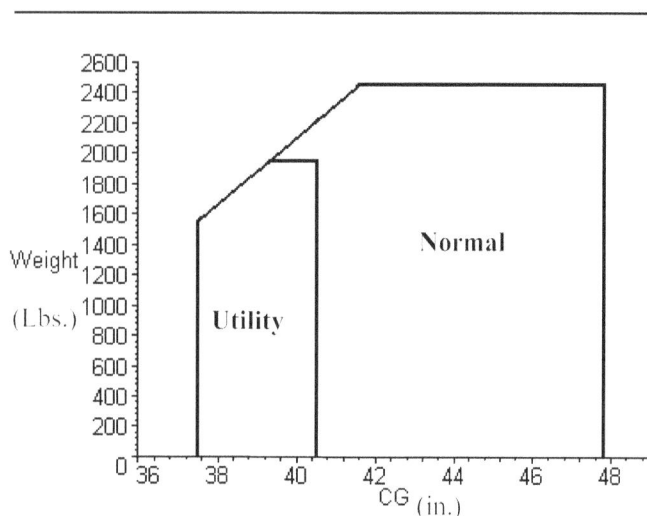

Fig 9-2. - *Normal/Utility Weight and Balance Envelope.*

Aerobatic – Aerobatic aircraft are certified to a limit load factor of at least +6.0 g's and -3.0 g's. This expanded load envelope allows the aircraft to be used for more aggressive maneuvers which would be unsafe in other aircraft.

Turbulence

Turbulence in the atmosphere can impose significant loads on an aircraft. The reason for this is the variation of the angle of attack without an initial variation in the airspeed. We know already that if the angle of attack is increased, the lift coefficient also increases. If the airspeed is held constant, an increase in the lift coefficient will result in an increase in the lift being produced. This lift increase will create a load factor on the aircraft. The loads due to turbulence are normally very short lived. This is partly due to the rapid fluctuations of the relative airflow, and partly due to airspeed and flight path fluctuations resulting from the loads. As well, the longitudinal stability of the aircraft as discussed in Chapter 7 will contribute to the dampening of loads in turbulence.

Aircraft with low wing loading will be more susceptible to turbulence than aircraft with higher wing loading. Wing loading is the amount of aircraft weight being supported per unit area of the wing,

$$WL = W/S \qquad (9.2),$$

where (WL) is the wing loading of the aircraft, (W) is the weight of the aircraft, and (S) is the surface area of the wing.

To understand the effect of wing loading, we can compare what happens to two aircraft of different weights flying in turbulence. The first aircraft is loaded to a weight of 3000 lbs, and the second aircraft is loaded to a weight of 1500 lbs. Both aircraft are of the same type and are flying at the same airspeed. This means that the first aircraft will have twice the absolute angle of attack of the second. The absolute angle of attack is the angle between the relative airflow and the "zero lift line".

Now let's consider what happens to these aircraft as they encounter the vertical gusts associated with turbulence. If a gust from below is encountered, the angle of attack will be increased since the relative airflow will now come from a direction more below the aircraft (Fig 9-3). If the angle of attack is increased, the lift being produced will be increased as well. Since the weight hasn't changed, this increase in lift will cause the aircraft to experience an acceleration upwards which will involve some positive load factor greater than one.

If the second aircraft (1500 lbs) is flown at a speed which requires an absolute angle of attack of 6°, then the first aircraft (3000 lbs), which is flown at the same speed, will require an absolute angle of attack of 12° (the lift being produced by a wing is proportional to the absolute angle of attack). Now consider a vertical gust which increases the angle of attack by 6°. The second aircraft will now be brought to an AOA of 12°, and the first will be brought to 18°.

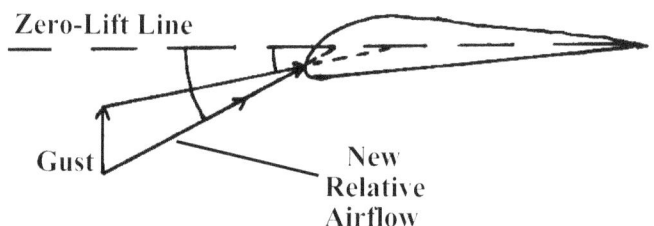

Fig 9-3. - *Gust Effect on AOA* - A vertical gust will increase the AOA of the wing. The actual increase will be the same for two wings travelling at the same speed, but a heavier aircraft will be operating at a higher AOA to begin with.

So with a doubling of AOA, the aircraft weighing 1500 lbs will experience a doubling of lift, and thus a load factor of 2. Meanwhile, the aircraft weighing 3000 lbs will only have it's AOA increased by a factor of 1.5, so the load factor will be only 1.5.

These two aircraft are both being flown at the same speed and they are both being exposed to the same turbulence. The difference between the two is in the wing loading.

Speed Limits

Along with acceleration loads imposed on the aircraft, there are pressure loads exerted by the airflow over the aircraft. These air loads combined with the limit load factor and the stalling speed of the aircraft are used to determine several limiting speeds pertaining to the aircraft's operations. These speeds are often referred to as "V speeds" since they all have "V" notations. Each speed can be notated by a V with one or two subscript letters following it.

Stall Speeds (V_s and V_{so})

The stalling speed of an aircraft is the minimum speed at which flight can be maintained. The published stall speeds are usually for the clean configuration (i.e. – flaps up) and for the landing configuration (flaps and gear extended, etc.) at maximum gross weight. As we know from Chapter 6, the stall speed will vary with weight and C of G, load factor, power setting, and configuration. The stall speed becomes the lower end speed limit for flight operations, and this limit will vary according to these factors.

Since a full chapter has already been dedicated to the quirks of the stall, we will end this discussion here and move on to speeds which we have not previously encountered.

Flap Extended Speed (V_{fe})

When the flaps are extended on an aircraft, they project into the airflow in such a way that they experience higher loads than when they are retracted. These loads act at some distance from the hinge line and as a result create a bending moment about the hinge line (Fig 9-4). The flap structure must be able to support this moment or it will fail. Because of this, most if not all aircraft have a maximum speed at which the flaps can be extended. This speed will be lower than the maximum speed at which the aircraft can operate with the flaps retracted. If this limiting speed is exceeded with the flaps extended, structural

damage will most likely occur, and structural failure may occur as well.

Fig 9-4. - *Flap Loads* - When flaps are extended, they experience loads that do not occur in the retracted position.

On some aircraft, more than one flap operating speed is published. This takes into consideration the fact that the flaps can be partially extended and can therefore take a higher speed before the loads become excessive. Do not, however, assume that this is the case. If more than one speed is not published by the manufacturer, assume that the V_{fe} for full flaps applies to partially extended flaps as well.

Landing Gear Speeds (V_{le} and V_{lo})

Just as loads on the flaps can be damaging, air loads can cause the landing gear to be damaged as well. Two limiting airspeeds are used for the landing gear. They are the gear extension speed and the gear extended speed. On some retractable aircraft, the two speeds are the same, on others the gear extended speed can be higher.

The gear extended speed (V_{le}) is the speed at which damage or failure may occur with the landing gear locked in the extended position. The gear extension speed (V_{lo} – sometimes called the gear operation speed) is the speed at which damage or failure will occur with the landing gear in transition between the retracted and extended positions. This speed will be equal to or lower than the gear extended speed.

V_{lo} takes into consideration joints in the landing gear that may be in a weakened position during gear cycling. As well, hydraulic and electrical actuators may not be able to withstand the air loads that the "down and locked" landing gear can. Another consideration for V_{lo} is that on some aircraft, the gear doors are closed when the gear is down, but they must open during cycling. These doors may not be able to withstand the loads that the gear itself can.

On some aircraft, the landing gear can be used as an emergency speed brake during unusual attitudes. In this case, the gear could be extended at speeds above V_{le}/V_{lo}, but an inspection would be necessary before any more flights were attempted. As well, if the gear was to be used as an emergency brake, ATC should be advised in case the gear was to collapse on the runway. After overspeeding the landing gear, it may not be reliable once you are on the ground.

Maneuvering Speed (V_a)

The maneuvering speed of an aircraft is the maximum speed at which the controls can be placed in the fully deflected position. This is the common definition of maneuvering speed, and is accurate enough, but there are some specific details which we need to be aware of.

In order for an aircraft to meet airworthiness requirements, a maneuvering speed must be established by the manufacturer. At this maneuvering speed: the ailerons must be strong enough to withstand the twisting moments and other stresses imposed by full deflection; the rudder must be able to withstand the moments and loads imposed by full deflection; and the elevator or stabilator must be able to withstand the moments and loads imposed by full deflection. The *minimum* allowable speed for V_a is the stall speed at the limit load factor. So we can determine what the maneuvering speed should be by converting equation (6.2) from:

$$\text{Stall Speed} = V_s(\sqrt{n}) \qquad (6.2),$$

to:

$$V_a = V_s\sqrt{LLF} \qquad (9.3).$$

The speed determined by equation (9.3) is often called the "corner speed" and is the maximum speed at which the aircraft will stall before the wings are overstressed in the positive direction. That is, at the minimum maneuvering speed it is impossible to overstress the aircraft in the positive direction. On many aircraft, this speed is the published V_a.

However, note that wing strength is not required to be taken into consideration in determining the actual maneuvering speed. The speed determined with equation (9.3) is just the *minimum*. It is possible to have a V_a that is higher than this speed. If this is the case, full deflection of the controls will not damage the control surfaces themselves, but full deflection of the elevator could lead to structural damage of the wings or other aircraft components due to overloading.

As the weight of the aircraft changes, so does the maneuvering speed. This is because at lighter weights, the aerodynamic loads become more pronounced. So deflection of the controls creates larger accelerations and therefore larger loads on the airframe. As a result of this, a lower gross weight means a lower maneuvering speed. Since the minimum maneuvering speed is a stall speed, the equation for determining stall speeds at different weights (6.1) can be applied as follows:

$$\text{New } V_a = V_a\sqrt{(GW/MGW)} \qquad (9.4).$$

Once again, be careful of aircraft on which the published maneuvering speed is higher than the minimum. On these aircraft, it would be wiser to apply equation (9.4) to the *minimum* maneuvering speed, or the "corner speed", rather than the actual published V_a. This will give a speed that is lower than te actual maneuvering speed for the weight, but the error is on the side of caution.

Beyond the maneuvering speed, the aircraft can still be operated safely. However, full control deflection must be avoided. Some may wonder exactly how far the controls can be deflected safely. An airworthiness requirement for light aircraft is that at V_{ne} the control surfaces must be able to withstand a $1/3$ scale deflection. Approximately linear interpolation between V_a and V_{ne} is acceptable.

Normal Operation (V_{no})

The normal operations speed (V_{no}) is the speed which is at the top of the green arc and the bottom of the yellow arc on the airspeed indicator. This speed is intended to limit the loads imposed on the airframe by turbulence. V_{no} is based on a 50 fps vertical gust applied suddenly to the wing. The effect of this gust is to increase the angle of attack as we discussed previously. That increase in angle of attack will cause a corresponding increase in lift.

The highest speed which will allow the aircraft to withstand a 50 fps vertical gust applied suddenly to the wing without overstressing the airframe is the V_{no}. This speed is designated to give pilots an idea of how high the airspeed can be allowed to get in rough or turbulent air.

Never Exceed (V_{ne})

As the airspeed of an aircraft increases, the loads exerted on the airframe by the airflow will increase as well. There comes a point where these loads become unacceptable in their effect(s) on the aircraft. As a result of this, there must be a defined speed beyond which the aircraft may not be operated under any circumstances. This speed is the never exceed speed.

The never exceed speed (V_{ne}) of an aircraft is, as the name implies, the airspeed which should never be exceeded. V_{ne} is indicated on the airspeed indicator by a red radial line and is determined as 90% of the dive speed (V_d). The dive speed, in turn, is the speed at which any of several undesirable phenomenon can occur.

The dive speed is the first speed at which control reversal, flutter, or divergence occur. These phenomenon may occur separately or together. The first to be encountered defines the dive speed. All of these problems fall under the umbrella of aeroelasticity. Aeroelasticity is the study of how the aerodynamic properties of an aircraft change as the aircraft flexes under aerodynamic loads.

Control Reversal – Control reversal is a dangerous phenomenon which can occur at high speeds. The twisting moments induced by the deflection of a control surface can cause the surface to act as a trim tab. For example, if an aileron is deflected downwards, it should increase the lift being produced by the wing. If instead the moments being produced caused the wing to twist forward, the angle of attack could be sufficiently reduced to cause a _reduction_ in lift. Combined with a similar effect on the opposite wing, an aileron input to the left could result in a roll to the right.

Control reversal can occur on all three control surfaces. With the ailerons, the wing twists as the ailerons are deflected. With the elevator and the rudder, the fuselage bends. Control reversal is more likely to occur on the ailerons first— especially on aircraft with swept wings.

Flutter – Flutter occurs when the oscillations of a component begin to amplify themselves. If allowed to continue, flutter can cause the component(s) to literally rip themselves apart. In some cases, flutter develops so quickly that the resulting structural failure seems to be instantaneous.

All objects have a natural resonating frequency which is determined by the stiffness and the mass of the object. Vibrations at this frequency will normally dampen out over time—that is, the energy in the vibration will be dissipated by the material of the wing (or tail, or control surface, etc). At high airspeeds, however, the air loads acting on a surface can discourage the dampening of vibrations. As speed increases towards the critical flutter speed, it takes a longer period of time for the vibrations to dampen out. Eventually a speed is reached where the vibrations will not dampen at all—as the energy of the vibration dissipates into the structure it is replaced by the air loads. At even higher airspeeds, energy will be added by the airflow faster than it can dissipate into the structure. When this happens the amplitude of the vibrations will become continuously larger—that is, _negative_ dampening will occur—until the structure fails.

Flutter can affect any surface, but tends to occur in control surfaces at lower speeds than in other surfaces (such as the wing). For this reason, control surfaces are usually stiffened and balanced to counteract flutter.

Divergence – Airfoil divergence occurs when an increase in lift causes the wing to twist upward. This twist increases the AOA and therefore increases the lift further. The increase in lift causes the wing to twist further, thus increasing the AOA again. This cycle continues until the wing fails.

Divergence is caused by the twisting axis (torsional axis) of the wing being in a position that causes an increase in lift to lead to a twist that increases the AOA. Put simply, this means that the torsional axis is somewhere behind the AC—so the effect of lift is to push the forward portion of the wing up.

The wing structure can be designed to make divergence impossible (on many aircraft, this is done). However, when divergence is a possibility, it becomes a criterion for determining the top speed of the aircraft.

Don't be fooled by the fact that these 3 undesirable effects occur at the dive speed and V_{ne} is only 90% of the dive speed. This 10% margin is not just an "idiot factor" and should not be taken lightly. The margin takes into consideration factors such as aging of aircraft components, manufacturing tolerances, etc. So be sure to pay close attention to the never exceed speed. On most training aircraft, the V_{ne} is well above the cruising speed range, but on some more advanced aircraft, cruising at speeds closer to "redline" is more common.

Maneuvering Envelope

We have enough information now to determine the maneuvering envelope of an aircraft. This envelope is basically a graphical representation of the allowable airspeeds and load factors that an aircraft can be flown at. It is a chart with two axis' of representation. The horizontal axis represents the airspeed, and the vertical axis represents the load factor.

We can begin by placing a box around the graph axis (Fig 9-5). The right side of this box represents the never exceed speed. Flight beyond this speed, as we know, can result in structural damage or failure. The top and bottom of the box represent the positive and negative limit load factors, respectively. Flight beyond these limits, also as we already know, will result in structural damage.

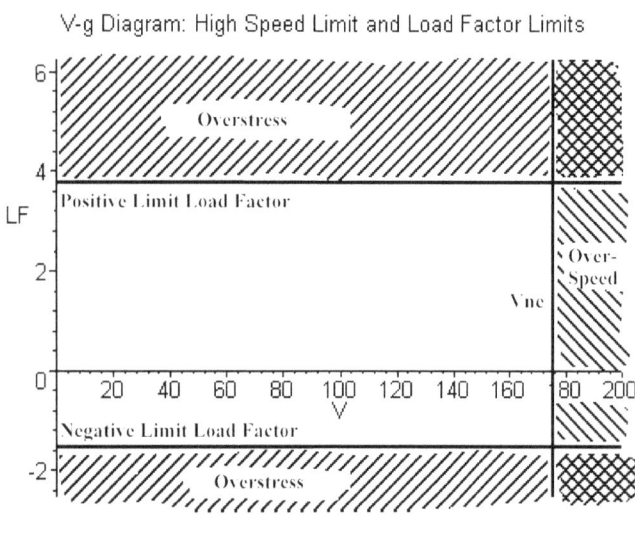

Fig 9-5. - *Speed and Load Factor Envelope.*

The next step is to put into place the stall limits (Fig 9-6). At a load factor of zero, the stall speed is zero. From there, the stall speed increases with the square root of the load factor. We can see from the diagram what the regular and inverted stall speeds are by checking the stall speed curve against one positive g and one negative g. Flight at speeds slower than the stall curve is impossible since the aircraft is stalled in this region. The point where the positive load stall curve meets the positive limit load factor is the corner speed, or the minimum allowable speed for V_a.

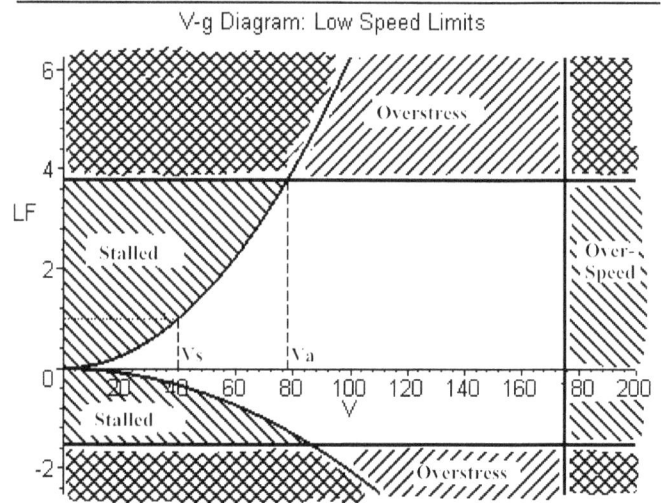

Fig 9-6. - *Adding the Stall Curve.*

We now essentially have the normal maneuver envelope for an aircraft. It is safe to operate the aircraft at any speed/load factor combination that fits within the limits of this envelope. We could also add a few things to the envelope to make it more complete (Fig 9-7). These items include the dive speed, the ultimate load factors, and V_{no}.

Note that V_{no} occurs at the speed indicated by the intersection of the limit load factor and the gust load factor (dashed line). This particular gust load factor line is for a 50 fps vertical gust.

As well, we could add a superimposed envelope which takes into consideration different configurations (Fig 9-8). In this example, the superimposed envelope is for the aircraft with flaps extended. Notice that the stall speeds are lower, V_{ne} has been replaced by V_{fe}, and in this example the limit load factors are lower (the LLF doesn't always change with configuration, but on some aircraft it does—check the POH).

V-g Diagram: Ultimate Limits and Turbulence Limits

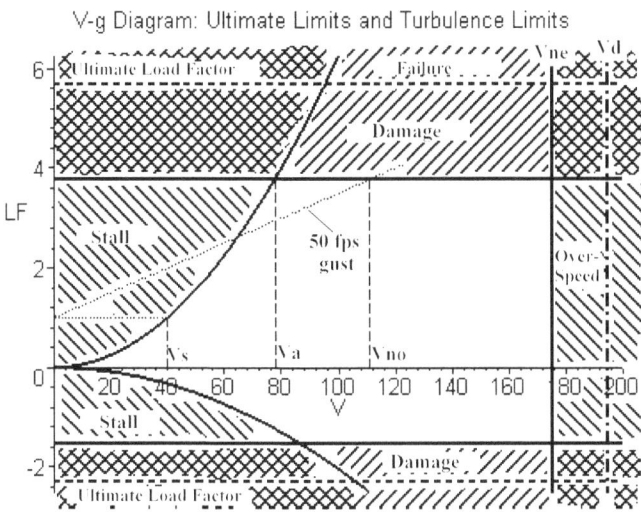

Fig 9-7. - The Maneuvering Envelope - This completed envelope gives us the speed and load factor limits that we must operate within to maintain structural integrity.

V-g Diagram: Configuration Changes

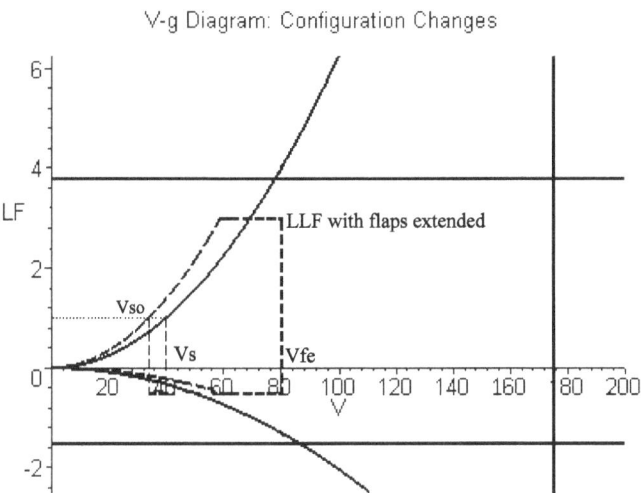

Fig 9-8. - Maneuvering Envelopes for Different Configurations - Load and speed limits will usually change with a configuration change. This example illustrates a flap extension.

Turning

As we discussed in the previous chapter, turning performance can be measured in terms of rate of turn and radius of turn. Generally speaking, a "maximum performance turn" is considered to be a minimum radius turn. We already know from (Eq. 8.12) that as the angle of bank increases and the airspeed decreases, the radius of turn decreases:

$$r = \frac{v^2}{g \tan AOB} \qquad (8.12).$$

So it would appear that we simply need to go to the maximum angle of bank with the lowest possible airspeed. This statement is reasonably accurate, but it is really not detailed enough. We know that as the angle of bank increases, so does the stall speed as well as the load factor. Both of these will be limiting to us.

Minimum Radius Turns

In performing a minimum radius turn, we want the minimum airspeed/maximum angle of bank combination discussed above. The problem is, for every airspeed there is a different maximum angle of bank, and for every angle of bank there is a different minimum airspeed. We need to find the perfect combination.

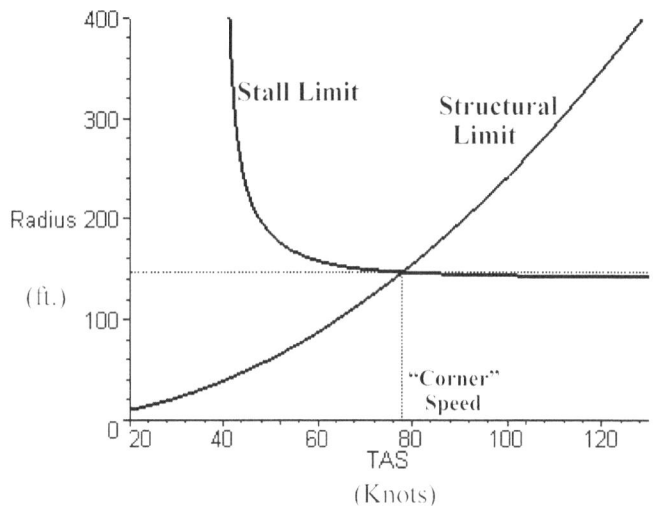

Fig 9-9. - Turn Radius vs. Airspeed and AOB.

Let's consider Fig 9-9 for a moment. The two curves represent angles of bank, and the points on the curves represent combinations of airspeed and turn radius. The "structural limit" curve represents the angle of bank that will bring the aircraft to it's limit load factor—any higher angle of bank will overstress the aircraft (this AOB is a constant for a given aircraft and configuration). Notice that as the airspeed increases, the turn radius increases as well (as per equation 8.12). The "stall limit" curve represents the maximum angle of bank that can be sustained at the given airspeed without stalling. Notice that as airspeed increases for the dashed curve, the turn radius decreases. Even though the increase in airspeed should increase the turn radius, the increase in angle of bank will decrease it even more.

The point where the two curves cross is the minimum turn radius that can be safely accomplished by this aircraft. This minimum turn radius can only be accomplished at the airspeed and bank angle indicated on the graph. The speed for a minimum radius turn is the corner speed (this is where the corner speed gets it's name). As well, on many aircraft this speed coincides with V_a (recall that the corner speed is the minimum allowable value for the maneuvering speed).

Note that this graph is only for one specific configuration. Varying the configuration may in some cases allow a reduction in turn radius due to a reduced stall speed, but a graph similar to Fig 9-10 would still apply to the new configuration.

Another consideration for minimum radius turns is power required and available. The above discussion assumes that the engine/prop can produce however much power is required by the aircraft. However, We've already seen in Chapters 3 and 8 that at higher load factors, more power is required due to induced drag. This means that in some aircraft, power available will restrict the aircraft's ability to perform a minimum radius turn.

Chapter Summary

Due to the stresses applied to an airframe by air loads, it is important to have some definitive limits which mustn't be exceeded during flight operations. These limits are expressed in terms of airspeed limits and load factor limits. Exceeding these limitations could lead to structural damage or failure.

Load factor limitations restrict the amount of lift that can be produced before airframe components begin to be damaged. The limit load factor for an aircraft is the maximum load factor that can be applied without permanent structural damage. The ultimate load factor, which is at least 1.5 times the limit load factor, is the point where structural failure becomes a possibility.

Turbulence and maneuvers will both subject an aircraft to load factors. In the case of turbulence, aircraft with higher wing loadings will be affected to a lesser extent.

Limiting airspeeds come in numerous forms. At the low end, we have stalls. Stalls are considered to be the aerodynamic limit to the maneuver envelope. At the high end, we have several speeds including the flaps extended speed (V_{fe}), the landing gear speeds (V_{le} and V_{lo}), the maneuvering speed (V_a), the normal operation speed (V_{no}), and the never exceed speed (V_{ne}).

V_{fe}, V_{le} and V_{lo} are based on the air loads and resulting moments that can be supported by the respective structures. V_a is the maximum speed which allows full control deflection without damage to the control surfaces. V_{no} is the maximum speed which allows the aircraft to fly through turbulence with 50 fps vertical gusts without structural damage. V_{ne} is the maximum speed at which the aircraft can be flown safely and is based on the design dive speed (V_d).

From the limiting speeds and load factors a maneuvering envelope can be constructed. This envelope is a graph of airspeed vs load factor and flight within the "box" is safe while flight outside the "box" is unsafe. The sides of the box are formed by the stall speeds, limit load factors, and V_{ne}. The envelope can also include the ultimate load factors and V_d.

Minimum radius turns are accomplished at the corner speed for a given configuration. This speed is the aircraft's stall speed at the positive limit load factor. If power restrictions are considered, some aircraft cannot perform minimum radius turns as they are described in this chapter. Instead, they must use a lower AOB and airspeed to allow for the power restrictions. This means that these aircraft are forced to perform wider turns than if they had sufficient power.

List of Formulae

Load Factor limits

$$LWL = W \times LLF \qquad (9.1)$$

Turbulence

$$WL = W/S \qquad (9.2)$$

Speed Limits
Maneuvering Speed

$$V_a = V_s \sqrt{LLF} \qquad (9.3)$$

$$New\ V_a = V_a \sqrt{(GW/MGW)} \qquad (9.4)$$

Questions and Problems

1) What is the difference between the limit load factor and the ultimate load factor?

2) Why doesn't the limit load factor normally increase with a decrease in weight?

3) If an aircraft is required to be stressed for negative g's in order to be airworthy, why are some aircraft prohibited from performing inverted maneuvers?

4) How does wing loading affect an aircraft's sensitivity to turbulence? Why?

5) What is the difference between V_{le} and V_{lo}? Why does this difference exist?

6) How can the stall speed be used to determine V_a? Is this always accurate? Why/Why not?

7) How is V_{ne} determined? Why is it important?

8) What is flutter?

9) How would you determine the speed to fly for a minimum radius turn?

10) Once exceeding V_a, how would you determine how much control input can be safely used?

11) An aircraft has a stall speed of 45 kts, a limit load factor of 4 g's and a published V_a of 105 kts. Would it be safe to use full aileron input at 100 kts? Full rudder? Full elevator? Why/Why not?

Introduction to Part II
Application

In part 1 of *Applied Aerodynamics for Private and Commercial Pilots* we examined the forces that act on an aircraft, along with their origins and their effects. The understanding of these forces and phenomenon are the background knowledge necessary to understand the control techniques used during flight maneuvers. However, these techniques have not been detailed yet.

Part 2 of this book is intended to take the theoretical information presented in Part 1 and use it to establish procedures and techniques to control the aircraft. These techniques will be founded mainly on the information already covered in previous chapters, but some new topics will also be introduced. As well, items previously covered will at times be covered in more detail or from a slightly different perspective.

Keep in mind when operating aircraft that some variables and design considerations are beyond the scope of this book. The techniques detailed herein are generally correct for most general aviation aircraft. However, due to individual design considerations some techniques may be inappropriate for specific aircraft. For accurate information on the operating procedures relevant to your aircraft, be sure to refer to the POH or the Flight Manual as appropriate. As well, discuss the individual flight characteristics of your aircraft with a qualified flight instructor who is current on type.

Another thing to bear in mind is that this book is written strictly from a "hands and feet" perspective. No attempt is made to consider pilot decision making or airmanship issues. These issues, such as *when* or *where* to fly the airplane, are just as important as *how* to fly the airplane. For more information on these subjects, refer to other aviation texts and to other pilots—particularly flight instructors.

Chapter 10
Weight and Balance

Throughout this text, numerous effects of aircraft weight have been discussed. Along with these effects, we have looked at the effects of the C of G location. Clearly, both weight and C of G will vary from flight to flight since we will not always have the same passengers, the same baggage, or the same amount of fuel. For that matter, weight and C of G will vary even throughout a given flight due to the burning of fuel. Burned fuel is dumped overboard in the form of exhaust gases, thus the weight will decrease and in most cases the C of G will shift.

Since the weight and C of G have such a significant effect of the aircraft's handling and performance, it is critical that we remain within the limits certified by the manufacturer. These limits will be published in the form of maximum weights and limiting C of G positions. In this chapter, we will be discussing these limits and how to insure that we remain within them.

Effects of Weight

The weight of the aircraft has several effects on handling and performance. These effects have already been detailed in previous chapters, but we will briefly review them here. Primarily, the effect of weight is due to the direct relationship between weight and mass.

An increase in weight means an increase in the mass of the aircraft. We know from previous chapters that this increase in mass will have one fundamental effect—an increase in the force required to achieve a given acceleration. All weight related effects can be traced back to this fact—directly or indirectly.

To illustrate the point, consider the effect weight has on the takeoff roll. An increase in weight means that the thrust produced by the engine at takeoff power (a constant for a given altitude/airspeed combination) will accelerate the aircraft more slowly down the runway. This loss in acceleration increases the distance required to achieve the same speed. This is combined with the change in lift-off speed. An increase in weight means that the wings need to produce more lift to offset the acceleration due to gravity. Once again this is a result of the increased mass of the aircraft (remember that weight is simply the mass multiplied by the acceleration due to gravity).

Weight will also increase our landing distance—once again due to the change in speed resulting from the need for more lift. Essentially, all of our performance characteristics will deteriorate with an increase in weight. Climb rate decreases because the excess power is divided over more weight. Climb angle is decreased because the climb rate is decreased, or because the excess thrust is divided over more weight (remember from Chapter 8 that these two statements mean the same thing). Cruising speed is reduced due to an increase in induced drag. Range and endurance both are decreased for the same reason. The stall speeds of the aircraft will increase with weight due to the increase in lift required even to maintain equilibrium.

The only performance characteristic that does not deteriorate is the glide range. However, even best glide range will occur at a different speed due to the weight change.

Clearly, there must be some practical limit on the allowable weight of the aircraft. This weight is determined by a number of factors including performance characteristics of the aircraft at different weights. As well, handling characteristics, stability, and structural strength will be considered when determining the maximum gross weight for an aircraft.

Effects of C of G

The center of gravity (or center of mass) is another factor which can have a huge influence on the handling and stability of the aircraft. As we know, the C of G is the "focal point" of the weight—it is the point at which we can generally assume that all of the weight acts. Each of the axis' of the aircraft intersect at this point, and as a result, all the movements (pitch, roll, and yaw) are about this point.

The C of G has a huge impact on the handling and stability of the aircraft. A forward C of G will result in a longitudinally and directionally stable aircraft, while an aft C of G can result in a longitudinally—and possibly directionally—unstable aircraft.

Control of the aircraft is also directly influenced by the C of G position. The elevator/stabilator and the rudder lose effectiveness as the C of G moves aft. This is because the arm that the control surfaces act on is shortened as the vertical and lateral axis' move aft (Fig 10-1).

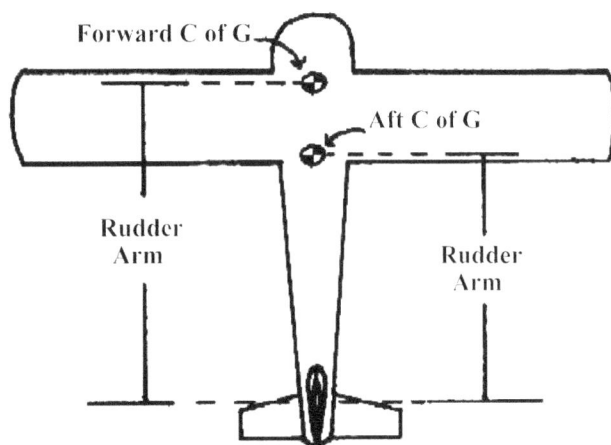

Fig 10-1. - *Control Arm Based on C of G Position* - As the C of G moves aft, the arms of the rudder and the elevator are shortened.

It appears so far that getting the C of G as far forward as possible is desirable so that we can improve the stability and control of the aircraft. This is in fact the case—up to a point. As the C of G moves forward, the tail down force must be increased to maintain equilibrium. This means that the total lift being produced by the aircraft is increased—and as a result so is the induced drag. So performance will suffer as the C of G is shifted

forward.

Another problem encountered with a forward C of G is the effectiveness of the elevator/stabilator. As the C of G moves forward and the tail is required to produce more and more down force to maintain equilibrium, we will eventually run out of aft movement of the control column—and thus tail down force. This phenomenon is usually the limiting factor on the forward position of the C of G. In ground effect, downwash is reduced and as a result, the elevator becomes less effective. When landing, the airspeed is constantly decreasing—which also reduces the effectiveness of the elevator. Since we must be in ground effect when landing, the two of these effects combine to significantly reduce the usefulness of the elevator. With a C of G too far forward, it would difficult (if not impossible) to flare properly.

So having the C of G too far aft can be a serious problem. As well, having the C of G too far forward can be a serious problem. It is important that we keep the C of G within the prescribed limits to insure the proper operation of our aircraft.

Terminology

Before discussing the weight and balance envelopes and calculations that are applicable to aircraft, it is important to understand some of the important terms that will be used.

The first set of terms refer to the weight of the aircraft.

Gross Weight – The total (or "all-up") weight of the aircraft—this includes fuel, oil, any other operating fluids (e.g. – hydraulic), flight crew, passengers, baggage, etc.

Maximum Gross Weight – The maximum allowable weight of the aircraft. For many light aircraft, this weight is the maximum weight for taxiing, takeoff, and landing. For some larger and/or more complicated aircraft, there can be several maximum gross weights including *maximum ramp weight, maximum zero fuel weight, maximum takeoff weight,* and, *maximum landing weight.*

Licensed Empty Weight – The weight of the aircraft itself with all of the minimum required equipment installed and the unusable fuel and oil.

Basic Empty Weight – The weight of an aircraft with all of the required and optional equipment installed and the unusable fuel and oil. Many manufacturers will also include full oil in the basic empty weight.

Useful Load – This is the difference between the maximum weight and the empty weight. It is essentially the amount of weight that can be carried on board the aircraft.

Payload – The payload is the weight that can be carried aside from operating fluids (fuel, oil, etc.) and the flight crew. This is an important weight for commercial operators, because it is basically the weight that they can be paid to carry.

After the operating weights of the aircraft, we need to know some terms relating to the C of G.

Datum – The datum is a reference point from which measurements are taken (Fig 10-2). The position of the datum does not matter, provided all measurements are taken from the same point. The datum location is chosen arbitrarily by the aircraft manufacturer. If weight and balance calculations were to be done on the same aircraft using different datum locations, it would be found that the numbers would look different, but in sketching those numbers on an aircraft schematic, the C of G would be in the same location.

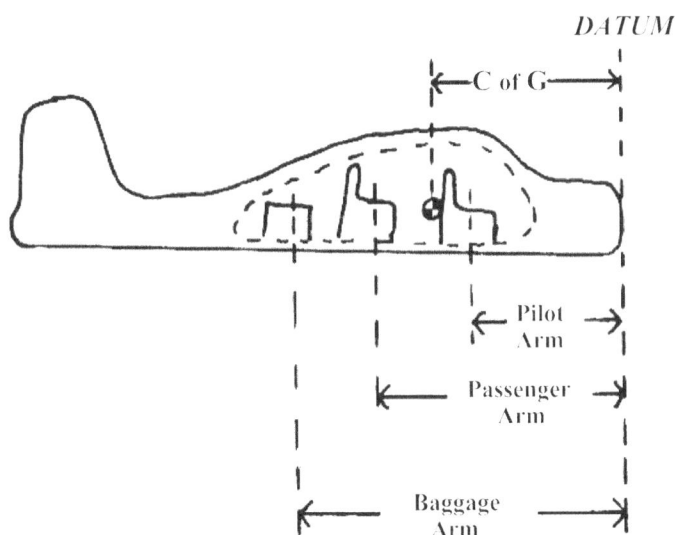

Fig 10-2. - The Datum - A reference point used for weight and balance calculations. Moment arms are measured from this point.

Arm – The arm is a measurement from the datum (reference) to a specified location on the aircraft. The arm is normally measured in inches and defines the position of an object to determine it's moment. As a sign convention, a positive arm normally indicates a position aft of the datum and a negative arm normally indicates a position forward of the datum.

Moment – Moments are determined from equation (1.25),

$$M = FA \qquad (1.25).$$

In the case of a weight and balance calculation, the force referred to is the weight of a given object. So to determine the moment of an object about the datum point, we can use the following,

$$M = W \times A \qquad (10.1).$$

Using these terms, we will now move on and consider the methods for keeping the aircraft within it's certified weight and balance limits.

Weight and Balance Envelopes

The weight and balance limitations of the aircraft are contained in the _POH_ or the _Flight Manual_. This information will normally be in graph form, but will sometimes be included in chart form as well. The graphs will either be in the form of "Weight vs. C of G" (Fig 10-3a), or "Weight vs. Moment" (Fig 10-3b). Both of the graphs give the same information in a different, but equivalent, format.

The weight vs C of G graph is very straightforward. The maximum weight of the aircraft is defined by the top horizontal line. The forward and aft C of G limits are defined by the vertical lines on the left and the right. When determining our position on this graph, we simply draw a line straight across from our weight and draw another line straight up from our C of G. If the lines cross inside the envelope, the aircraft is safe to fly.

The weight vs moment envelope is essentially the same thing, except the presentation looks slightly different. The envelope is skewed to the right significantly. The reason for this is that as the weight increases, a given C of G position will give a higher moment. Use of the graph is the

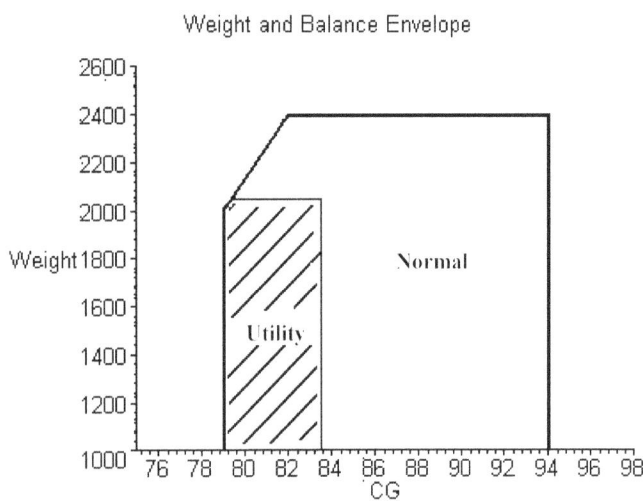

(a) A weight vs C of G envelope for a fictional aircraft.

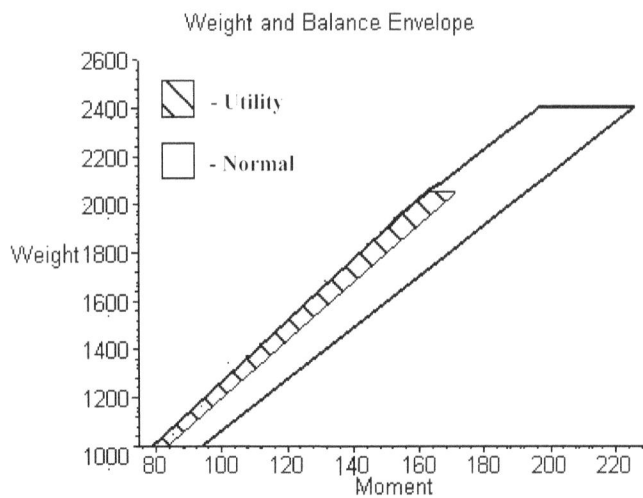

(b) A weight vs moment envelope for the same (fictional) aircraft as in (a).

Fig 10-3. - *Weight and Balance Envelopes.*

same as the weight vs C of G. A line drawn straight across from the weight will intersect with a line drawn straight up from the moment. Provided this intersection occurs within the envelope, the aircraft is safe to fly.

Calculations

The calculation method for weight and balance is fairly straightforward. We simply add up all of the weights (empty weight, fuel, oil, pilot, passengers, baggage, etc.) to determine the gross weight of the aircraft. Then we multiply the weight of each object by it's arm to determine each moment, which we add to get the total moment for

the aircraft. If we are working with a "weight vs moment" envelope, the calculation is now complete. If, on the other hand, we are working with a "weight vs C of G" envelope, the total moment must be divided by the gross weight to get the arm of the loaded aircraft—this is the C of G.

To determine the arm of a certain position in the aircraft, simply refer to the *POH* or the *Flight Manual.* The information will normally be presented graphically in the weight and balance section of the book (Fig 10-4a). Sometimes the same information will be presented in the format of a weight vs moment chart for each station (Fig 10-4b).

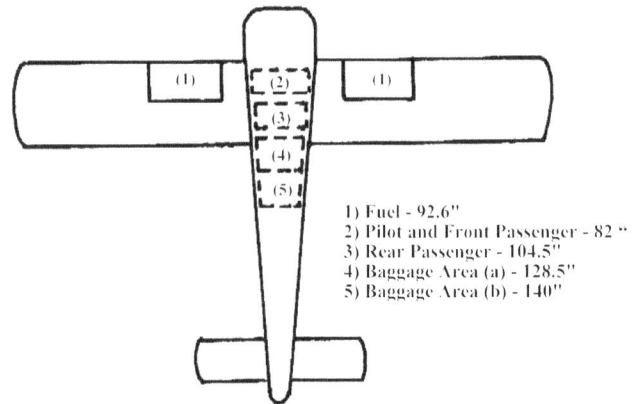

1) Fuel - 92.6"
2) Pilot and Front Passenger - 82 "
3) Rear Passenger - 104.5"
4) Baggage Area (a) - 128.5"
5) Baggage Area (b) - 140"

(a) Graphical representation of aircraft stations for a fictional aircraft.

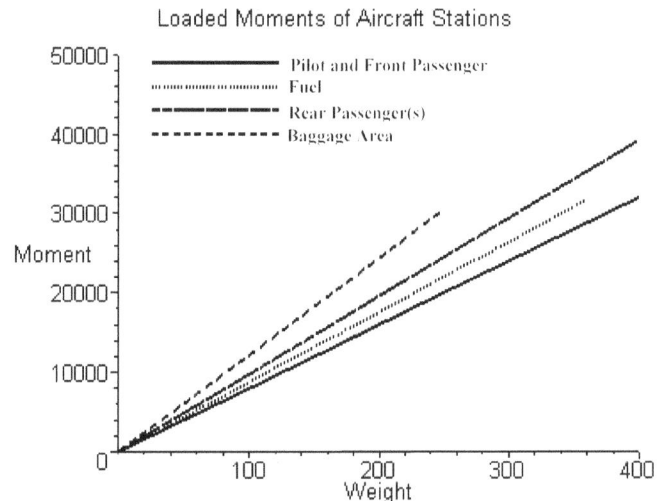

(b) Graph of weight vs moment for different aircraft stations

Fig 10-4. - *Typical Arm Representations.*

To do an example we can start with a weight and balance document. This document is developed by AME's (aircraft maintenance engineers) for an individual aircraft and provides us with information on the empty weight, C of G, and moment of the aircraft (Fig 10-5).

Now working with the position arms provided in Fig 10-4, we can perform the weight and balance calculation illustrated in Fig 10-6. Notice that three weight/C of G combinations are of importance to us. The first is the subtotal. This provides us with weight and balance information on the loaded aircraft without fuel. The second is the landing total. This one provides us with weight and balance information for the landing—after burning the planned amount of fuel. The third is the takeoff total. This provides us with takeoff weight and balance information based on the amount of fuel we intend to carry on departure.

ACME Aircraft Maintenance Inc.

Weight and Balance Report

Aircraft: C-GXYZ
Maximum Gross Weight: 2650 lbs.
Forward C of G: 82"
Aft C of G: 100"

Note: Empty weight values include full oil and unuseable fuel.

Removed:

Airspeed Indicator	-2.7 lbs	71"	-191.7"lbs
Rear Seats	-22.4 lbs	95"	-2128"lbs

Added:

Airspeed Indicator	+2.5 lbs	71"	+177.5lbs

TOTAL CHANGE:

	-22.6 lbs	-2142.2"lbs

Original Empty Weight:

1604.3 lbs	87.7"	140697.11"lbs

New Empty Weight:	1581.7 lbs
New Empty Moment:	138554.91 "lbs
New Empty C of G:	87.6"

Fig 10-5. - *A Typical Weight and Balance Document.*

Notice that the weight decreases as we burn fuel. As well, the C of G will shift as the fuel is burned. The reason for this will be discussed in detail in the next section. For now we simply need to consider the importance of this fact. If we were to complete a weight and balance calculation for takeoff and—finding that we are within the prescribed limits—leave it at that, we may very well find that part way through the flight the aircraft is no longer within limits. To avoid this, we must complete calculations for takeoff *and* landing. It is also a good idea to complete a calculation for a

zero fuel weight and balance since it is possible that the flight will end up being longer than planned (therefore we will burn more fuel), and thus a new landing weight and balance will be in effect.

	Weight	Arm	Moment
Empty	1335.8	84.1	112340.78
Pilot	*175.0*	82.0	*14350.0*
F. Pass	*200.0*	83.5	*16700.0*
R. Pass	*170.0*	104.5	*17765.0*
Baggage	*60.0*	128.5	*7710.0*
Subtotal	*1940.8*	*87.01*	*168865.78*
Unused Fuel	*90.0*	92.6	*8334.0*
Landing	*2030.8*	*87.26*	*177199.78*
Used Fuel	*120.0*	92.6	*11112.0*
Takeoff	*2150.8*	*87.55*	*188311.78*

Fig 10-6. - *Weight and Balance Calculation* - Straight numbers are published either by the manufacturer or on the weight and balance document. Weights that are *scripted* are weighed, and moments and arms that are *scripted* are calculated.

The Effect of Fuel Burn

As the fuel burns off, the weight/C of G shift will be *approximately* linear (as long as the weight of fuel burned is small in comparison with the weight of the aircraft, this approximation is sufficiently accurate). This means that if we were to plot the full fuel (or takeoff fuel) weight and C of G with the zero fuel weight and C of G, the C of G travel would be along a roughly straight line between these two points. The weight and balance envelope in Fig 10-7 illustrates this point for the calculations done in Fig 10-6.

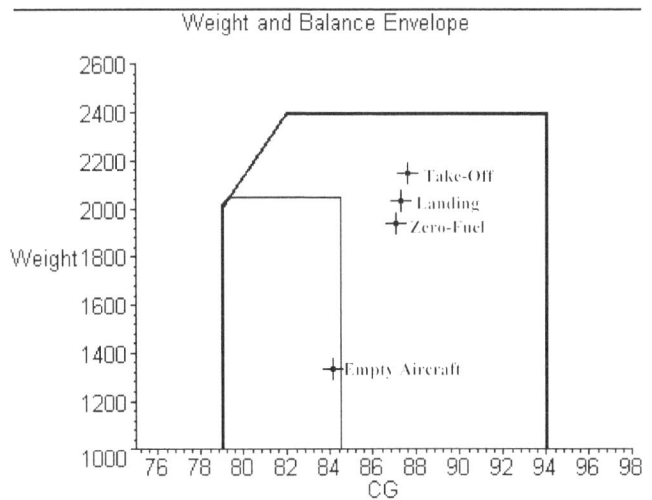

Fig 10-7. - *Envelope Placement of the Aircraft in Fig 10-6.*

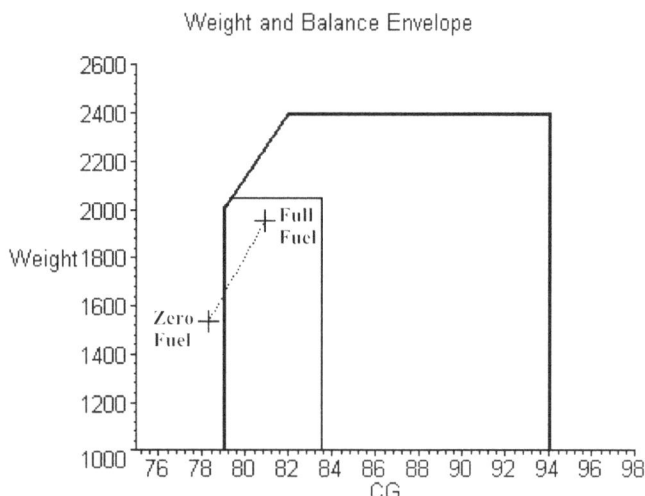

Fig 10-8. - *Fuel Burn Effect* - In this example, the burning of fuel brings the aircraft outside the weight and balance envelope. When this is the case, the useful fuel on board is restricted by the weight and balance limitations.

It is important to realize that if fuel is used as ballast to bring the aircraft into the certified limits, these limits can be exceeded as the fuel is burned off. For example, we can consider an aircraft that has a total fuel capacity of 40 gal US. If this aircraft is loaded in such a way that it has a C of G that is too far aft with less than 10 gal US of fuel, the effective fuel capacity is only 30 gal US. Departing with full tanks, the pilot only has 30 gal US of fuel to work with—burning more than this will cause undesirable handling characteristics. This is illustrated in Fig 10-8.

A commonly asked question is whether fuel will cause the C of G to shift forward or aft. The answer will depend on the aircraft, and may even vary for different loadings of the same aircraft. The issue to consider is whether the arm of the fuel tank places it forward or aft of the aircraft's loaded C of G. If the fuel arm is aft of the loaded C of G (Fig 10-9a), the C of G will shift forward as the fuel is burned off. This is due to a reduction of weight in the aft section of the aircraft. On the other hand, if the fuel arm is forward of the loaded C of G (Fig 10-9b), the C of G will shift aft as the fuel is burned off. Similar to the forward shift, this is due to a reduction in weight in the forward portion of the aircraft.

Load Shifting

We can sometimes find ourselves in a situation where the aircraft is loaded so that we are well within the weight limits, but we are not within the C of G limits. In this case, we have three options. We can load ballast to move the C of G, we can unload the aircraft until the C of G is moved to within the limits, or we can shift the load around until the C of G is within limits. 99 times out of 100, shifting the load will be the more desirable option—particularly for revenue operations that make money by carrying passengers and/or cargo. Loading ballast means carrying more weight and degrading performance, while unloading the aircraft means leaving passengers or cargo behind.

The questions that need answering here are how much weight should we move, and how far should we move it. The answer lies in a simple relationship which can be derived from the standard weight and balance calculations. Consider this—if an object inside the aircraft is moved, the change in the objects moment will also be the change in the total moment for the aircraft. From this, we get,

$$D_O \times W_O = \Delta M_{AC}. \tag{10.2}$$

Where D_O is the distance the object moved, W_O is the weight of the object, and M_{AC} is the moment of the whole aircraft. Remember from Chapter 1 that Δ is the letter delta used to represent a change. If we are calculating a weight and balance for an aircraft with a weight vs. moment graph, this is the equation for us to use in load shifting.

On the other hand, if our aircraft has a weight vs. C of G graph, we need to make some changes for this equation to be a bit easier to work

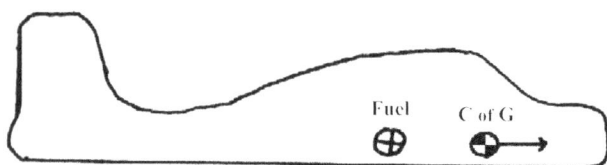

(a) With the fuel arm aft of the C of G, the C of G shifts forward as the fuel burns off.

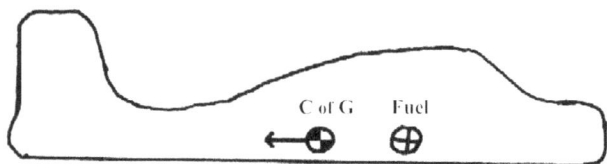

(b) With the fuel arm forward of the G of G, the C of G shifts aft as the fuel is burned off.

Fig 10-9. - *Fuel Effect of C of G Travel.*

with. Knowing how to get the moment of the aircraft, equation (10.2) can be changed to,

$$D_O \times W_O = W_{AC} \times D_{CG}$$

Where WAC is the weight of the aircraft and DCG is the distance that the C of G moved. From this, we can get,

$$\frac{W_{AC}}{W_O} = \frac{D_O}{D_{CG}} \qquad (10.3).$$

Knowing any two of the variables in equation (10.2) or any three of the variables in equation (10.3) will allow us to determine the other. For example. We can consider an aircraft with a loaded gross weight of 2500 lbs, in that aircraft we have a bag which weighs in at 250 lbs. How far would we have to move that bag forward to shift the C of G forward 1.5 inches? Using equation (10.3),

$$\frac{2500}{250} = \frac{D_O}{1.5}$$

therefore,

$$D_O = 15 \text{ inches}.$$

We could also complete the calculation in a different way. If moving the bag to the next forward compartment meant a 30 inch movement, how far would the C of G shift?

$$\frac{2500}{250} = \frac{30}{D_{CG}}$$

therefore,

$$D_{CG} = 3 \text{ inches}.$$

This equation could also be used to determine how much baggage to move given distances for the C of G and the baggage movement.

Chapter Summary

The weight and C of G location of an aircraft have an extremely large influence on the aircraft's performance, controllability, and stability. For this reason, aircraft designers will establish limitations on the weight of the aircraft as well as the range of positions for the C of G. These restrictions must be adhered to if one expects to experience the advertised (and safe) flight characteristics of the aircraft.

The acceptable weight and balance envelope is often presented graphically and there are two common methods of representation. The first is a graph of weight vs C of G. With this method, the allowable weights are plotted against one axis of the graph while the allowable C of G positions are plotted against the other. If the aircraft's weight and balance falls within the envelope, it is safe to fly. The second method is to plot the weight against one axis again while the total moment about the datum point is plotted against the other axis. Once again, if the aircraft's weight and balance falls within the envelope the aircraft is safe to fly.

The calculation of a weight and balance is fairly simple. We start with the empty weight of the aircraft and add the weights of anything that is carried. This includes fuel, oil, flight crew, passengers, baggage, etc. Then the moment of each item (including the aircraft) must be added up to a total. Dividing the total moment by the total weight gives us the C of G.

As fuel is burned during flight, the weight of the aircraft will decrease, and the C of G will shift. What direction the C of G moves in will depend on where the fuel tank is relative to the current C of G. As fuel burns off, the C of G will move further away from the fuel tank. For this reason, weight and balance calculations should include the weight and C of G for takeoff, landing, and zero fuel. Most aircraft are designed with fuel storage as close as practical to normal C of G positions, but some small amount of movement is inevitable.

Sometimes the initial weight and balance calculation will indicate that the aircraft is outside of acceptable C of G limits even though the maximum weight has not been exceeded. When this is the case, shifting the weight around is often the easiest solution. The new C of G or moment, along with the shifts that must be made, can be determined using equation (10.2) or (10.3).

List of Formulae

Terminology

$$M = W \times A \qquad (10.1).$$

Load Shifting

$$D_O \times W_O = \Delta M_{AC}. \qquad (10.2)$$

$$\frac{W_{AC}}{W_O} = \frac{D_O}{D_{CG}} \qquad (10.3).$$

Questions and Problems

1) How does weight and balance affect an aircraft's flight characteristics?

2) Based on the following data, calculate a weight and balance:

Item	Weight	Arm	Moment
Empty aircraft	1272.7 lbs	81.2"	103343.24"lbs
Fuel	240.0 lbs	83.9"	
Oil	15.0 lbs	18.0"	
Pilot	185.0 lbs	82.3"	
Front Passenger	200.0 lbs	83.8"	
Rear Passenger	130.0 lbs	117.4"	
Baggage	105.0 lbs	131.5"	

Fuel Consumption = 8 gal US/h (1 gal US weighs approximately 6 lbs at 15°C)
Planned Duration of Flight = 2h30m

3) Based on the weight and balance envelope in Fig 10-3a, is the aircraft in Q2 safe to fly?

4) If the front and rear passengers traded places, where would the new takeoff C of G be? Landing C of G? Zero Fuel C of G?

Chapter 11
Taxiing

The first activity in the aircraft that makes the knowledge of flight theory and useful is taxiing. When taxiing the aircraft to or from the runway, or anywhere else on the aerodrome, proper use of the controls is important and sometimes critical. Methods for starting and stopping, controlling taxi speed, and directional control—particularly in high winds—will come in handy on every flight.

These methods are not complex or difficult, but they are important for the successful completion of a flight. After a few training flights, taxi methods will be second nature.

Starting and Stopping

To start the aircraft rolling, simply release the brake. Often, this is all that is required to get moving. Sometimes, however, extra power may be needed to initiate the motion. Simply open the throttle a little, allow the aircraft to start rolling, and reset the throttle to it's initial setting (around 1000 RPM on most training aircraft). Be careful not to leave the power on for too long, as this will cause too much speed to build up—simply use the extra power to start the motion and then let the aircraft maintain that motion with a lower power setting.

To stop the aircraft, remove all power first—that is, close the throttle to bring the engine to full idle. Once the power has been removed, apply the brakes as necessary. Be sure to remain one step ahead of the airplane at all times so that a sudden application of the brakes will not be needed. As well, be sure to remove all power prior to applying brakes.

Power must be removed to stop for several reasons. Generally, the main reason is simply to allow the aircraft to stop shorter and to avoid wear on the brakes. However, there are times when it can be dangerous to brake with power on. In particular, consider taxiing with a strong tailwind and having to stop suddenly. The sudden stop creates a tendency for the aircraft to nose over, the tailwind doesn't make things any better, and if power is being carried, it may be enough to cause a loss of control. This is especially the case for tail-draggers, since a nose over movement will often bring the C of G ahead of the main gear.

Speed Control

Speed control when taxiing is extremely important. Airplanes are ungainly creatures on the ground, they can't maneuver nearly as well as a car can. Because of this, excessive speed can easily lead to an accident.

On level ground with calm winds, the aircraft will usually maintain an appropriate taxi speed on it's own provided the power is not set too high (once again, about 1000 RPM is a good taxi power for most training aircraft). It is important, however, to be careful of the effects of uneven ground or moderate to high winds.

If the aircraft is travelling uphill with a headwind, no problems should occur. The aft component of weight and the headwind will mean that an increase in power will be necessary just to maintain a normal taxi speed. The problems start when taxiing downhill and/or with a tailwind. These situations will tend to accelerate the aircraft to higher taxi speeds—sometimes very gradually so that the pilot doesn't notice unless he or she is really on top of the situation. Keep in mind at these times that it is easier to remain slow than to slow down—don't let the speed get out of hand. The best thing to do is to pull the power all the way back to idle and then use the brakes only as necessary.

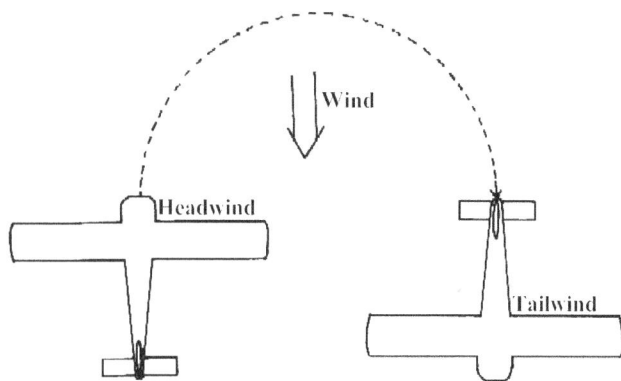

Fig 11-1. *- Wind Direction Relative to the Aircraft Changes After a Turn. - Taxiing with a headwind requires extra power, but after a turn, the resulting tailwind requires a reduction in power*

Beware of the windshift relative to the aircraft as a turn is made (Fig 11-1). We may be battling a headwind and carrying extra power to maintain taxi speed. But after a turn, that headwind may become a tailwind. If the power is not adjusted immediately, the speed will come up very quickly.

Directional Control

There are three main methods for directional control on the ground. They are aerodynamic steering with the rudder, steerable nosewheel/tailwheel, and differential braking.

Aerodynamic steering is very inefficient and difficult, it requires constant vigilance on the part of the pilot. Due to the ineffectiveness of the rudder at such low speeds as taxi, this method is practically unheard of on modern aircraft. Manufacturers tend to lean towards wheel steering and differential braking.

Wheel steering is simply a mechanical linkage between the rudder pedals and the nosewheel/tailwheel. Step on the left rudder pedal and the aircraft will turn left, step on the right rudder pedal and the aircraft will turn right. On some aircraft, this mechanical connection is made with bungees instead of steel cables. When this is the case, a delayed reaction to control application must be anticipated.

Differential braking is available on some aircraft to allow for a much tighter turn radius on the ground. Independent brakes are supplied for each main wheel, and they can be applied separately to effect the steering of the aircraft. In most if not all

cases, differential braking is controlled by toe pads immediately above the rudder pedals. This arrangement is referred to as "toe brakes" since most non-differential brake systems are operated by a hand brake. Some manufacturers design their aircraft with a free-swinging nosewheel, and rely entirely on differential braking for the steering of the aircraft on the ground.

Effects of Wind

Strong winds have two separate effects on a taxiing aircraft other than the effect on speed which we have already discussed. The first is the directional effect of taxiing in a strong crosswind (Fig 11-2). The second is the tendency to lift a wing and/or the tail during taxi.

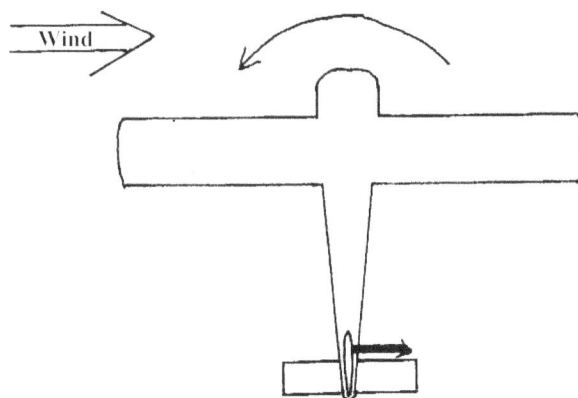

Fig 11-2. *- Taxiing With a Crosswind Causes the Aircraft to Weathervane - This is a result of the aircraft's directional stability.*

Taxiing in a strong crosswind, the aircraft is effectively in a sideslip. Due to directional stability, a yawing moment is created which will tend to turn the aircraft into the wind. This is known as "weathervaning" and must be compensated for by the use of rudder (or any other directional control device, e.g. – differential braking).

As for the lifting tendency, it is due to the fact that air is flowing over the wing and tail surfaces, and this causes them to create lift. If the airflow is at a high enough speed, the lift could be sufficient to upset the aircraft. With the wind head-on to the aircraft, all controls should be neutral to minimize this effect. As the wind moves around relative to the aircraft (or vise versa), some control inputs are necessary.

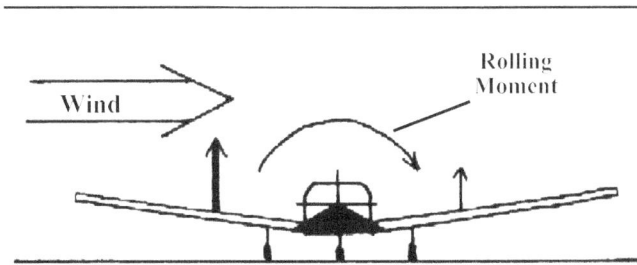

(a) Crosswind without corrective control inputs

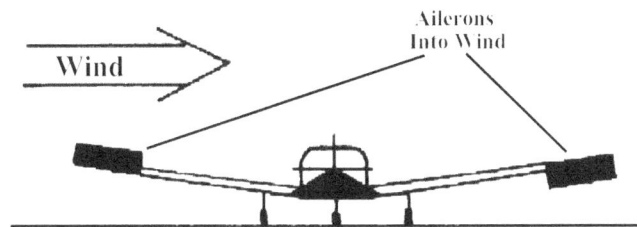

(b) Aileron correction for the crosswind

Fig 11-3. - *The Effect of a Crosswind During Taxi* - With a quartering headwind or a direct crosswind, the aircraft will tend to roll away from the wind. Counteract this with an appropriate aileron input.

A quartering headwind (Fig 11-3) or a direct crosswind will tend to roll the aircraft away from the wind. To minimize the problem, the ailerons should be deflected into the wind. This reduces the lifting tendency on the windward side and increases it on the leeward side. The net effect is to reduce or eliminate the rolling moment. The elevator/stabilator should still be left neutral.

With a quartering tailwind, the aircraft still wants to roll, plus the tail tends to get lifted. To compensate, we still need to create a rolling moment into the wind, plus we now need to create a pitching moment to prevent the tail from being lifted. However, the control inputs used to accomplish this may not be what one would initially expect.

First, consider the ailerons. The wind is now from behind, so the airflow over the wings is essentially backwards. An aileron deflection into the wind (Fig 11-4a) in this case will actually increase the rolling tendency. So a "backwards" input is used to counteract the roll (Fig 11-4b). In other words, the aileron deflection used is one that would seem to aggravate the effect of the wind. A tailwind from the left would roll the aircraft to the right, and a right aileron input is required to counteract it.

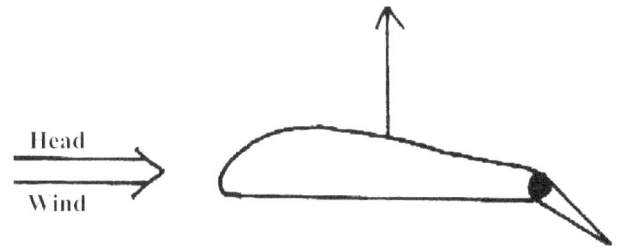

(a) With a headwind, the control surfaces behave as expected.

(b) With a tailwind, the effect of the control surface is reversed due to the reversed airflow.

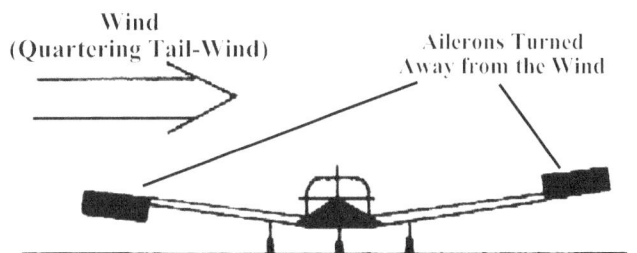

(c) To decrease the rolling moment, a reversed aileron input should be used.

Fig 11-4. - *Tailwind Effect During Taxi.*

Now consider the elevator/stabilator. The airflow is "backwards" over the tail as well, so once again the control effect will be reversed. If we want to hold the tail down, it will be necessary to place the elevator/stabilator in the down position (Fig 11-5).

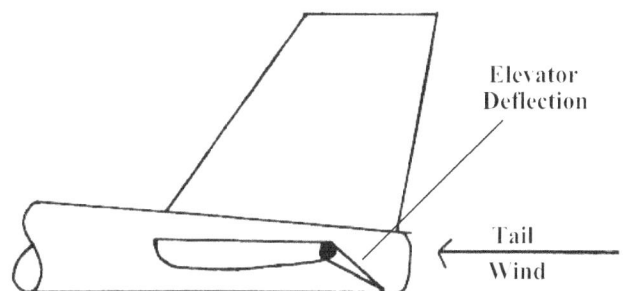

Fig 11-5. - *Elevator Position to Correct for a Tailwind.*

Chapter Summary

In order to taxi an aircraft safely, particularly in high winds, some thought on the mechanics involved is necessary. Speed control is extremely important under all circumstances since aircraft don't handle well on the ground.

Directional control on most aircraft is maintained with the steerable nosewheel/tailwheel and/or differential braking. In high winds, the aircraft will tend to weathervane and directional control may become more difficult. As well, aileron and elevator control inputs may be necessary in high winds.

Questions

1) While taxiing along runway 33 at an airport, the winds are reported as 020° magnetic at 25 knots gusting to 35 knots. What control input will be necessary? (Note for readers new to aviation—runways in southern domestic airspace are numbered according to their magnetic direction.)

2) After turning 90° left to exit on a taxiway, what control inputs will be necessary for the aircraft in Q1?

3) When starting to taxi from rest, it is often necessary to add extra power and then remove it. Why?

Chapter 12
Attitudes and Movements

In order to pilot an aircraft competently, it's critical that we understand the basic control movements and their effects on the aircraft. Without a full understanding of these fundamentals, it is impossible to progress onto the key maneuvers of flying the aircraft such as straight and level, climbs, descents, and turns.

The fundamental control functions and their bearing on aircraft operations are normally covered on the first training flight (sometimes along with other basics). This exercise is referred to as "Attitudes and Movements" and covers the basic use of all of the primary control surfaces along with some of the factors that affect these controls.

Attitude vs. Movement

Before discussing this topic any further, we need to discuss what an attitude is and what a movement is. These two terms are often confused, but they are distinctly different. The relationship between the two is very important as well.

Attitude – The attitude of the aircraft is the aircraft's position relative to the horizon. Attitude is defined by the pitch angle and the bank angle. Pitch attitude is a term which refers to how high (or low) the nose of the aircraft is relative to the horizon (Fig 12-1a). The pitch attitude could also be considered to be the angle between the longitudinal axis and the horizon as seen from the side of the aircraft. Bank attitude refers to the angle between the lateral axis and the horizon (Fig 12-1b). When an aircraft is established in a constant attitude, it is stationary relative to the horizon—that is, it is not rotating about the lateral or longitudinal axis.

Movement – A movement is a rotation about one of the three axis'. The three movements

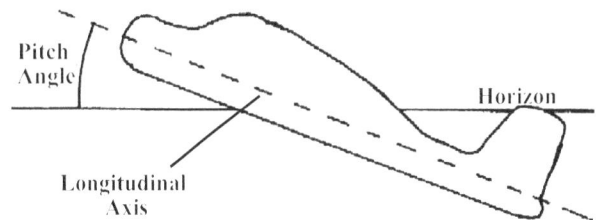

(a) Pitch Attitude – Determined by how high (or low) the nose is relative to the horizon.

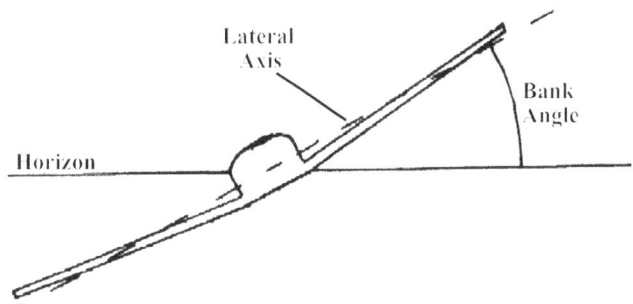

(b) Bank Attitude – Determined by the angle between the wings (or the lateral axis) and the horizon.

Fig 12-1. - *Attitude* - The position of the aircraft relative to the horizon. Attitude is measured as a pitch angle and a bank angle.

available to us are pitch (movement about the lateral axis – Fig 12-2a), roll (movement about the longitudinal axis – Fig 12-2b), and yaw (movement about the normal/vertical axis – Fig 12-2c).

The movements of pitch and roll are used to transition between different pitch and bank attitudes, respectively, and are controlled by the elevator and the ailerons, respectively. Uncoordinated yaw is generally undesirable, the reasons for which have been discussed previously. Yaw is controlled by the rudder.

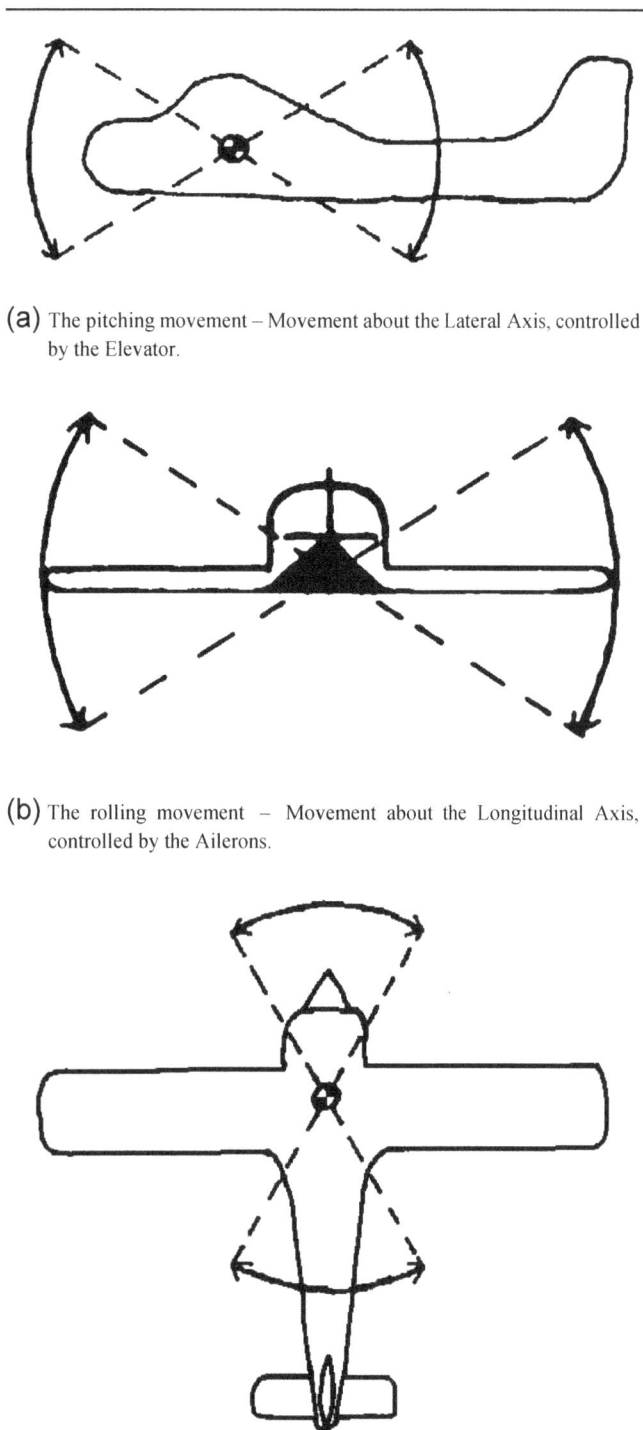

(a) The pitching movement – Movement about the Lateral Axis, controlled by the Elevator.

(b) The rolling movement – Movement about the Longitudinal Axis, controlled by the Ailerons.

(c) The yawing movement – Movement about the Vertical/Normal Axis, controlled by the Rudder.

Fig 12-2. - *Movements* - A movement is a rotation about one of the aircraft's three axis'.

Pitch

As stated above, the pitch attitude is the position of the nose (high or low) relative to the horizon. It can also be considered to be the angle between the longitudinal axis and the horizon—as seen from the side. Pitch is controlled by the elevator, which is in turn controlled by the position of the control column (or stick) in the cockpit. If we hold the control column steady, the pitch attitude will remain constant (ignoring for the moment air currents and the effect of dynamic stability). If we move the control column back, the elevator will pivot upwards, and the aircraft will pitch up into a new pitch attitude. In this case, the control column is not returned to it's original position, but is held in the new aft position. On the other hand, moving the control column forward will lower the elevator and pitch the aircraft down.

The effect that pitch has on the aircraft is an important element in the control of the aircraft. For the purpose of this chapter we will be considering the effect of pitch without accompanying power changes. This is usually demonstrated in the aircraft with the power set to a cruise setting.

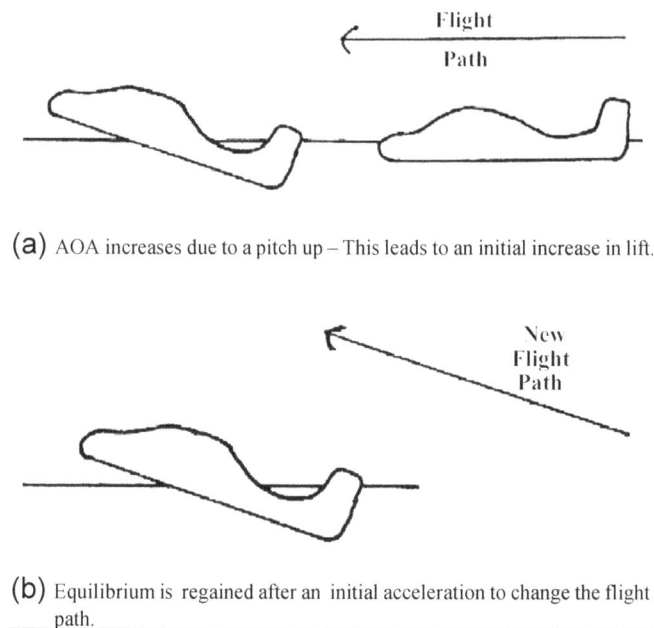

(a) AOA increases due to a pitch up – This leads to an initial increase in lift.

(b) Equilibrium is regained after an initial acceleration to change the flight path.

Fig 12-3. - *The Effect of Pitching Upwards.*

As the aircraft pitches up, the angle of attack is increased (Fig 12-3a). This, of course leads to an increase in lift which will initially accelerate the aircraft upwards. As the flight path changes, the angle of attack will be reduced due to the relative airflow coming from above (remember that the relative airflow is always opposite the flight path). This reduction in the angle of attack will eventually reach the point where lift once again equals weight and equilibrium is regained (Fig 12-3b). The aircraft will now be traveling upwards (assuming it

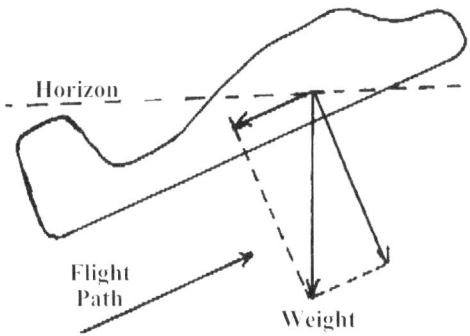

Fig 12-4. - *An Aft Component of Weight Results From a Pitch-Up* - This aft component of weight will cause a drop in airspeed until drag is reduced sufficiently and equilibrium is reestablished.

was previously level).

The upwards flight path means that now a component of the weight is acting opposite the flight path, so the aircraft will initially slow down (Fig 12-4). As the airspeed is reduced, the drag being produced is also diminishing—thus we eventually reach a point where the thrust being produced by

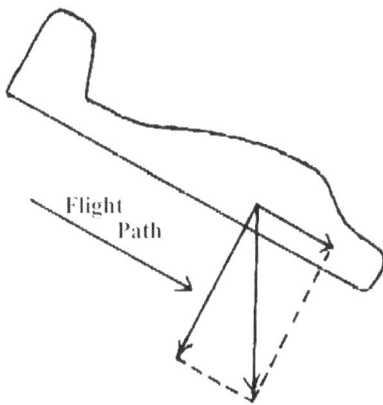

Fig 12-5. - *A Forward Component of Weight Results From a Pitch-Down* - The excess propulsive force causes an acceleration until drag increases sufficiently and equilibrium is restored.

the engine is again equal to the total retarding force. This means that the airspeed will stabilize at some value below the original one.

If instead we were to lower the nose, the exact opposite would occur. The angle of attack would be reduced initially, leading to a reduction of lift. The aircraft would assume a descending flight path (causing the AOA to increase and lift to be restored), and a component of weight would be acting to accelerate the aircraft (Fig 12-5). As the

aircraft accelerated, the drag would increase, and we would eventually reach a speed where the drag would be equal to the total propulsive force (the thrust and the forward component of weight).

So to sum up, the pitch attitude will have a direct impact on the vertical speed and the airspeed of the aircraft. Pitching the aircraft up will initiate an increase in altitude with a decrease in airspeed. Pitching the aircraft down will cause a decrease in altitude with an accompanying increase in airspeed.

Pitch vs. Angle of Attack

Pitch Attitude and *Angle of Attack* are often confused with one another—especially by beginning students. The two are decidedly *not* the same, they are however related to one another.

(a) Level flight – The relative airflow coincides with the orientation of the horizon, therefore the AOA coincides with the pitch angle.

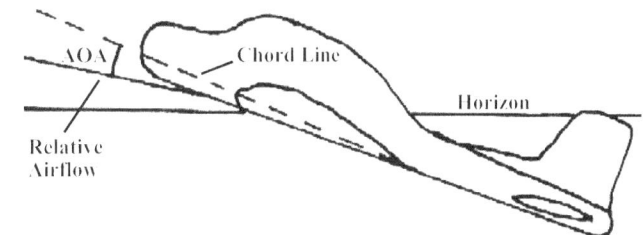

(b) Climbing – The relative airflow is from above, so the AOA will be less than the pitch angle.

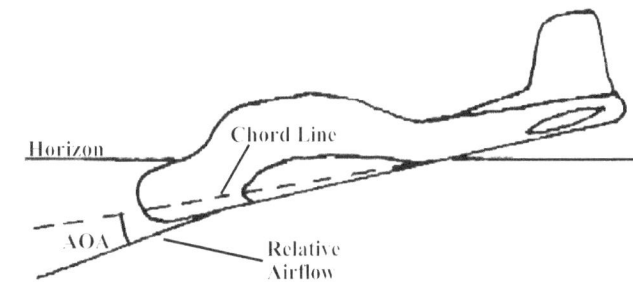

(c) Descending – The relative airflow is from below, so the AOA will be greater than the pitch angle.

Fig 12-6. - *Attitude vs. Angle of Attack* - The AOA is determined by a combination of pitch attitude and flight path.

The angle of attack is determined by combining the pitch attitude with the flight path. In level flight, the pitch attitude and the AOA will be the same (ignoring the angle of incidence), but in a climb or descent, they will not be equal (Fig 12-6).

Roll and Bank

The bank attitude of the aircraft, as stated previously, is the angle between the lateral axis and the horizon. We use the rolling movement to transition the aircraft from one banked attitude to another. The rolling movement, in turn, is controlled by the ailerons.

To establish the aircraft in a banked attitude from wings level, we use the control column (or stick). Rotating the control column to the right will cause the right aileron to be deflected upwards and the left aileron to be deflected downwards (Fig 12-7). As we've seen from Chapter 5, this will cause the aircraft to roll to the right. As long as the ailerons are deflected, the aircraft will continue to roll.

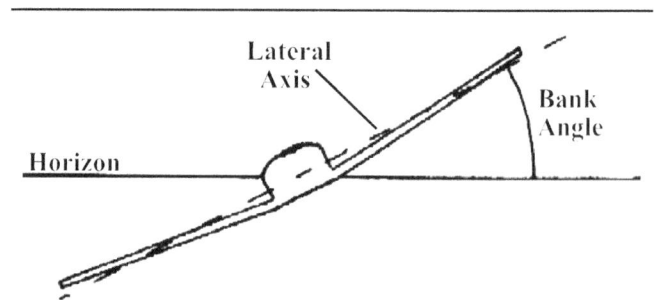

(a) Outside view

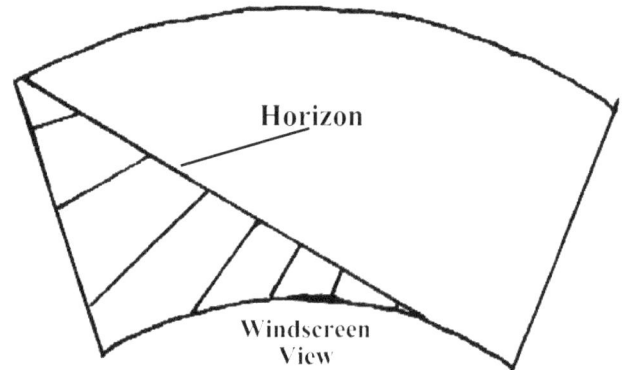

(b) Inside view

Fig 12-8. - A Banked Attitude.

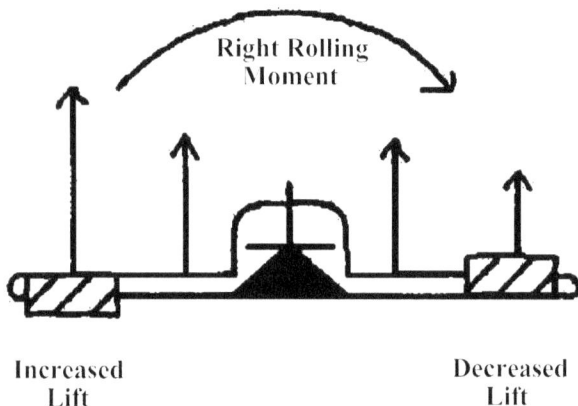

Fig 12-7. - Aileron Deflection Causes a Roll. - This deflection is controlled from the cockpit with the control column (or stick).

Once the aircraft is established in the desired bank attitude, the ailerons must be neutralized to stop the roll. This means returning the control column to center. The aircraft will now be established in a steady banked attitude (Fig 12-8). In a perfect world, this bank attitude would remain constant. However, in practice, we need to make constant minor corrections with the ailerons to maintain the desired bank attitude (that applies to wings level as well).

When we want to return to wings level from

the banked attitude, we simply initiate a roll in the opposite direction. For example, starting from a banked attitude to the right, we move the control column to the left to cause a roll to the left. Once the aircraft is returned to wings level, we center the control column once again to stop the roll and maintain wings level.

On some aircraft, adverse yaw is a problem during the rolling movement, this problem will be discussed in more detail in the section on yaw.

While the aircraft is in a banked attitude, some effects of the attitude will be apparent. The most notable of these is the change in direction of the aircraft. We discussed in Chapter 8 why this occurs and we will discuss the details some more in Chapter 15. To sum it up, the aircraft will change it's direction of flight towards the low wing. So when the aircraft is banked to the right it will turn to the right, and when the aircraft is banked to the left it will turn to the left.

Another phenomenon—which will be more noticeable at higher angles of bank, but negligible at low angles—is the reduction in airspeed which occurs when a constant pitch attitude is maintained. This speed reduction is a result of the increase in induced drag caused by the increased lift being produced by the wings. This increase in drag, along with it's causes and results, have been discussed in detail in previous chapters.

Bank Effect on Trim

In Chapter 5, we discussed the trim of an aircraft. The trimmed angle of attack is the angle at which no pitching moments are present. This means that the aircraft will naturally tend to remain at this angle of attack. If something causes the aircraft to be displaced from it's trimmed position, moments occur which will return it to the trimmed angle (assuming positive stability). If we want to fly at a higher or lower angle of attack, pressure is required on the control column—or we can readjust the trim tab to change the trimmed angle.

Recall also from Chapter 8 that in a banked attitude the wings must produce lift which exceeds the weight of the aircraft. If the airspeed does not vary, this means the angle of attack must be increased. The stability of the aircraft will oppose this and cause the nose to drop in a banked attitude. Because of this, if the aircraft is left alone it will enter a nose low attitude, gain some airspeed, and begin descending. To prevent this, the pilot must maintain some aft pressure on the control column to hold the nose up. As the angle of bank becomes steeper, this pressure becomes more pronounced.

Combinations

An aircraft cannot be said to be just in a pitch attitude or just in a bank attitude. At any given moment, the total attitude of the aircraft is a combination of the two. When discussing pitch attitudes above, we assumed that the wings were laterally level with the horizon. We also assumed when discussing bank attitudes that the pitch attitude remained constant at the "cruise" attitude.

It is possible, common in fact, to place the aircraft into an attitude which is a combination of both pitch (nose high, nose low, or cruise) and bank (banked right, banked left, or wings level). When establishing these combinations, the order of the control inputs (bank then pitch, or pitch then bank) is not critical and will vary according to the objective.

The key to controlling attitude combinations is in understanding the effect that bank has on the

pilots pitch reference. When the wings are level, the pitch attitude is determined from the cockpit by referencing the distance between the engine cowling (or the dash) and the horizon (Fig 12-9a). However, when the aircraft is banked, this pitch reference will almost always cross the horizon (Fig 12-9b). In this condition, it is difficult or impossible to control pitch by using the same reference as in a wing level attitude.

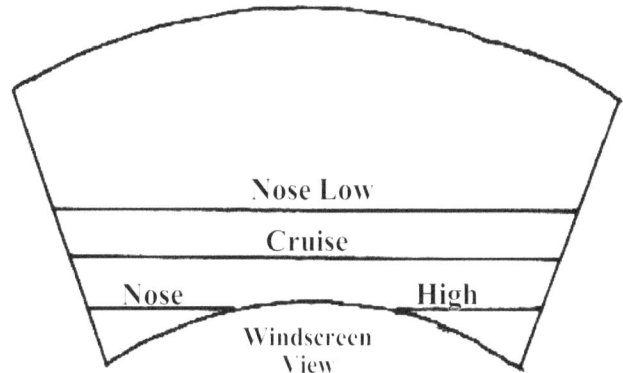

(a) Pitch Reference with the Wings Level (Inside View)

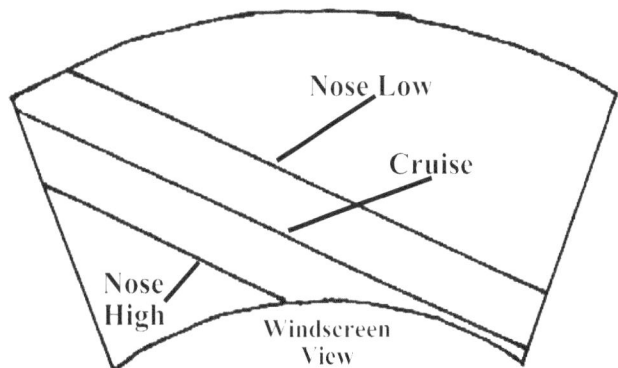

(b) Pitch Reference in a Banked Attitude (Inside View)

Fig 12-9. - *Pitch References in Various Attitude.*

Initially, it would seem that if we simply use the centerpoint of the cowling, pitch would still be easy to judge. In some aircraft this is true. However, in most aircraft the pilot does not sit on the centerline of the machine. This means that looking at the center of the cowling will lead to a skewed perspective while banked (Fig 12-10a). Instead, we should pick a reference point on the cowl which is directly to the front. This will allow a perspective which is unaltered from one bank attitude to the next (Fig 12-10b).

Understanding this concept will allow pilots to fly much more accurately when in a banked attitude. As well, pilots who make the conversion to the right seat (instructor students, commercial first officers, or pilots who just wish to expand their horizons) will find the switch much easier with this idea.

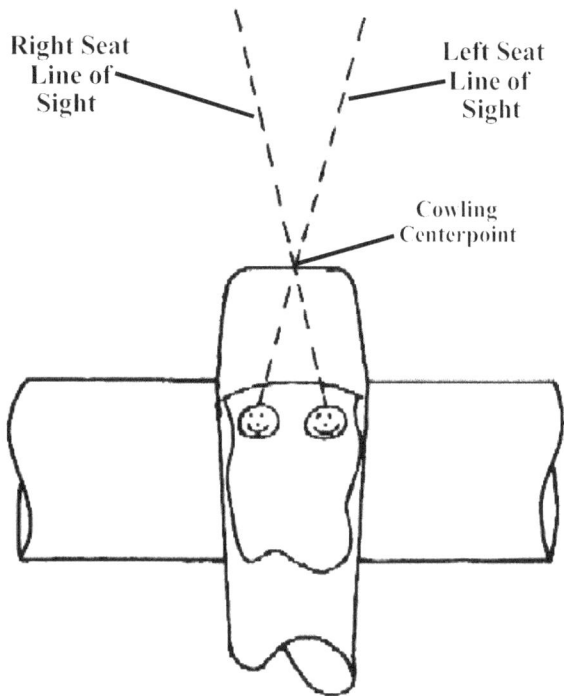

(a) Using the cowling center point as a pitch reference in a banked attitude will lead to inaccurate attitude control. In a left banked attitude, attempting to place the horizon in the "proper" position will lead to a nose low (lower than intended) pitch attitude. A right banked attitude will lead to a nose high pitch attitude. This effect is reversed in the right seat.

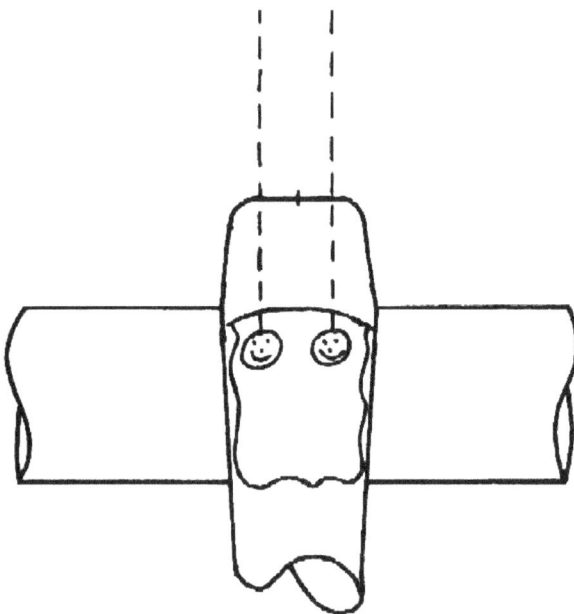

(b) Using a point directly in front of the pilot instead of the center of the cowling/dash solves this problem.

Fig 12-10. - *Pitch Reference in a Banked Attitude.*

Yaw

Yaw is the movement of the aircraft around the vertical axis. This movement is controlled, but generally not produced, by the rudder. The rudder in turn is controlled by the pedals on the floor of the aircraft. Stepping on the right pedal will deflect the rudder to the right and therefore yaw the aircraft to the right. Stepping on the left pedal will deflect the rudder to the left and therefore yaw the aircraft to the left.

It has been stated previously in this book that yaw is undesirable—or more correctly, *uncoordinated* yaw is undesirable. Uncoordinated flight means that the rate of yaw is inappropriate for the angle of bank/airspeed combination, and as a result the aircraft is flying in a sideslip (Fig 12-11). Ideally, this sideslip is eliminated by the vertical fin and coordinated flight is maintained. Under some circumstances, however, the vertical fin needs to be augmented by the use of rudder.

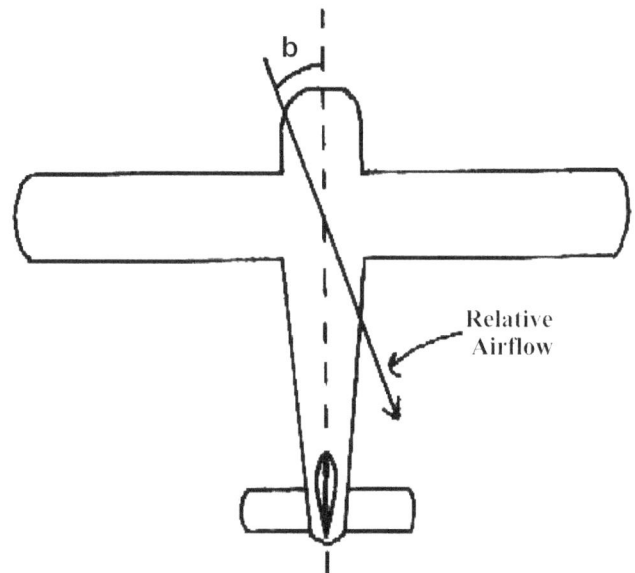

Fig 12-11. - *Uncoordinated Flight Means that a Sideslip is Present.*

The most common source of unwanted yaw is the engine. As we discussed in Chapter 4, the reciprocating engine/propeller combination creates a tendency for the aircraft to yaw to the left. This tendency is partly corrected by the offsetting of the fin or some other correction. Unfortunately, this correction is optimized for cruise power and speed. At low airspeeds and high power settings, the correction will be insufficient and the aircraft will yaw to the left—mandating a correction with right rudder. At high airspeeds and low power settings, the correction will overcompensate and the aircraft will yaw to the right—requiring a correction with left rudder.

Adverse yaw resulting from an aileron deflection will also require a correction with rudder. On most modern aircraft, this is unnecessary due to the effects of either differential or frise ailerons. However, these corrections, too, are intended for cruise speed and will not be as effective when operating at reduced airspeeds.

Allowing a yaw to develop and/or continue will cause problems with flying the plane. Due to the secondary effect of yaw, the aircraft will roll in the direction of the yaw. This will accelerate the rate of direction change and will lead to problems maintaining straight flight or a constant attitude. Other problems include performance degradation and under some circumstances, such as low airspeed operations, the possibility of a departure from controlled flight (i.e. – a spin).

Trim

As we know, changing the pitch attitude of the aircraft will also change the airspeed—pitching up will reduce the airspeed and pitching down will increase the airspeed. The weight of the aircraft remains constant, therefore the lift being produced must remain constant. This means that as the airspeed changes, the required angle of attack will also change.

The inherent stability of the aircraft will produce pitching moments that will tend to return the aircraft to it's original AOA. In order to maintain the new AOA (and pitch attitude), pressure will be required on the control column. This pressure holds the elevator in the necessary position to maintain the new AOA. Unfortunately, this pressure can be very fatiguing to the pilot, and can lead to inaccurate and imprecise flying.

The solution for this control column pressure is the trim tab. Once the aircraft is established in the desired attitude, the trim tab can be adjusted from the cockpit to eliminate aircraft pitching moments. This means that the trimmed angle of attack is readjusted by the trim tab and control column pressure will no longer be necessary to maintain the new attitude and AOA.

Effects of Power

In later exercises, changes in power setting will become frequent. The power being produced by the engine does more than provide us with propulsion—it also affects the trim of the aircraft. The trimmed *angle of attack* does not change appreciably with power changes (although it does change some due to the Induced airflow over the tailplane), but the attitude of the aircraft can vary substantially.

An increase in power will tend to accelerate the aircraft along it's flight path. An increase in airspeed, however, will increase the lift being produced and thus accelerate the aircraft upward. As the flight path changes and the aircraft begins to climb, the relative airflow will come from above and reduce the angle of attack. The aircraft, however, will tend to return to the trimmed angle of attack—thus it will pitch up (if preventive control inputs are not made). This will result in a pitch up which prevents a change in airspeed.

In fact, the increased downwash over the tail resulting from the increased slipstream—which in turn results from the increase in power—will tend to increase the pitch up so that airspeed is actually reduced by an increase in power. Of course, this is assuming that the controls are left alone and the aircraft is allowed to move freely. To counteract this pitch up tendency, the pilot must apply forward pressure to the control column. Then the trim tab can be used to eliminate the need for pressure.

A decrease in power will have the exact opposite effect. The nose will tend to drop due to the (initial) loss of airspeed so that a constant angle of attack can be maintained. Once again, the effect of slipstream over the tail will be apparent—the aircraft will actually end up at a higher airspeed after a power reduction.

The other effect of power has already been discussed—yaw. An increase in power will result in a yaw to the left. A decrease in power will result in a yaw to the right.

Configuration Changes

The configuration of the aircraft can also have a significant effect on the trim of the aircraft. As an example we will discuss flaps since flaps are the most common method of configuration change—particularly on training aircraft. The flaps can in some cases create large pitch changes as they are extended or retracted.

On a low winged aircraft, extending the flaps will *generally* cause the aircraft to pitch down. This is desirable since it allows the aircraft to maintain approximately the same speed even though the drag has been increased. The pitch down is caused by the increased pitching moment of the wing which occurs with the increased camber resulting from the flaps.

On a high winged aircraft, things are *usually* a little bit different. The flaps will change the angle at which the downwash leaves the wings (Fig 12-12). This will result in a change of the horizontal tail's angle of attack—causing the aircraft to pitch up. It is important to be aware of this, since the combination of the pitch up and the increased drag from the flaps can cause the airspeed to become dangerously low. As the flaps are extended, a control input must be made to keep the nose down.

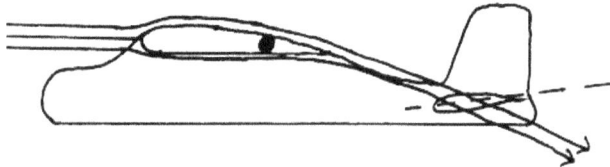

(a) With the flaps retracted, the downwash leaving the wing has a reduced influence on the horizontal stabilizer.

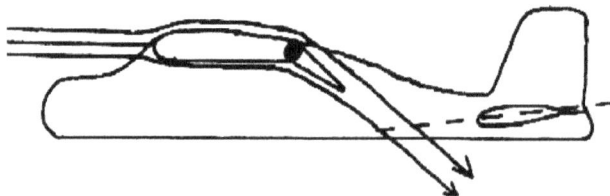

(b) Extending the flaps increases the downwash and therefore increases the down force on the tail. This leads to a tendency for high winged aircraft to pitch up when flaps are extended.

Fig 12-12. - The Effect of Flaps on a High Winged Aircraft.

Chapter Summary

The attitude of an aircraft is it's position relative to the horizon. Movements are defined in relation to the three axis of the aircraft and can be considered to be transitions between different attitudes.

The pitching movement is around the lateral axis and is controlled with the elevator or stabilator. The pitch attitude will have a direct effect on the airspeed and vertical speed of the aircraft. A nose high attitude will lead to a lower airspeed, and generally lead to a higher climb rate. A nose low attitude will lead to a higher airspeed and a higher descent rate.

There is often some confusion between the pitch attitude and the angle of attack. The difference is very important. The pitch attitude is the aircraft's position relative to the horizon while the AOA is the angle between the wing chord and the relative airflow. For a fixed flight path, an increase in pitch attitude will mean an increase in AOA. However, if the flight path changes, this will affect eh AOA since the relative airflow will change.

The bank attitude of an aircraft is the angle between the lateral axis and the horizon. During a banked attitude, the aircraft will turn in the direction of the low wing. The movement used to transition between various banked attitudes is the rolling movement. Roll is a rotation around the longitudinal axis and is controlled with the ailerons.

The movement of yaw is around the vertical axis and is controlled with the rudder. To keep the aircraft in coordinated flight (zero sideslip angle), it is necessary to maintain a rate of yaw which is appropriate to the angle of bank and airspeed. This means that with the wings level, yaw must be eliminated. Undesirable yaw can be caused by a number of factors, including power, aileron drag, turbulence, and incorrect control inputs.

Trim is used to eliminate pressure on the controls once a desired attitude is established. Pitch trim adjustments will normally be required after any change in power, airspeed, and/or configuration. As well, when a new bank attitude is established, the pitch trim will be affected. As a general rule, however, bank attitudes are not maintained for long enough to make trim adjustments practical. For this reason, control column pressure is maintained while in the banked attitude.

Questions

1) How is an aircraft transitioned from the cruise attitude to a nose high attitude? How will this affect the airspeed and altitude if no power adjustment is made? What will the outside reference look like?

2) What is the difference between an attitude and a movement?

3) What are the three movements? What axis are they associated with? What control surfaces control these movements?

4) How is the aircraft returned to cruise flight from a banked attitude?

5) How do pitch attitude and AOA relate to one another?

6) What is the function of the trim tab?

7) How and why does the power setting affect the trim of the aircraft?

8) How and why would a configuration change affect the trim of the aircraft?

9) What are some causes of undesirable yaw? How do you correct for them?

Chapter 13
Cruise Flight

In this chapter, we will build on Attitudes and Movements to develop a technique for maintaining Straight and Level Flight. Straight and Level is often referred to as cruise flight, and it involves maintaining a constant altitude and direction. Cruise can be accomplished at a variety of airspeeds, but the selected airspeed is generally held constant once it is established.

Straight Flight

To begin with, we need to establish the aircraft in straight flight. This, of course, means maintaining a constant direction. To do this, we need to select a directional reference outside the aircraft. The best reference to use is a point on the horizon—the further away the point is, the better it will be (Fig 13-1).

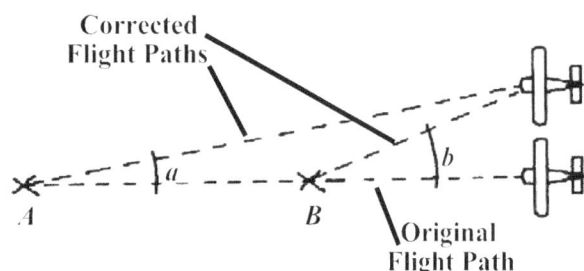

Fig 13-1. - *Directional Reference* - The further away a reference point is, the more accurate it will be. Nearby points lead to large directional changes when small amounts of drift are present. This problem becomes more pronounced as the crosswind component gets stronger (and therefore the lateral drift is increased).

Maintaining flight towards this point is fairly simple. Keep the wings level with the ailerons and control yaw with the rudder. That is, maintain direction with the rudder—if the wings are level, a constant direction will mean the aircraft is in coordinated flight.

Note here that the rudder is being used to *maintain* direction, not to change the direction or to make corrections. The difference is subtle but important. If the direction drifts, it must be re-established with a small turn—and we *do not* turn with rudder. This turn can be accomplished by rolling slightly into a banked attitude. The banked attitude will initiate the necessary turn and the aircraft should be returned to wings level very quickly.

When flying straight, the goal should be to maintain heading +/- 5°. With directional tolerances this tight, the bank required for corrections should be unnoticeable to passengers. As we will see in Chapter 15, the angle of bank used should not exceed half of the heading change. Try to get used to judging these angles with outside reference instead of using the heading indicator and attitude indicator.

When maintaining a compass heading, a visual reference can still be used. Simply establish the aircraft on the compass heading by the directional gyro or the magnetic compass, and then select a reference point on the horizon. This technique allows us to keep looking outside to control the attitude of the aircraft.

Level Flight

While maintaining straight flight, we also want to maintain level flight. This means holding a constant altitude. We do this by manipulating the pitch attitude to hold the altitude and correct for minor deviations. At altitude it is difficult or impossible to determine height visually, so occasional reference to the altimeter will be necessary. This having been said, pitch is used to control the altitude, so constant outside reference is still needed.

Establishing the cruise attitude appropriate

to the airspeed should keep the altitude fairly constant, but some correction will always be necessary. This means that if the aircraft climbs slightly above the target altitude, we lower the nose slightly. If the aircraft descends slightly below the target altitude, we raise the nose slightly. We know from the previous chapter that these corrections will also cause variations in the airspeed. These speed deviations are acceptable if they are slight, and they will be slight as long as we do not need to make any large altitude corrections.

In practice, it should be our goal to keep the altitude within a +/- 50 ft tolerance. Deviations larger than 100 ft will normally require more than simply a pitch correction—a power adjustment will also be necessary (this is covered in the next chapter).

Airspeed Control

As was stated previously, airspeed is generally held constant in cruise once it is established. If the power setting is kept constant, holding altitude will allow the airspeed to remain constant as well. However, sometimes, we wish to alter our airspeed for some reason (i.e. – circuit spacing, fuel economy, etc). Changing airspeeds and maintaining altitude requires a power change.

To change airspeed in cruise, first adjust the power. This power change will initially result in a speed change—but that is not all. As the speed changes, so will the amount of lift being produced by the wings. This means that the aircraft will accelerate into a climb or a descent (depending on whether the power was increased or reduced) until the angle of attack is adjusted to regain equilibrium. So all in all, the aircraft will change speed, but it will also climb or descend.

To prevent this, we need to adjust the pitch attitude so that the angle of attack will change without a climb or descent. Increasing the airspeed means increasing the power and lowering the nose. Decreasing the airspeed means decreasing the power and raising the nose. The pitch attitude can't be adjusted too quickly or we will find that we are over-correcting—the attitude must be changed as the airspeed changes so that a constant altitude can be maintained.

Last but not least, we need to adjust the trim. After a speed change, the aircraft will be flying at an angle of attack other than the one it is trimmed

for. Sustained flight in this condition is difficult and tiring. So we need to adjust the elevator trim tab to relieve the pressure on the control column.

Normally, speed changes involve a specific target airspeed. It is useful, then, to anticipate how much to adjust the power by. A commonly used rule of thumb is to use 100 RPM for each 5 knots of airspeed adjustment:

$$100 \text{ RPM} = \sim 5 \text{ knots.}$$

In an aircraft with a variable pitch prop, a similar rule of thumb can be used:

$$1"\text{Hg} = \sim 5 \text{ knots.}$$

These rules are fair approximations and should get us close to the target airspeed. From there we simply make smaller power changes in the appropriate direction.

So to sum it up for speed changes, we adjust the power to accelerate/decelerate, adjust the pitch attitude to maintain a constant altitude, and trim to eliminate the control pressure.

P – Power

A – Attitude

T – Trim.

Cruise Performance

The term 'Cruise Performance' usually refers to both cruise speed and fuel economy. For this chapter, we will look only at the issue of cruise speed. The speed available will depend on the power required and the power available. The theory behind cruise performance has already been discussed, so right now we want to look at the information provided in the POH.

Most operating handbooks will contain charts similar to the ones in Figure 13-2. These charts compare cruising altitude to cruise speeds at various power settings. Note here that the altitude used in performance considerations is *always* density altitude. This means that the actual conditions must be reduced to the standard atmospheric conditions.

The power setting provided in Figure 13-2 are given in percentages. We need to convert them into something useful—this means either an RPM

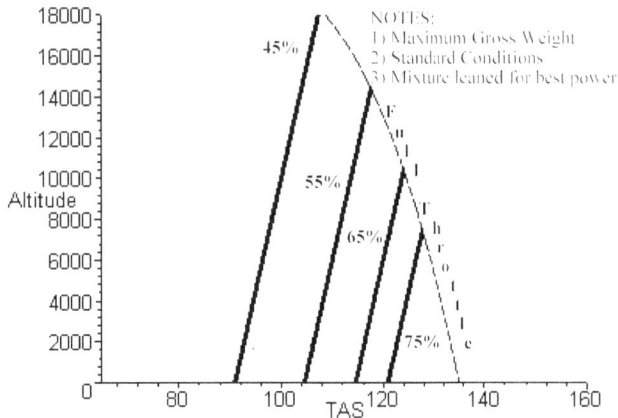

Fig 13-2. - *Typical Cruise Performance Graph* - Cruise speed is plotted against density altitude and power setting.

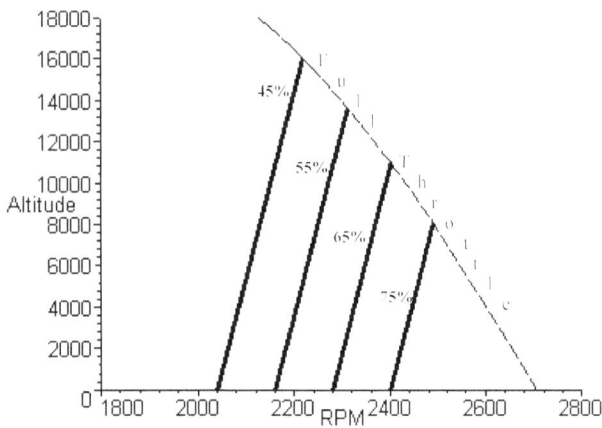

Fig 13-3. - *Typical Power Setting Graph* - Power Setting is plotted against density altitude and RPM or RPM/Manifold Pressure.

setting or an RPM/Manifold Pressure combination, depending on what type of prop the aircraft has (fixed pitch or variable pitch). Fig 13-3 is an example a chart which may be provided in the POH. Once again, the altitudes provided are density altitudes.

Chapter Summary

Cruise flight is often referred to as straight and level and involves maintaining a constant direction, a constant altitude, and often a constant airspeed.

Straight flight is maintained by using the ailerons to keep the wings level and eliminating yaw with the rudder. If a correction becomes necessary, it should be made with a slight bank attitude in the appropriate direction.

Level flight is maintained by establishing and holding the appropriate cruise attitude. When altitude corrections need to be made, they can be made by pitching up or down as necessary to reestablish the target altitude. Large corrections may require a power adjustment as well.

Airspeed is controlled by the power setting in level flight. To change the airspeed, the power is first reset, then the pitch attitude is adjusted to maintain a constant altitude. The excess drag or thrust, as the case may be, will cause an acceleration until drag and thrust are once again equal and equilibrium is reestablished.

The cruise performance of an aircraft can be determined in advance from the POH or Flight Manual. This information will always be based on density altitude.

Questions and Problems

1) How is straight flight maintained? Why aren't heading corrections made with the rudder?

2) What procedure would be followed to slow down from 130 knots to 100 knots while remaining in straight and level flight?

3) Based on the performance charts in Fig's 13-2 and 13-3, What RPM would be required to fly at 55% power at 5000 ft.? What TAS would be the result?

Before we can establish the aircraft in cruise flight, we have to climb to our cruise altitude. As well, before landing, we have to descend down from our cruising altitude. These and many other circumstances will dictate changes in altitude throughout a flight—this of course means climbing or descending.

As we've seen in previous chapters, excess power will make the aircraft climb, and a power deficiency will cause the aircraft to descend. This is assuming, of course, that the airspeed is held constant—as opposed to allowing an acceleration or deceleration.

Climbing

A climb, as the name implies, is an increase in altitude. This increase is caused by an excess in power from the engine/prop for the airspeed being maintained. This airspeed will be held constant by controlling the pitch attitude of the aircraft as it was detailed in Chapter 12. Pitching the aircraft up will provide for a decrease in airspeed, and pitching the aircraft down will cause an increase in airspeed.

Entry and Recovery

If we start from cruise and increase the power, the aircraft will tend to accelerate. However, if we allow the aircraft to pitch up to maintain the cruise airspeed, the aircraft will climb instead. Notice that we *allow* the aircraft to pitch up—we do not need to actively pull back on the control column in this case. This is because of the trim change resulting from the power change—we may in fact have to push forward to establish the pitch attitude necessary.

Most climbs, however, are performed at some airspeed below cruise. This means that instead of adding power and allowing the aircraft to

pitch up, we can pitch the aircraft up first and then add power. It is important to keep in mind the trim changes that will occur with the power and speed changes. Once the climb attitude is selected, we will maintain it with the necessary pressure on the control column until the trim has been adjusted (whether this pressure is forward or aft will depend on the combined effect of and the speed and power changes).

When trimming, we have to maintain a constant attitude—otherwise the airspeed will vary and the resulting trim forces will vary as well. If we have a specific target airspeed (as we usually will), the initial attitude change may not get the exact speed we are looking for. When this is the case, small pitch adjustments can be made prior to trimming. Adjusting the trim before these corrections are made will mean trimming all over again after the speed has been readjusted.

So the entry to the climb is conducted by adjusting the pitch attitude to the climb attitude, applying power, and trimming:

A – Attitude

P – Power

T – Trim.

In practice, once a student has become thoroughly familiar with the climbing procedure, the attitude and power are generally adjusted simultaneously.

The recovery back to cruise flight follows a similar procedure. First, we lower the pitch attitude back to cruise. Once this is done, the excess power provided by the engine will accelerate the aircraft instead of causing a climb. After re-establishing our target cruise airspeed the power can be reduced back to the appropriate cruise setting, and we can trim out the control forces. So the pneumonic that we can use for a climb recovery is the same one we used for the climb entry:

A – Attitude

P – Power

T – Trim.

In the case of the climb recovery, we won't get to the point where the attitude and power are adjusted simultaneously since some time must be allowed for the aircraft to accelerate. Some pilots do use the procedure of setting cruise power immediately, but this will lead to a sluggish acceleration and a long period of time before the trim can be adjusted properly.

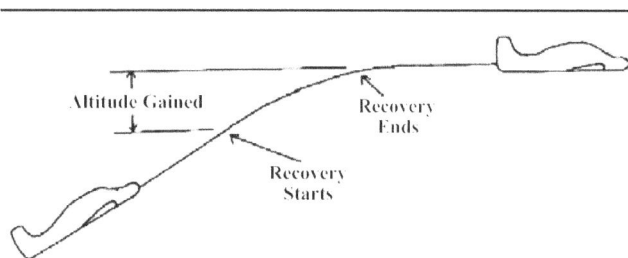

Fig 14-1. - *Climb Recovery Flight Path.*

One important point about the climb recovery is that we must begin the recovery prior to our target altitude. This is because the transition from the climb to cruise is not instantaneous (unless we wish to experience partial or negative g's), and a slight climb will continue as we recover (Fig 14-1). So the actual altitude which we recover at will be somewhat higher than the altitude which we begin the recovery at. A rule of thumb which works well is to begin the climb recovery prior to the target altitude by 10% of the vertical speed. This vertical speed can be read off of the vertical speed indicator (VSI) and will be in feet per minute (FPM). We simply divide the number by 10 and lead the climb recovery by that much.

For example, consider an aircraft which is climbing at 500 FPM. If we want to return this aircraft to cruise at 4500 ft, we would want to begin the recovery at 4450 ft (4500 - 500/10).

Types of Climbs

When climbing, we always want to get the maximum possible performance out of our aircraft.

In order to do this we have to ask a question: What defines maximum performance? In other words, what *type* of performance are we looking for? The answer to this question will determine the type of climb that we use.

If we want to get to our cruising altitude as quickly as possible (to take advantage of a tailwind aloft perhaps), we climb for best rate (V_y). This climb would allow us to gain the most altitude per unit of time. Best rate is also used as a safety item. The farther we are from the ground if something were to go wrong (e.g. – an engine failure), the safer it is. Altitude provides us with more options in an emergency. On the other hand, if we need to clear an obstacle, climbing for best angle (V_x) would be more useful. Best angle allows us to gain the most altitude per unit of distance. Best rate and best angle are considered to be the two "performance" climb speeds.

Another climb speed (speed *range* actually) which can be useful is the cruise climb. The cruise climb is intended to cover ground at a reasonably high speed while climbing. The climb speed is higher than that of best rate, and as such it allows us to see better (the pitch attitude will be lower), cover ground faster, and keep the engine cooler (more air will be flowing over the engine). The cruise climb is often used on flights when the wind aloft does not provide a significant advantage.

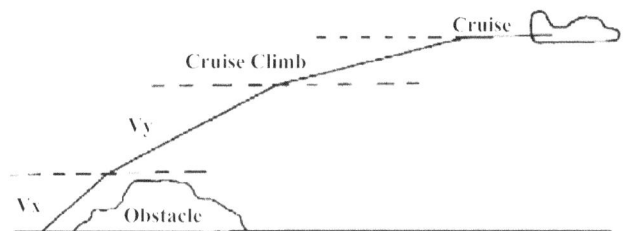

Typical Climb Profile

Considering all of these climb options, a typical climb profile is as follows. 1) Best angle to clear obstacles (if necessary), 2) Best rate to some predetermined altitude, 3) Cruise climb to cover ground.

Best Rate

The best rate of climb allows us to gain the most altitude per unit of time. We saw in Chapter 8 that the speed for best rate is the speed which allows us the most *excess* thrust horsepower. This speed should be published in the POH. Establishing a pitch attitude which maintains this airspeed will give us the maximum *sustainable* vertical speed.

Notice the word "sustainable" here. It is possible to climb faster than the best rate for a short period of time. We can do this by pitching up rapidly and converting airspeed to vertical speed. This airspeed will quickly bleed off due to the effect of gravity however, and we will then be climbing slower. This technique is known as a "zoom climb" and in non-tactical aircraft has no useful application.

The speed that best rate of climb is accomplished at will vary according to several variables that were detailed in Chapter 8. They were weight, C of G, and altitude. In some operating handbooks, the different speeds for varying conditions will be published. Sometimes, however, V_y is only published for maximum gross weight and standard sea level. In this case, some pilots prefer to stick with the same speed regardless of the conditions. However, if we wish to get the maximum possible performance out of the aircraft, some speed adjustment will be necessary.

We know that with an increase in altitude, V_y will decrease slightly. A fair rule of thumb which can be used for aircraft with no published speeds for different altitudes is to maintain a constant *true* airspeed. Holding a constant TAS means that at higher density altitudes the calibrated airspeed will be lower (recall the examples given in Chapter 2). This allows us to compensate for the altitude change and maintain a rate of climb which is close to the maximum possible.

As for weight changes, V_y will decrease as weight decreases. Once again, without the data from the manufacturers power curves, exact speeds cannot be determined for various weights. A rule of thumb which can be used is to apply equation (6.1) to V_y:

$$\text{Stall Speed} = V_s \sqrt{(GW/MGW)} \qquad (6.1)$$

becomes:

$$\text{New } V_y = V_y \sqrt{(GW/MGW)}$$

Keep in mind that V_y is not a constant AOA maneuver with weight variations, so this is just a

rule of thumb, and as such may not be very accurate for a particular aircraft. With experience on a particular type of aircraft, it may be found that another rule is more effective.

When checking the operating handbook for airspeed information, we don't want to just check sections on general information or normal operating procedures. The performance charts or graphs can also contain some useful information (Fig 14-2). If we refer to the rate of climb graph, speeds *may* be supplied for different weights. Even if they aren't, knowing what performance should be expected can help identify problems if they occur.

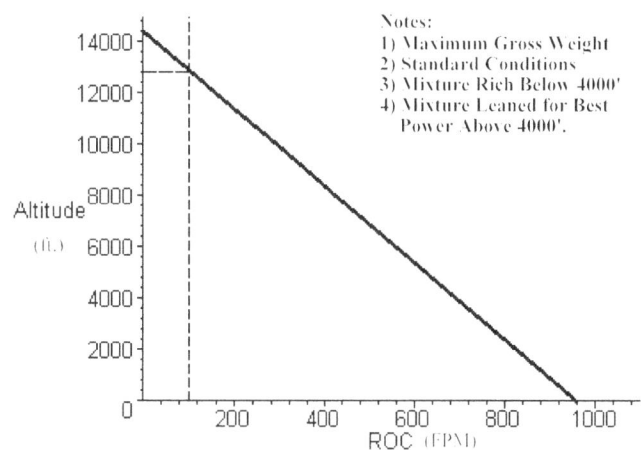

Fig 14-2. *- Rate of Climb vs Density Altitude. -* Note that the service ceiling is found by locating the altitude that provides a rate of climb of 100 FPM.

Best Angle

The best angle of climb speed (V_x) is the speed which gives us the most altitude for the least distance over the ground and will invariably be lower than the best rate of climb speed. This type of climb is normally only used to clear obstructions along the aircrafts flight path. Once the obstruction is cleared, the aircraft is accelerated to V_y for the next segment of the climb. V_x gives a lower rate of climb than V_y, but the lower forward speed allows for a greater climb angle (Fig 14-3).

Like V_y, V_x is affected by the weight and the altitude of the aircraft. As well, V_x will be affected by the wind just as we discussed in Chapter 8. Once again, some operating handbooks will contain information for different conditions while others will just contain V_x for maximum gross weight and

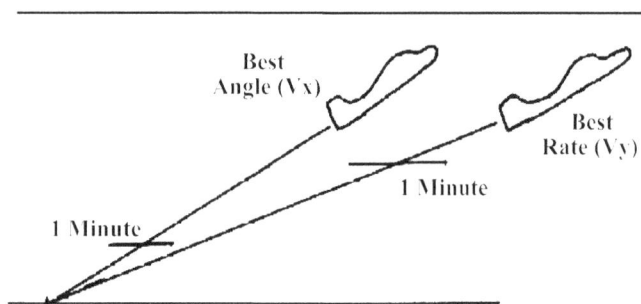

Fig 14-3. - V_y vs V_x.

standard sea level.

For a decrease in weight, V_x can be decreased as well. The rule of thumb used for V_y can be used here also, but keep in mind that it is just a rule of thumb. Like V_y, V_x is not a constant AOA maneuver at different weights, so this rule is approximate at best.

As for altitude, the best thing to do with V_x is to keep the calibrated airspeed constant for altitude changes unless the operating handbook states otherwise. Remember to check the performance charts for this information as well as the sections on normal operations. The chart to check for V_x is the takeoff chart, it will usually contain information on obstacle clearance. Be careful, however, using this information for en-route obstacle clearance. Remember that the takeoff obstacle climb takes into consideration the fact that the aircraft is still accelerating. As such, the climb speeds listed on takeoff charts may in fact be lower than the actual V_x.

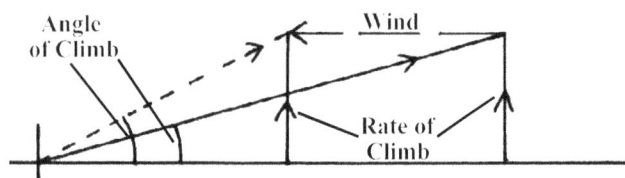

Fig 14-4. - *Wind's Effect on Angle of Climb.*

Wind, as we know, will affect the angle of climb due to the variation in groundspeed (Fig 14-4). Normally, takeoffs are conducted as into the wind as possible—especially when taking off over an obstruction. This wind will also affect the speed to be used as V_x. Recall from Chapter 8 that with a headwind we can improve the angle of climb even further by decreasing the climb speed slightly, and

with a tailwind we can reduce the winds effect by increasing the climb speed slightly.

As a repeat of the example we looked at in Chapter 8, let's consider an aircraft with a V_x of 65 knots and a V_s of 45 knots. If this aircraft were to take-off into a 50 knot headwind and climb at V_x, the climb would be extremely steep. However, if instead the speed was reduced to 50 knots, the groundspeed would be zero and the climb would be vertical. A rule of thumb which can be applied here is to reduce the climb speed by half of a headwind or increase it by half of a tailwind. The limitations to this rule are that the speed should not be lower than the stall speed or higher than V_y.

En-Route

The en-route climb is used to climb to altitude while covering ground at the same time. There is no specific speed to use for an en-route climb, but a speed *range* is designated. If the aircraft is flying faster than V_y but has climb power set and is climbing, it is established in an en-route climb. So the speed range for an en-route climb is between V_y and the maximum level cruise speed. Which speed we select will depend on how fast we wish to climb and how fast we wish to travel. The speed we choose will be the a compromise between a rapid climb and rapid travel. If we wish to *climb* fast, we would select a speed closer to V_y. On the other hand, if we wish to *travel* fast, we would select a higher speed—perhaps at or about cruise speed.

Climb Power Settings

On most training aircraft, the climb power setting is full power. With a fixed pitch prop, this means full throttle and whatever RPM setting results from this. In an aircraft with a variable pitch prop, full power means having full throttle and having the props set to full fine (maximum RPM). Generally, however, in an aircraft with a variable pitch prop, full power is only used for the take-off and initial climb.

For a high powered aircraft, good climb performance can still be accomplished with less than full power. So power is often reduced to the climb setting shortly after take-off. This allows for less wear and tear on the engine, as well as good cooling since the engine is not producing as much heat.

For an obstacle clearance climb (V_x) the power should normally be left at full power at least until the obstacle is cleared.

Descending

During a descent, the aircraft is losing altitude. This altitude loss is a result of a power deficiency for the airspeed being maintained. Like the climb, the airspeed in a descent is maintained by control of the pitch attitude. Lowering the nose will increase the airspeed and raising the nose will decrease the airspeed. Power is used to control the vertical speed. For a constant airspeed, an increase in power will decrease the rate of descent and a decrease in the power setting will increase the rate of descent.

Entry and Recovery

Entering a descent from cruise is straightforward. If we reduce the power from the cruise setting and hold a constant attitude, the aircraft will decelerate and initiate a descent. If, on the other hand, we lower the nose to maintain the airspeed we will reach equilibrium at a higher rate of descent. Like the climb, power and speed changes will necessitate an adjustment of the trim. So the procedure for entering a descent is,

P – Power

A – Attitude

T – Trim.

This procedure allows us to establish a descent with minimum control manipulations since most descents are carried out at speeds equal to or below cruise speed.

Recovery from the descent is also straightforward. Return power to the cruise setting, adjust the pitch attitude for cruise, and retrim the aircraft. So, once again,

P – Power

A – Attitude

T – Trim.

If we were to adjust the attitude first, the airspeed would drop (recovery from a descent requires a pitch up) and we would have to wait longer to return to the cruise airspeed. This means a longer period before being able to trim properly.

With practice, descent proficiency can be brought to the point where the power and the attitude can be adjusted at the same time.

The descent recovery must be started prior to the target altitude. This is because, like the climb, the aircraft is still descending during the transition to cruise. Once again, the amount of lead can be determined by checking the VSI and leading the recovery by 10% of the vertical speed. So if we want to level off at 3000 ft, and we are descending at 800 FPM, the level off can be initiated at 3080 ft (3000 + 800/10).

Types of Descents

We basically have four types of descents to choose from—three of which will be discussed in this chapter. They are 1) the en-route descent, 2) the glide descent, 3) the emergency descent, and 4) the approach descent. Each of these types of descents can really be considered to be variations of the en-route descent. Approaches will be discussed in detail in Chapter 23.

En-Route Descent

An en-route descent is intended to bring the aircraft down to the target altitude gradually while maintaining a reasonably high groundspeed. Generally, en-route descents are conducted at or about the cruise airspeed. Power is reduced and the aircraft is pitched down to maintain the cruise speed. This results in a descent from altitude due to the power deficiency. The vertical speed is controlled by the power while the airspeed is controlled with the pitch attitude. En-route descents are normally conducted in the clean configuration (flaps, gear, etc. retracted). On the power curve, this technique places us below the power required at the selected speed (Fig 14-5).

In many commercial operations, power is left at the cruise setting and pitch alone is used to initiate the descent. This results in a descent airspeed which is higher than cruise while descending to the target altitude. The rationale behind this technique is that for most scheduled

Gliding

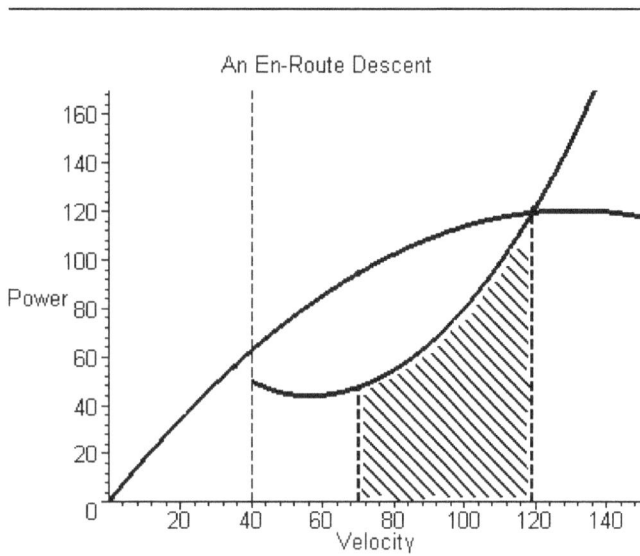

Fig 14-5. - *En-Route Descent* - Any descent will place us below the power curve. An en-route descent places us below the power curve at a relatively high speed.

The glide is the simplest type of descent. With the power reduced to idle, the airspeed is controlled by the pitch attitude to achieve the desired performance. Generally, glides are performed at one of two key speeds—either best range or best endurance. As the names imply, best range will allow us to glide the greatest distance per unit of altitude, and best endurance will allow us to glide for the greatest amount of time per unit of altitude.

The locations of these speeds on the power curve have been determined previously in Chapter 8. Best range is on the tangent line from the origin to the power-required curve, and best endurance is at the lowest power required (Fig 14-7). These speeds are often published in the operating handbook for the aircraft, particularly best range since this is the speed normally used during an emergency glide (i.e. – after an engine failure).

Glides can also be accomplished at speeds other than best range and best glide. To select different speeds, we simply use the pitch attitude to accelerate or decelerate. On the power curve, this will place us at various points along the dashed line in Fig 14-8. As we can see there, different power deficiencies will result from different airspeeds. Therefore different rates of descent will also be the result.

and charter operations, time is of essence. Slowing the aircraft down for the descent or even maintaining a constant cruise speed can't be justified when safety is not jeopardized by speeding up. On the power curve, we end up increasing the speed from cruise and as a result being below the power required curve (Fig 14-6). This is a legitimate technique that can be used by private operators as well, but as a general rule it isn't used often.

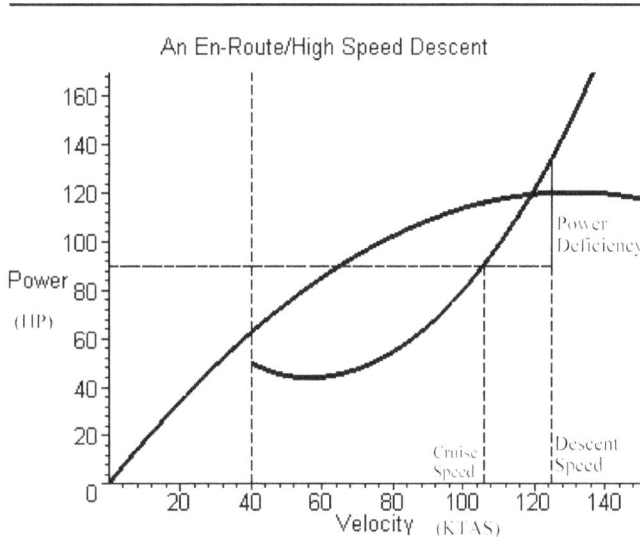

Fig 14-6. - *High Speed Descent on the Power Curve.* - Both the cruise and the descent are at the same power setting. However, the descent is conduced at a higher airspeed. This places the aircraft below the power curve.

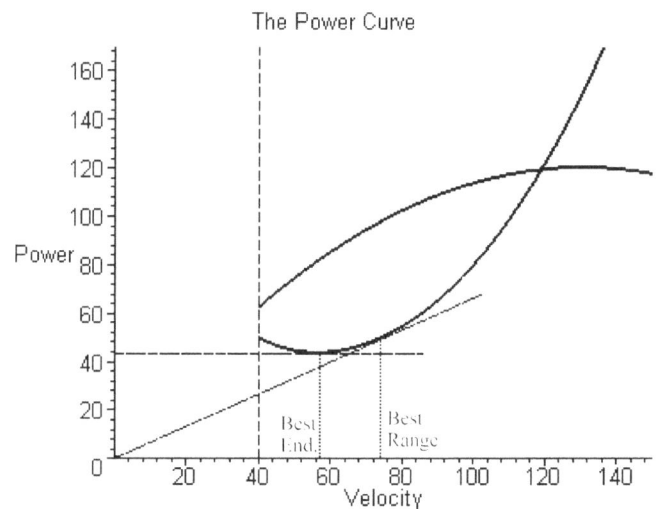

Fig 14-7. - *Best Range and Best Endurance* - Descending at these speeds will have the same effect as cruising at these speeds. The difference is that now, instead of burning fuel to provide the power required, we are giving up altitude.

A common tendency among many pilots (especially low time pilots) is to try to "stretch" the glide range of the aircraft by raising the nose. We've seen already that a speed lower than best range will always result in a steeper descent angle. Initially, however, a reduction in speed from best range will seem to have improved the glide performance. This is due to the fact that the aircraft has to lose airspeed after a pitch change. The excess airspeed is initially converted into glide range—so much so that in many cases an aircraft can be leveled off all together to lose airspeed prior to recommencing the descent. The resulting flight

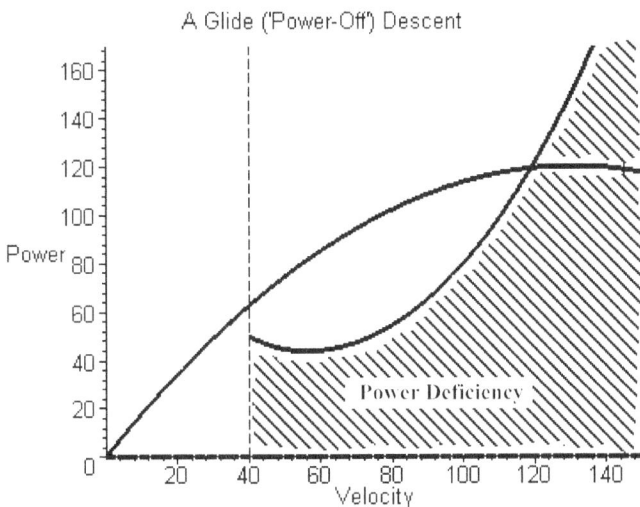

Fig 14-8. - *Gliding at Different Airspeeds* - Each airspeed has a corresponding power required and therefore power deficiency in a glide. This results in different rates of descent at different airspeeds.

path will be similar to the one in Fig 14-9. As the glide angle is reduced, the visual cues which indicate range (discussed in two sections time) will indicate an improvement in range. Unfortunately, this improvement quickly disappears and range will in fact be reduced.

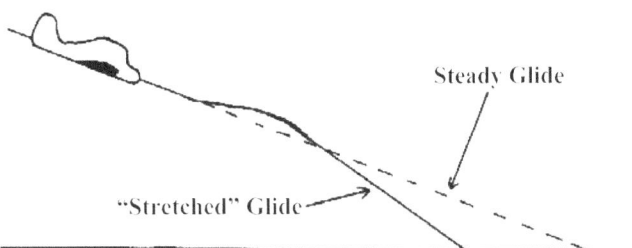

Fig 14-9. - *"Stretching" the Glide* - Pitching up to increase glide range from the best glide speed will not work—although it will initially appear that it did.

For a pilot trying to "stretch" the glide with pitch, this reduction in range will lead to a further raising of the nose. We will see in Chapter 19 that this is one of the danger areas for stalls and spins and should therefore be avoided at all costs.

Emergency Descent

The emergency descent is not a commonly practiced maneuver in general aviation aircraft. It is usually reserved for use in pressurized aircraft operating at high altitude after a depressurization. However, it is also useful when (if) we experience an in-flight fire which cannot be extinguished. The objective is to get down with the highest safely sustainable vertical speed.

The procedure can vary significantly from one aircraft to another due to different design characteristics and speed limitations. There are two common procedures, one of which will apply to most aircraft.

The first option is to create as much drag as possible and then, with the power to idle (or in some cases with the engine(s) shut down), to lower the nose to achieve the highest allowable airspeed. The speed limit will usually be V_{fe} since the flaps will be extended. To create all of the drag that we want, we extend all secondary control surfaces that are available (i.e. flaps, speed brakes, spoilers). As well, an intentional slip will create even more drag and increase the descent rate. Slips are discussed in Chapter 20.

The second option to consider is to leave the aircraft in the clean configuration and, with the power to idle (or the engine(s) shut down), lower the nose to achieve the highest allowable airspeed. This is similar to the first procedure, however the aircraft is in the clean configuration. This allows a higher limiting speed—which means a great deal of parasite drag. Depending on the aircraft design (such as which drag producers are available and what the limiting speeds are) this procedure may or may not be better than the first.

Judging Range

While in a descent, it is often useful to determine the glide range of the aircraft—that is, where the aircraft would touch down if the descent was to be continued right to the ground. This is important for any descent which will be continued right to the ground—a forced approach, an emergency descent, or just a regular landing. The aiming point can be identified with a simple visual indication.

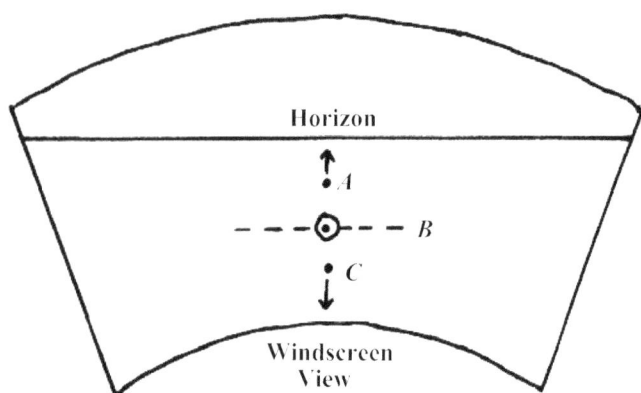

Fig 14-10. - *Visual Indications in a Glide* - (a) Points moving up on the windscreen are beyond the current glide range, (b) Points moving down on the windscreen will be flown past prior to reaching the ground, and (c) The point that is stationary on the windscreen is where ground contact would occur if a flare was not conducted.

Looking out the windscreen of the aircraft, three indications can be seen (Fig 14-10). At the top of the windscreen, points on the ground will appear to be moving upwards—these points are beyond the range of the aircraft with it's current glidepath. At the bottom of the screen, points on the ground will appear to be moving downward—these points will be passed in the glide prior to reaching the ground. The point of importance is the one which will be somewhere in the middle. This point appears to be stationary on the windscreen. The stationary point will be the touchdown point if the descent is continued right to the ground and is called the aiming point. The aiming point and the *actual* touchdown point are not quite at the same spot. As we will see in Chapter 23, the touchdown point will be slightly further than the aiming point due to the flare maneuver.

Chapter Summary

The climbing procedure is used to increase the altitude of the aircraft. From level flight, the climb is entered by first establishing a climb attitude, then applying climb power, and trimming for the new configuration. The climb attitude used will depend on the type of climb being carried out and therefore the airspeed being used. Best angle of climb is the lowest speed used and therefore involves the highest pitch attitude. A lower pitch attitude (and higher airspeed) would be used for best rate. An en-route climb, or cruise climb, would have an even lower pitch attitude and higher airspeed.

Recovering from the climb to level flight is simply a reversed entry. The pitch attitude is reduced to the cruise attitude and airspeed is allowed to build. Once the airspeed reaches the desired cruise value, the power is reduced to an appropriate setting and the trim is readjusted.

For a descent, the entry and recovery procedures are somewhat different. The power is adjusted first, and the pitch attitude is adjusted next to establish the desired descent airspeed. Once the aircraft is in the descent, the trim is adjusted to relieve control pressure. The level-off follows a similar order. Power is increased to the cruise setting, the pitch attitude is returned to cruise, and the aircraft is re-trimmed.

Descents can be en-route, glide, or emergency descents. As well, approach descents are discussed later in Chapter 23. En-route descents are often referred to as power assisted descents since some power is left on and the descent is made at a relatively high airspeed and low vertical speed. Glide descents are conducted with the power off or at idle. Glides are normally conducted at the aircraft's best glide speed, but they can also be conducted at other speeds. Emergency descents are used to lose as much altitude as possible as quickly as possible.

During a descent, the descent range can be determined visually by finding a point of land that remains stationary on the windscreen. If the descent was to be continued to the ground without a flare, this point would be the initial contact point.

Questions

1) What is the difference between best rate and best angle of climb?

2) Why isn't a climb recovery initiated with a power reduction?

3) What is(are) the advantage(s) of a cruise climb over best rate or best angle? Disadvantages?

4) What are the differences between a glide descent, an en-route descent, and an emergency descent?

5) Descending at 600 FPM to 1700 feet in order to join a circuit pattern, when should the level-off procedure be initiated?

Chapter 15
Turning

In turning an aircraft, the intention (obviously) is to change the direction of flight. This change comes about as a result of the effects of directional and pitch stability when the aircraft is placed into a banked attitude. Contrary to some popular misunderstandings, the rudder does not turn the aircraft, nor do the ailerons. Rudder is used to control yaw, while the ailerons are used to control roll. As we will see shortly, however, this control of yaw and roll (along with pitch) will have a direct influence on turns.

Effect of Bank

We saw in Chapter 8 that the aircraft turns because of bank. The lift being produced by the wings acts in the direction perpendicular to the relative airflow and the wingspan. So, as seen from behind, an aircraft in a banked attitude is producing lift inclined to the side (Fig 15-1a). The horizontal component of lift accelerates the aircraft to the side, which produces a sideslip (Fig 15-1b). Due to directional stability, the aircraft's tendency to eliminate a sideslip causes a yaw in the direction of the bank.

In actual fact, the aircraft is yawing and simultaneously pitching in the turn. To see this, we can consider two extremes. First, we look at an aircraft with it's wings level that is yawing (Fig 15-2a). This aircraft is in fact turning in the direction of the yaw, but the turn is uncoordinated since the rate of yaw does not match the angle of bank. Next, consider an aircraft at an angle of bank of 90° (Fig 15-2b). This is impossible since no vertical component of lift is present to counteract weight, but we can consider it hypothetically. For this aircraft to turn, the movement involved would be pure pitch. In reality, when an aircraft turns, the bank angle will be somewhere between these two

examples. So the actual movement will be a combination of pitch and yaw.

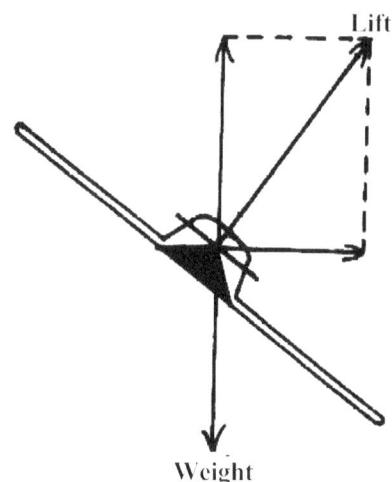

(a) The Effect of Bank – Inclining the lift creates an unbalanced force to the side.

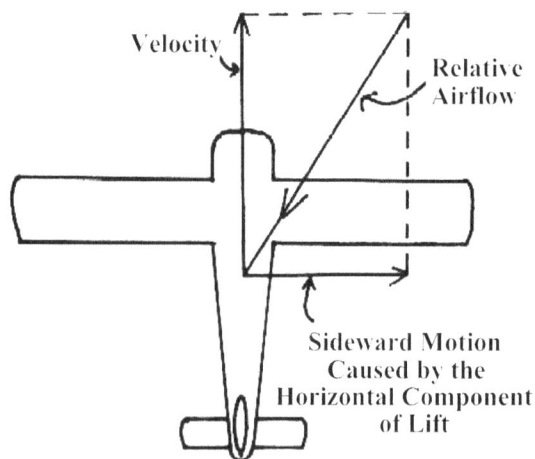

(b) Sideslip produced by the banked attitude

Fig 15-1. - *Forces in a Turn* - The horizontal component of lift accelerates the aircraft to the side and produces a sideslip. The sideslip causes yaw into the turn.

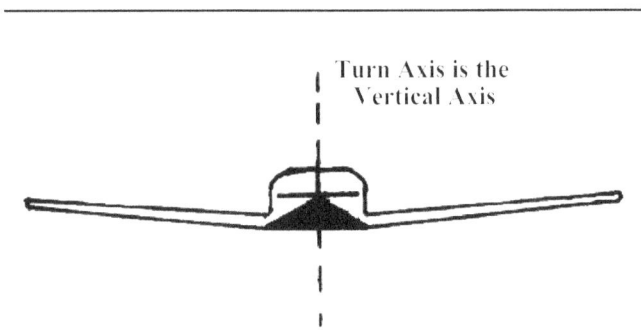

(a) Pure Yaw

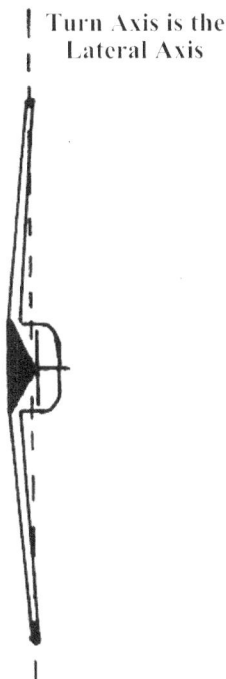

(b) Pure Pitch

Fig 15-2 - *"Perfect" Turns -* In reality, turns are at some point between these two extremes, and as a result involve both pitch and yaw.

Gentle and Medium

Gentle and medium turns are the most common type of turns executed. Gentle turns range up to 15° angle of bank, and medium turns range from 15° up to 30° of bank. The 15° and 30° angles don't hold any significance other than arbitrary definition. Executing a gentle or medium turn simply involves establishing the desired bank attitude and maintaining the cruise pitch attitude.

As we discussed in Chapter 12, entering a banked attitude means that we have to roll. This means using the ailerons in the direction of the desired turn. But the ailerons are not turning the aircraft—they are rolling the aircraft into the desired bank attitude. From there, the ailerons should be neutralized and the aircraft will turn of it's own accord. Some rearward pressure will be required on the control column to prevent the aircraft from pitching downwards, but at shallow angles of bank this pressure will barely be noticeable.

Once established in the turn, the use of rudder and aileron will normally be necessary to a very small degree. As the aircraft travels around in a circle, the outside wing is further from the center than the inside wing (Fig 15-3). As a result of this, the outside wing travels faster and produces slightly more lift than the inside wing, this tends to roll the aircraft into the turn. To compensate, an aileron input to the outside of the turn may be necessary. Meanwhile, the increased speed of the _outside wing will also result in more drag being produced on that side. The imbalance in drag will cause the aircraft to yaw to the outside of the turn, requiring inside rudder to maintain coordination.

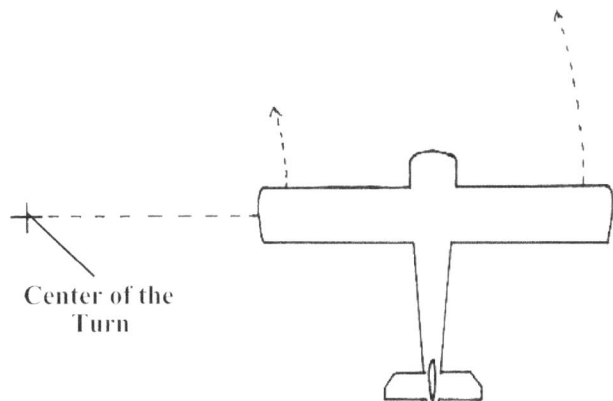

Fig 15-3. - *Effect of a Turn on Wing Speed.*

These control inputs are very small. Overdoing them will result in uncoordinated flight just as not doing them will. If the aircraft is equipped with a yaw string (unfortunately this is not an option on most single engine aircraft due to the slipstream of the prop), "step on the space" to coordinate the aircraft. A yaw string will indicate the direction of a slip by aligning itself with the relative airflow. Applying pressure to the rudder opposite the string will yaw the aircraft into the airflow and eliminate the slip.

Most aircraft will have an inclinometer (often referred to as "the ball") instead of a yaw string. The inclinometer (Fig 15-4) works on a combination of gravity and centrifugal force. When the wings are level and no yaw is present, gravity keeps the ball at the low point in the tube—this means the ball will be centered. If the aircraft were to be placed in a banked attitude and no turn occurred, the ball would still be held low by gravity, but the low point of the tube would no longer be the center, it would be towards the low wing. On the other hand, if the aircraft were to be yawed without being banked, the flight path of the aircraft would be changing and centrifugal force would bring the ball to the outside of the yaw. Ideally, in a turn the yaw rate and angle of bank should match up so that the resultant of gravity and centrifugal force would keep the ball centered. If the ball is off center, we can "step on the ball" to eliminate the undesirable yaw (or create the necessary yaw) and coordinate the aircraft.

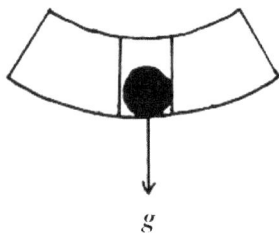

(a) With the wings level and no yaw present (and therefore no sideloads due to centrifugal force), the ball will remain centered due to gravity.

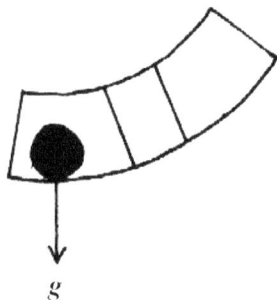

(b) With the aircraft banked and no yaw present (i.e. – no turn), the ball will still be positioned by gravity—this time toward the low wing.

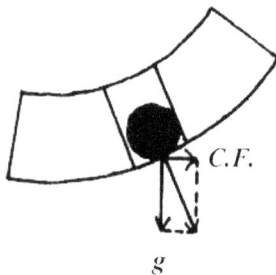

(c) In a banked attitude with an appropriate amount of yaw present, centrifugal force and gravity will combine to give a resultant that centers the ball. This is a coordinated turn.

Fig 15-4. - An Inclinometer or "The Ball".

Recovering to straight and level flight from a gentle or medium turn simply means returning to a wings level attitude. To commence recovery, the ailerons are deflected to the outside of the turn to roll the wings level. Once the wings are back to the level attitude, the ailerons must be neutralized to stop the roll.

Like climbs and descents, the recovery from a turn must be initiated prior to the point where we want to be recovered. The reason is that during the rollout, the aircraft is still turning. So if we wait until we are at the heading we wish to roll out at, we will turn right through it as we roll out. A rule of thumb which works well for turn recoveries is to initiate the rollout half of the angle of bank prior to the target heading.

Steep Turns

Steep turns are turns with an angle of bank exceeding 30°. Once again, the defining angle is arbitrary, but the turning procedure does change somewhat at higher angles of bank. The exact angle at which these changes occur will vary from one aircraft to the next as will the significance of the changes.

The two fundamental differences between gentle/medium turns and steep turns are the amount of back pressure required to maintain the proper pitch attitude and the amount of power required to maintain the entry airspeed. Both of these differences originate from the same cause—an increased angle of attack to increase the amount of lift being produced. Recall from Chapter 8 that in a turn the aircraft must produce more lift than in straight and level flight. This is illustrated for various angles of bank in Fig 15-5. Increased lift means an increased angle of attack, but due to pitch stability the aircraft has a tendency to maintain it's trimmed angle of attack. This tendency must be overcome by a pitch input from the pilot.

The increase in lift also means an increase in induced power required. As the aircraft rolls to higher angles of bank, the induced power required will increase and as a result the total power required will be increased. The increase in power required will vary between different aircraft types, but it will always be reduced at higher airspeeds. That is to say that at a lower airspeed, a given angle of bank will require a certain amount of power above that required for cruise. However, at a higher airspeed the same aircraft at the same angle of bank will not require as much of an increase. As we saw in Chapter 8, this is because at higher airspeeds parasite power required is predominant and is not affected by changes in lift. At lower airspeeds, induced power required is prevalent and *is* affected by the lift changes.

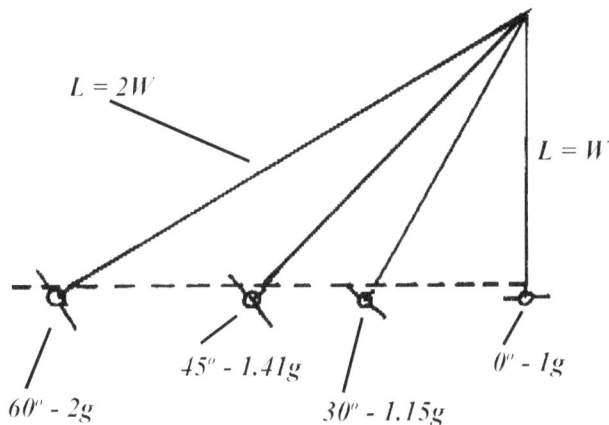

Fig 15-5. - The Extra Lift Needed at Various AOB's - As the AOB is increased, the amount of lift needed is increased. This is because the vertical component of lift must be equal to weight in order to maintain vertical equilibrium.

One more important item to consider in the steep turn is the pitch reference. This is a concern in gentle and medium turns as well, but the effect becomes much more pronounced in steep turns. The effect referred to is the illusion of being nose high or nose low due to the pilots seat position during a turn. This was discussed in Chapter 12 and it was seen that using a reference directly in front of the pilot instead of on the center of the cowling (or dash) will reduce or eliminate the problem.

The procedure for entering a steep turn is similar to gentle and medium turns, except the back pressure required on the control column will be greater and an increase in power will be required (for aircraft with high cruise speeds, such as twins or higher performance singles, this power change may not be required). On the rollout, the back pressure must be released and the power must be reduced back to cruise. Like the gentle and medium turns, the rollout must lead the target heading. The rule of leading the rollout by half of the angle of bank is still effective. So for example, in a 60° bank turn, the rollout should begin approximately 30° prior to the desired heading.

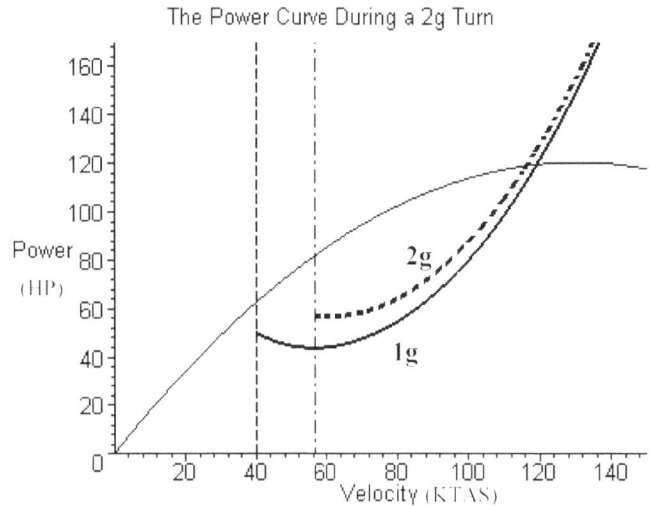

Fig 15-6. - Change in the Power Curve in a Turn - In a turn, the increase in induced drag and the increase in stall speed will restrict the airspeeds that are available to fly at. This example is for a 60° banked turn (2g).

Steep turns can only be accomplished at higher airspeeds. This is because of the effect of induced drag at lower airspeeds and the effect that load factor has on the stall speed. As an example, Fig 15-6 indicates the change in the power curve for a fictitious aircraft when it enters a 60° bank turn. Notice the change in available airspeeds for this aircraft. The turn could not be accomplished at low speeds because of a power deficiency—instead, the speed must be increased prior to entering the turn.

Climbing and Descending

It is a common practice in flight operations to perform a turn while climbing or descending. This can be done by entering the climb or descent first, or by entering the turn first. After proficiency is gained it is possible to enter both at the same time. The control inputs for the entry and the recovery are simply a combination of a turn and a climb or descent. However, things are slightly different in maintaining the turn.

We've seen in Chapter 6 that for a dihedral or anhedral aircraft, the two wings are at different angles of attack in a climbing or descending turn. Because of this, the aircraft will tend to roll. This rolling tendency needs to be eliminated with the ailerons. Almost all training aircraft are dihedral, so they will roll into a climbing turn and out of a descending turn. So in a climbing turn, the ailerons need to be held out of the turn more than in a level turn and in a descending turn the ailerons may actually need to be held into the turn.

Climbing during a steep turn may not be possible in some aircraft due to power restrictions. As the power required increases in the turn, the *excess* power available decreases. *Excess* power is what makes the aircraft climb, so as a result climb performance will always be reduced in a turn. Steep turns in an "underpowered" aircraft (like many trainers) may eliminate the aircraft's ability to sustain a climb at all.

Rate and Radius

In Chapter 8, we discussed the rate and the radius of a turn. It was seen that the *rate* of turn increases with the angle of bank and decreases with an increase in airspeed. It was also seen that the *radius* of turn decreases with an increase in the angle of bank and increases as the airspeed increases (by the square of the airspeed in fact). In practice, these are the exact tendencies which we see in the aircraft.

If we tried to fly a perfect circle over the ground, a given airspeed would require a certain angle of bank. If on every turn we were to increase the airspeed, a corresponding increase in bank angle would be required to maintain the same flight path. This phenomenon is very noticeable to pilots trying to upgrade to higher performance aircraft. For example, turning from base leg to final in the circuit in a faster aircraft requires significantly more lead than in the slower trainers given the same angle of bank in the turn (Fig 15-7).

Rate of turn is a similar issue. To increase the turn rate (rate of heading change) at a given airspeed, the angle of bank must be increased. Likewise, to maintain the same turn rate at higher airspeeds, the angle of bank must be increased. Rate of turn is not usually an issue of concern except during instrument flying. The "rate one turn" is often used when flying by instrument. A rate one turn means that the aircraft's heading is changing at 3° per second, or 180° per minute. This rate of turn (like any other) will require a greater angle of bank at higher airspeeds. A rule of thumb which is commonly used for rate one turns is to take 10% of the TAS and add 7 for knots or 5 for MPH. The

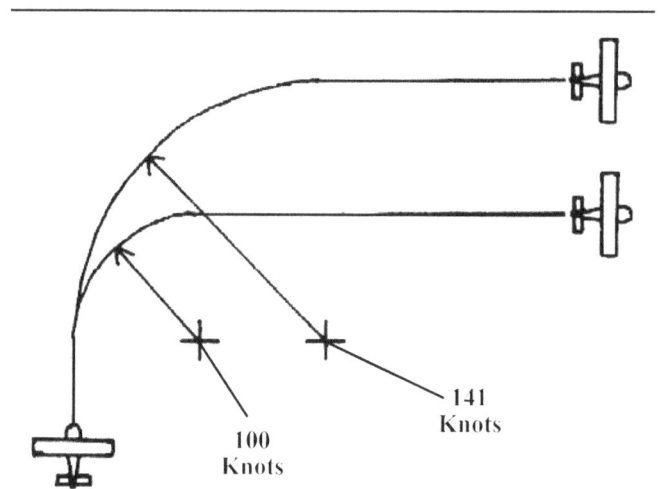

Fig 15-7. - *Effect of Speed on Turn Radius* - The turn radius increases with the square of the airspeed, so at higher speeds, significantly more lead is required on turns.

angle determined here will be a good approximation of the angle of bank required for a rate one turn. A more precise calculation could be done using equation (8.13), however this rule of thumb is sufficient for practical purposes.

Canyon Turns (Minimum Radius)

The concept of a minimum radius turn has already been discussed in Chapter 9. The most common practical application for this maneuver is to turn around after inadvertently flying into a box canyon during mountain flying operations. This means that the canyon we are flying in is a dead end and the aircraft cannot out climb the terrain. The procedure used will vary from one aircraft to the next depending on limit load factors and airspeeds. The variation in procedure will generally be more in the configuration used as opposed to the actual control inputs.

The procedure for a canyon turn will usually be one of the following two:

1) Enter the turn at V_a (or corner speed if it is not the same) and roll to the angle of bank appropriate to the aircraft's limit load factor. Apply power as necessary to maintain V_a and roll out after 180° of the turn has been completed. We saw in Chapter 8 why this procedure works, but some other considerations may make the second procedure more appropriate.

2) Enter the turn at V_{fe} (or, if the aircraft has power to spare, the flaps extended corner speed—this speed will not be published, you have to figure it out) and extend flaps while rolling to the angle of bank appropriate for the aircraft's limit load factor with flaps extended. Apply power as necessary to maintain V_{fe} *or* allow the airspeed to decrease as much as possible without stalling (turning around to exit a box canyon is *not* a good time to stall) and roll out after 180° of the turn have been completed.

These two procedures are fairly general and may vary from aircraft to aircraft depending on the limiting speeds and load factors of the aircraft. An interesting exercise is to use the equations given in Chapters 8 and 9 and the speeds and loads appropriate to a particular aircraft to determine the best procedure.

Chapter Summary

In order for an aircraft to execute a coordinated turn, a banked attitude must be established. The banked attitude inclines lift to the side and produces a sideslip. The directional and pitch stability of the aircraft react to the sideslip by producing a turn in the direction of the low wing.

Turns are classified as gentle, medium, and steep. Gentle turns involve an angle of bank of up to 15°, medium turns go as far as 30° of bank, and steep turns are greater than 30° of bank. As well, turns can be climbing or descending.

Rolling into and out of a turn is done with aileron, but sometimes requires rudder to coordinate the roll. Once established in the turn, small corrections with the ailerons and rudder will normally be necessary to maintain the desired angle of bank and coordination. In the case of a steep turn, many aircraft will require an increase in power setting due to the extra induced drag created by the turn.

The rate and radius of a turn can both be important considerations for a pilot. The rate of turn will increase with an increase in bank angle and decrease with an increase in airspeed. The radius of turn will decrease with an increase in bank angle and increase with an increase in airspeed.

To conduct a minimum radius turn (AKA – canyon turn), we need to fly at the lowest possible airspeed with the highest possible angle of bank. These are contradicting requirements due to the increase in stall speed with bank angle, so a proper combination must be found. On many aircraft, the perfect combination means flying at the aircraft's corner speed (usually the same as V_a) and establishing the angle of bank that brings the aircraft to it's limit load factor.

Questions

1) What are the primary differences between gentle/medium turns and steep turns? Why do these differences occur?

2) How does a banked attitude make an aircraft turn?

3) Given the following information from an aircraft's POH, what would be the most appropriate method for a canyon turn (assume there is enough THP available for either method):

   a) V_s = 65 knots   d) LLF (no flaps) = 4.4 g's
   b) V_{so} = 52 knots   e) LLF (full flaps) = 3.5 g's
   c) V_{fe} = 110 knots

4) While setting up for an IFR approach in a light twin, a rate one turn is conducted. Approximately what angle of bank must be used if the airspeed is 120 knots? What load factor will this angle of bank create?

Chapter 16
Aircraft Control Considerations

Now that we've looked at the basic flight maneuvers, we can consider the use of the flight controls with the perspective of hindsight. The proper use of the controls is often confusing, especially to the beginner. Several apparent contradictions occur when trying to learn procedures by rote, but if instead we try to *understand* the proper function of various controls these "contradictions" suddenly disappear. Knowing these proper applications will help to avoid confusing and even dangerous control use (or misuse) in the future.

Pitch vs. Power

Ever since the dawn of aviation, there has been an ongoing argument among pilots about the use of pitch and power. One side of the argument states that power controls the airspeed and pitch controls the vertical speed:

> Power   ➤   Airspeed
>
> Pitch   ➤   Vertical Speed

The other side of the argument states that pitch controls airspeed while power controls the vertical speed:

> Pitch   ➤   Airspeed
>
> Power   ➤   Vertical Speed

Both sides can cite examples to prove the point: "In straight and level flight, don't you hold your altitude with pitch?"; "In a descent, don't you hold your airspeed with pitch?".

Often, pilots just try to memorize when to use pitch or power to control airspeed or vertical speed. Unfortunately, this can lead to some potentially dangerous control inputs under some circumstances. It's far more useful to understand how the controls are really meant to be used.

Which side of the argument is correct?. Both of them are partly right while neither of them is completely right.

The Role of Power

Power, as we know from Chapter 1, is the rate of doing work—that is, it is the rate of transferring energy. In an aircraft, we have four types of energy to consider. The first is chemical energy—stored in the fuel which we carry. The second is kinetic energy—airspeed. The third is potential energy—altitude. And the fourth is heat energy—the energy we always lose to the atmosphere due to drag. It is the function of the engine to convert fuel energy into useful mechanical energy to operate the aircraft. The rate at which this occurs is power.

So the power provided by the engine will either accelerate the aircraft to a higher airspeed, climb the aircraft to a higher altitude, or simply replace energy lost to the atmosphere via drag.

The big question is, which of these functions will the power fulfill?

The Role of Pitch

The answer to this question lies in pitch. Pitch divides up the energy of the aircraft. If we pitch the aircraft up we will lose airspeed and gain altitude. This means that we are trading kinetic energy for potential energy. On the other hand, we can pitch the aircraft down and gain airspeed by trading off altitude. This ignores the effects of power and drag which complicate things somewhat.

With no power, the energy lost to the atmosphere must be constantly replaced by giving up altitude—otherwise we lose all of our airspeed and eventually stall (stalls, of course, occur when we exceed our critical angle of attack, but in this case, the angle was exceeded due to a low airspeed). With an engine, this energy is maintained by the burning of fuel—this way we can maintain both airspeed *and* altitude.

An excess of power (fuel energy being used at a higher rate than necessary) will cause an acceleration or a climb. Which of these two occurs depends on the pitch input that accompanies the power increase. Pitching the aircraft up with a power increase will lead to a climb, while pitching the aircraft down with a power increase will lead to an acceleration. Maintaining a constant pitch attitude with a power increase will lead to a small acceleration with a small climb.

Likewise, a power deficiency (not enough fuel energy being used) will lead to a deceleration or a descent. Once again, which of the two occur will depend on the pitch input accompanying the power change. Pitching the aircraft down with a power reduction will lead to a descent, while pitching the aircraft up with a power reduction will lead to a deceleration. Like the power increase, maintaining a constant pitch attitude with a power reduction will lead to a deceleration *and* a descent.

Utilizing Pitch and Power

So we can see here that the engine (power) provides the aircraft with the necessary energy to maintain operation, and pitch control is used to divide that mechanical energy up into useful forms—speed (kinetic energy) or altitude (potential energy). So the two statements made about the roles of pitch and power can both be true if they are qualified by certain conditions. The purposes of pitch and power can never be really separated, the proper way to define their combined roles would be to say that pitch and power combined control airspeed and vertical speed combined:

Pitch + Power ➤ Airspeed + Vertical Speed.

This statement leads to the commonly used flight training maxim:

Attitude + Power = Performance.

For simplicity, it is a common (and effective) practice to separate the roles of pitch and power. However, in order for this method to work, the variables need to be reduced—that is, either airspeed or vertical speed must remain constant while the other is varied. If both airspeed and vertical speed are varied simultaneously, it will quickly be found that since one has a direct effect on the other, effective and consistent control is lost to both. Using two co-dependant controls to manipulate two co-dependant performance features is too complicated to be effective in practice (although in theory it is entirely possible).

To solve this problem, a "critical constant" should be selected during a maneuver. This means that either airspeed or vertical speed/altitude should be chosen to remain constant while the other is varied as necessary. The selected constant can be held with the use of pitch, while the variable is controlled with power. This method allows the energy from the engine to be dedicated to controlling the variable of the maneuver.

In the case of straight and level flight, the desired constant is normally altitude. This means maintaining a constant altitude with pitch control and controlling airspeed with the power. In normal cruise flight, the power can be set and left alone since the airspeed is usually a constant as well. Small altitude deviations can be corrected with pitch inputs, and the altitude can be maintained by maintaining the appropriate pitch attitude. Varying the airspeed can be accomplished by changing the power setting and adjusting the pitch attitude to maintain altitude as the airspeed changes.

As for climbs and descents, it is usually more important to maintain a constant airspeed (e.g. – best rate of climb, best glide, etc.). So in this case, pitch would be used to control the airspeed and power would be used to control the vertical speed. If we maintain a constant airspeed in a descent (using pitch), an increase in power will decrease the descent rate while a decrease in power will increase the descent rate. The same can also be said for climbs—the difference being that in a climb the power is usually set to "climb power" and a performance speed is maintained with the pitch attitude.

Keep in mind of course, that a change in power to adjust the vertical speed will require a corresponding change in pitch to maintain a constant airspeed.

Rudder and Ailerons

Another common misunderstanding in aviation—leading to frequent arguments during "hangar talk"—is in the roles of the ailerons and the rudder. Common disagreements are usually restricted to "Which one turns the airplane?" and "Which one coordinates the airplane?". Some individuals are of the opinion that the rudder turns the aircraft—although this faction seems to be constantly decreasing in size. Others believe that the ailerons turn the aircraft—this side of the argument seems to be getting larger. In fact *neither* statement is correct. As for coordination, *both* the rudder *and* the ailerons are used in conjunction with one another.

Individual Roles

Ailerons – Ailerons, as we discovered in Chapter 5, are the control surfaces at the outboard trailing edge of the wings. They pivot up and down to change the camber—and therefore the lifting properties—of the wings. The two ailerons move opposite to one another so that as one wing gains lift (the one with the down-going aileron) the other (with the up-going aileron) loses lift. However, this lift differential is *not* what makes the aircraft turn—it makes the aircraft *roll*. Roll is used to transition the aircraft from one bank attitude to another.

This is the only purpose for ailerons. The turn results from a banked attitude (the mechanics of which have been discussed previously) not from the aileron input.

Rudder – Like the ailerons, the rudder was discussed in detail in Chapter 5. The rudder is the control surface which is attached to the trailing edge of the vertical stabilizer. It pivots side to side to vary the camber of the stabilizer so that it can induce or eliminate yaw. As stated in Chapter 5, the rudder is intended to *control* (and not necessarily to produce) yaw.

Combining the Two

Aileron and rudder generally go together during control manipulations. For example, in an aircraft without differential or frise ailerons the rudder will need to be deflected in the same direction as the ailerons in order to compensate for aileron drag and the resulting adverse yaw. The rudder and ailerons work in conjunction with one another to coordinate the aircraft—the ailerons control roll which in turn controls the bank attitude, and the rudder controls the rate of yaw. We've already seen that the rate of yaw and the angle of bank must match for the aircraft to be coordinated.

To coordinate the aircraft after a sideslip has been introduced, the rudder can be used to *yaw* the aircraft into the slip or the ailerons can be used to *roll* the aircraft away from the slip. Both techniques are legitimate, and each has different applications in the aircraft.

In using rudder, the aircraft is rotated so that the relative airflow is once again parallel to the longitudinal axis. For example, if the aircraft is slipping to the left (the relative airflow is coming from the left), pressure can be applied to the left rudder pedal to yaw the aircraft to the left—this aligns the longitudinal axis to the relative airflow.

When using ailerons, the aircraft is transitioned (rolled) to a banked attitude (or perhaps a *different* banked attitude) which allows the rate of yaw to become appropriate to the bank angle. For example, if the aircraft was again slipping to the left, a roll to the right would correct the problem. Remember that if a slip is occurring, either the aircraft is yawing, or the aircraft is banked without an appropriate amount of yaw. If the aircraft was banked to the left and not yawing, a left slip must be occurring, the slip can be eliminated by adding left rudder (yawing into the slip) or by leveling the wings (rolling to the right—away from the slip).

When deciding which technique to use, the circumstances will dictate which is easier. In straight and level flight, the direction is constant—thus the rate of turn is zero. When this is the case, it is generally easiest to *maintain* direction with the rudder, and coordinate the aircraft with the ailerons. Theoretically, this should lead to a perfectly straight flight path with the wings always perfectly level. In reality, some minor corrections will always be necessary.

When turning the aircraft in VFR (Visual Flight Rules) flight, the rate of turn is not usually considered important, but the angle of bank generally is. This means that it would be easier to maintain a constant angle of bank using the ailerons for corrections, and to coordinate with the rudder. This means selecting an angle of bank and maintaining the appropriate rate of turn for the bank angle with the rudder—even though no *direct* effort is made to maintain a specific *rate* of turn.

During IFR (Instrument Flight Rules) flight, rate of turn becomes more important. An angle of bank is usually selected which provides for a standard rate turn (3° per second). Once the turn is established (using ailerons to roll in), it is easier to maintain the rate of turn with the rudder while coordinating with the ailerons. For the reader not familiar with instrument flying, the turn coordinator (normally at the bottom left of the instrument "six pack") is the instrument that is used to indicate a standard rate turn. The rate of turn can be maintained by applying rudder pressure in the appropriate direction, while coordination is maintained by rolling towards "the space" on the inclinometer.

So again, as in the case of pitch and power, a critical constant is selected—either rate of turn or angle of bank. However, this time instead of always controlling the constant with the same control, the constant determines which control to use. If rate of turn is the intended constant, the rudder is used to maintain the rate of turn and the ailerons are used to coordinate. On the other hand, if the angle of bank is the desired constant, ailerons maintain the bank angle while the rudder is used to coordinate.

Be careful not to get confused on the use of the rudder. The rudder controls the rate of yaw—which is also the rate of turn. Yawing with the wings level will produce a sideslip. If we wish to execute a coordinated (zero sideslip) turn, the aircraft must be banked. Establishing the banked attitude is accomplished with the ailerons—*not* the rudder. So rudder is used if necessary to *maintain* the turn, but not to *establish* the turn.

Ideally, rudder will be unnecessary *in* the turn as well—the vertical fin should yaw the aircraft sufficiently to eliminate the slip induced by the banked attitude. Unfortunately, other factors come into play which will often necessitate the use of rudder in the turn. One of these factors is the difference in the speed of the two wings. As the aircraft travels around in a circle, the inside wing is travelling slightly slower than the outside wing. This speed difference will lead to two problems—a lift differential and a drag differential between the two wings. The lift differential must be corrected by the ailerons—meaning that the ailerons will be deflected slightly out of the turn to maintain the appropriate bank angle. The drag differential must

be corrected by rudder—the outside wing has more drag, so rudder into the turn will be necessary. These phenomenon are more pronounced in lower speed aircraft with long wingspans. For this reason, gliders will normally require much more rudder input than a "power" pilot would be accustomed to.

The Use of Trim

At several points in this book, we have discussed the use of trim and it's importance. We know that trim eliminates the need to be constantly pulling or pushing on the controls. Almost all aircraft have a pitch trim device, and some also have roll and/or yaw trim. For this discussion, we will focus on the pitch trim, but the procedures described would apply equally well to roll and yaw trim.

First, we should look at what trim is *not*. Trim is not a primary control, therefore it is not used to control the aircraft's attitude or movements. Trim does not control the aircraft's airspeed, vertical speed, angle of attack, or pitch attitude (strictly speaking from an engineers perspective, trim actually *does* control the AOA and therefore the airspeed, but we will soon see why a pilot shouldn't look at it that way). Trim is not normally a direct means to control the aircraft (we will, however, see an exception to this rule in the next section of this chapter).

When trim is used to directly manipulate the pitch attitude, the result is usually a roller coaster ride. Consider what happens if the aircraft is in straight and level flight and trim is used to pitch up into a climb attitude. As the trim wheel (or lever or switch) is moved back to raise the nose, the aircraft is no longer flying at it's trimmed AOA—the trimmed angle is increased. Static pitch stability (see Chapter 7) dictates that the aircraft will pitch up to reduce it's AOA. Unfortunately, this reaction is normally slightly behind the movement of the trim tab, and by the time the aircraft reaches the desired pitch attitude the trim has been over-adjusted. This causes the aircraft to pitch up beyond the desired point and dictates a forward adjustment of the trim by the pilot. As the trim is adjusted forward the same thing happens, an over-adjustment results and the cycle starts all over again. This problem is magnified by the effects of dynamic pitch stability (Chapter 7 again). Even if the trim is adjusted just so, the aircraft will have to travel through several oscillations before settling on the proper attitude.

The problem with trim as a pitch control device is again amplified by the use of power. If a power change is made, the induced airflow over the tail will change the trimmed AOA even though the trim itself hasn't been adjusted. In the case of a power increase, the aircraft will pitch up further. With a power decrease, the aircraft will pitch down.

So if the trim will result in pitch oscillations, how do we use it? Simple. In Chapter 7, we discussed the ideas of stick-fixed and stick-free stability. Holding the elevator steady during pitch oscillations tends to reduce the size of the oscillations as well as the number of cycles. When adjusting trim, we will use the elevator to eliminate the oscillations altogether. First, instead of using the trim to pitch the aircraft, we will use the elevator. Once the aircraft is established in the desired attitude, the power should be adjusted (if it wasn't adjusted prior to the pitch change). At this point, pressure will be required on the control column to maintain a constant attitude because the aircraft is not flying at it's trimmed AOA. To correct this, the trim should be adjusted to eliminate the pressure on the control column while the control column (i.e. – elevator) is used to maintain the pitch attitude.

Remember to maintain a constant attitude while trimming. Allowing the attitude to vary with the power set will lead to airspeed variations. As a result, the required trim setting will change. This can lead to a false indication of proper trim. As well, always leave the trim until last. Adjusting the attitude *or* the power after the trim is set will create the need to trim all over again.

Primary Control Failures

In aviation, pilots are constantly preparing for emergencies. These emergencies are few and far between in reality, but when you do need to be prepared you *really* need to be prepared. In the training environment, emergencies are usually restricted to engine failures, instrument failures and fires since these are the most common ('common' being a relative term). Another emergency which can be very dire—although very unlikely—is a primary control failure.

There are three main types of control failures (for direct controls—we won't consider hydraulically or electrically boosted controls here). The first is a structural failure of the control surface.

This will sometimes lead to an unrecoverable loss of control, especially if the elevator and/or horizontal stabilizer have failed. However, this type of failure is also so unlikely as to be considered practically impossible—*provided that we keep the aircraft within it's certified limits (see Chapter 9).* The second possibility is a jammed control. A jammed control can be very serious depending on the position which the control is jammed in—however, in many cases the secondary effects of the other controls will be enough to get the aircraft down safely. The third type of failure is a cable break. This will lead to a control surface which is floating freely in the airflow. Control of the aircraft will be tedious, but manageable.

The main idea behind compensating for a control failure is to use the secondary effects of the other controls. This will allow the gaps in control of the aircraft to be filled in. Above all else, when dealing with a control failure, the pilot must be able to anticipate what the aircraft is going to do and stay *at least* one step ahead of the aircraft. This means understanding the effects of each of the other controls as well as the stability characteristics of the aircraft.

First, upon recognition of the failure, the aircraft should be returned to straight and level flight and then turned towards an airport for a landing. Straight and level, once established, is not too difficult. The approach and landing is the tricky part and must be planned well in advance. Large control inputs should be avoided if at all possible.

Rudder – If the rudder is the surface which fails, the problem is *relatively* easy to deal with. Unless landing with a strong crosswind is required, some easy precautions will reduce the problem of no rudder to easily manageable proportions. Generally, we want to control yaw with the rudder because of the problems associated with it. These problems include excess drag, a rolling tendency, and the possibility of leading to a spin.

As for the excess drag, the problem is (in this particular circumstance) negligible since control is much more important than performance, so it can usually be ignored. The roll induced by uncompensated yaw can be corrected with the ailerons. This leads to uncoordinated flight, but once again, the problem of the drag can be ignored. The spin potential created by uncoordinated flight can only be overcome in one way – don't stall. For this reason, an approach with no rudder control should be flown at a significantly higher airspeed than usual (we will see in Chapter 19 that the approach is one of the danger areas for stalls) and no sudden or excessive attitude changes should be made.

Ailerons – If the ailerons fail, the problem is slightly worse, but still easily manageable. Roll must now be controlled by the rudder. Recall that the secondary effect of yaw is roll, so application of right rudder will induce a yaw to the right which will in turn lead to a roll to the right. As with the rudder failure, this will lead to uncoordinated flight. However, if we are cautious to avoid the danger areas of uncoordinated flight (namely spins), this is a minor problem in comparison to the control failure.

Elevator – The most serious control failure would be an elevator failure. A loss of pitch control could be disastrous if allowed to progress. The controls which we will use to replace the elevator will be the trim and the power. When doing this, it is extremely important that we keep the effects of dynamic stability in mind. These effects will frustrate all but the most subtle control inputs.

Power will have a direct influence on the pitch of the aircraft for three reasons. First of all, the thrust/drag couple will assist in pitching the aircraft up (in most aircraft) while power is on. Second, induced airflow over the tail will increase the down force on the horizontal stabilizer—thus pitching the aircraft up—while power is on. Third, the pitch stability of the aircraft will lead to an indirect effect of power. Remember that with pitch stability, the aircraft tends to maintain a constant AOA. Adding power will initially accelerate the aircraft, but with a fixed AOA the aircraft will then be producing too much lift—causing an acceleration upward. The change in flight path will cause a reduction in AOA which will lead to the aircraft pitching up to retain the trimmed AOA.

If the effects of the thrust/drag couple and the induced airflow over the tail could be eliminated, the airspeed would not change with a power change. The aircraft would simply pitch up and initiate a climb at the original trimmed airspeed (remember that it is actually the AOA that is trimmed to remain constant, but for a given weight, a constant AOA will result in a constant airspeed). However, considering the amplifying influence that

the couple and the induced airflow have on the power effect, the pitch will be more pronounced and the airspeed *will* change. In fact, with an increase in power, the airspeed will *decrease* and with a decrease in power, the airspeed will *increase.* These airspeed changes will also be accompanied by corresponding vertical speeds.

Along with power changes, trim will allow us to control airspeed and vertical speed "independently" of one another. In the case of an elevator failure where the control surface is moving freely and "floating" with the relative airflow, trim functions normally. Rolling the trim forward in this case will lower the nose, while rolling the trim back will raise the nose. It can't be overemphasized here that these adjustments—power *and* trim—must be kept small or the pitch oscillations will make things *very* difficult.

If the elevator is jammed in place and the jam is at the elevator hinge or very near it in the control circuit, the trim will need to be used differently.   To understand what is happening, imagine that the elevator is now a fixed part of the stabilizer (which it *is* now) and the trim tab is functioning as an elevator. To pitch the aircraft down, instead of *raising* the trim tab to lower the elevator, we would *lower* the trim tab to act as an elevator. This means that the trim will be moved in the direction *opposite* to that which we would normally use.

If the jam is near the control column in the control circuit, the elevator cable(s) will usually stretch enough to allow the trim to be used normally. Trimming nose down will lower the nose because the elevator will still have some limited motion. This motion will not be directly affected by the control column, so the trim will still need to be used.   Faced with an elevator jam, some quick experimentation will allow us to identify which situation we are dealing with.

A Few Words on Rules of Thumb

At several places in this book, rules of thumb have been introduced. As we progress through the remaining chapters, several more rules of thumb will be looked at. When using rules of thumb in practice, caution must be used—we must not assume that they are absolutely reliable. They are not intended as *laws* governing the operation of the   aircraft—they are intended as guides in establishing a reasonable estimation of the aircraft's performance.   Rules of thumb are invariably intended to replace some physical law or set of laws by an oversimplified version of the same law(s). This simplification allows us to complete necessary calculations quickly and often mentally—meaning that these laws become useful to us in practical, "real world" situations.

Unfortunately, oversimplification means that some variables are not considered (for example, rules of thumb concerning wind effects on takeoff distance generally ignore prop efficiency and rolling friction). Ignoring variables will ultimately lead to inaccuracies in the calculations which must be considered. In some cases these inaccuracies are not important (e.g. – at a constant airspeed 100 RPM will change the vertical speed by 100 FPM), however in other cases, the inaccuracies could be critical (e.g. – determining takeoff distance over an obstacle).

For this reason, rules of thumb should _not_ be used when the information sought is contained in the POH or Flight Manual. This information has been determined by engineers and confirmed by test pilots, so it is fairly reliable. This having been said, published figures are often best case numbers and variations from the published conditions must be taken into account. These variations can be considered through the use of rules of thumb, but only if the manufacturer of the aircraft has not provided correction factors for the variations (such as wind, runway surface condition, runway gradient, etc.).

Most of the rules of thumb contained in this book do not include an extra conversion to account for the ignored variables. For critical performance factors, it would be a good idea to change the numbers an appropriate amount to account for the variables not considered by the thumb rule. This would include variables such as pilot experience (total _and_ on type), aircraft condition, aircraft handling characteristics, and of course, any factors which are ignored by the rule of thumb.

Chapter Summary

The pitch attitude and the power setting are inseparable control inputs and work closely in conjunction with one another. The power of the engine provides the aircraft with the energy that it needs to continue operating. The pitch attitude can be considered to be a method of designating the function of that energy, either to airspeed or altitude. In combining the two, pitch is normally used to maintain a "critical constant" while power is used to control a variable.

The rudder and ailerons also work closely in conjunction with one another. It is important to understand their relationship to each other in order to use them properly. The ailerons control roll while the rudder controls yaw. The coordinated use of these two controls is important to eliminate a sideslip (AKA – uncoordinated flight).

Trim is used to relieve control pressure. Pitch trim is the most common type, although trim can be applied around all three axis'. In using pitch trim, the attitude and airspeed should be established and maintained first since variations will cause "false" control pressure indications.

If a primary control surface fails in flight, the key to controlling the aircraft is in understanding the primary and secondary effects of the other controls as well as the stability characteristics of the aircraft. This is especially true in the case of a loss of direct pitch control.

Rules of thumb are presented throughout this text as well as many others. One must be cautious, however, not to simply accept rules of thumb on blind faith but to understand where they come from. This understanding will allow you to recognize the limitations of these rules and to avoid any misapplication. Always remember that rules of thumb are simplifications of more complicated physical laws or groups of laws. This simplification will inevitably lead to inaccuracies which can at times be critical.

Questions

1) Based on the principles discussed in this chapter, how would one transition from cruise at a given airspeed to a descent at a higher airspeed? Explain what is happening in this process.

2) What does someone mean when they say, "Attitude + Power = Performance"?

3) What are the four types of energy we manipulate when operating an aircraft? How do they relate to one another?

4) How can the rudder be used to maintain coordinated flight? How about the ailerons?

5) Why would the use of rudder be necessary once established in a turn?

6) Explain the proper procedure for eliminating control pressure with trim. Why is this procedure important?

7) How would you establish, maintain, and recover from a turn after an aileron failure?

8) How would you establish, maintain, and level off from a descent after the cable controlling the elevator snapped?

9) What exactly is a rule of thumb? What are the shortcomings of rules of thumb? How can they still be useful?

Chapter 17
Range and Endurance

In Chapter 8 we discussed the concept and the theory behind flight for maximum range and flight for maximum endurance. We saw that range allows us to fly the greatest distance per unit of fuel while endurance allows us to fly for the greatest period of time per unit of fuel. We also saw where range and endurance occur on the power curve. In this chapter, we will look at how range and endurance can be applied in normal flight operations and how to establish the aircraft in flight for maximum range or flight for maximum endurance.

The Power Curve

As we know, the power curve follows a "U" shape (Fig 17-1). Points on this curve relate various airspeeds to the power which is required at that speed to maintain level flight. Power required is high at lower speeds and decreases to a minimum point as the velocity increases. Beyond this point, the power required once again increases with an increase in velocity.

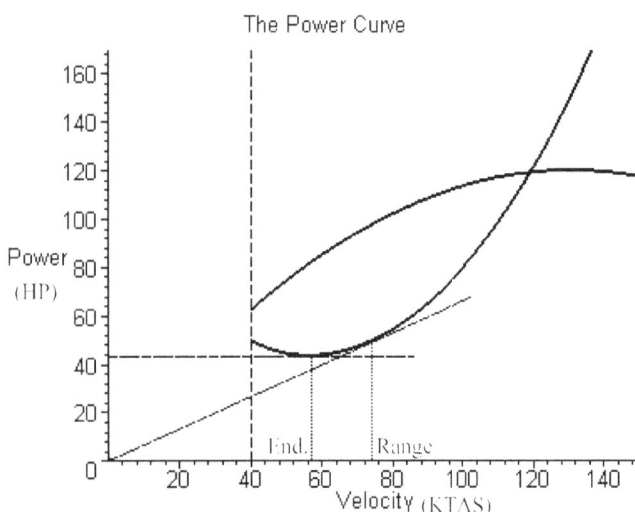

Fig 17-1. - *The Power Curve* - A tangent drawn from the origin determines best range speed, and the lowest point on the curve determines best endurance speed.

On this curve, drawing a line from the origin so that it is tangent to the curve will indicate the speed and power required for maximum range. This speed allows for the highest velocity to power ratio, and it also corresponds to the best lift to drag ratio. The lowest point on the curve indicates the speed for maximum endurance. This point allows for the least power required, and thus allows for the lowest fuel flow.

Range

Flight for maximum range allows the aircraft to cover the greatest possible distance per unit of fuel. This speed would clearly be used to maximize the range of the aircraft (hence the name of the speed). Applications would include reaching a destination in an unplanned diversion, improving fuel economy, minimizing fuel stops along an intended route, and other distance related goals. As we've seen above and in previous chapters, the objective in flight for range is to establish the aircraft at the speed which provides for the highest available speed to power ratio.

Factors Affecting Range

We've seen in Chapter 8 what the factors are that will affect the range of a given aircraft. To review and sum them up:

Altitude – We've seen that altitude will not affect the optimum range of the aircraft. This statement, however, ignores the fuel burned in a climb to altitude. Thus, when we try to achieve optimum range, lower altitudes are better to minimize fuel wasted in the climb.

Weight and C of G – Added weight will decrease range since with more lift, the aircraft will produce more induced drag—this increases the power required for a given airspeed. As for the C of G, we know that a forward C of G increases the required tail down force and thus increases the lift required to maintain equilibrium. This will also increase the induced drag and therefore decrease range.

Drag – If the aircraft has more drag it won't be able to travel as fast for a given power setting. This of course will decrease the range. The drag produced by a given aircraft can be directly affected by aircraft configuration (flaps, speed brakes, spoilers, etc.) and condition (new paint and wax vs. a dirty airplane).

Mixture – It was stated in Chapter 8 that fuel flow is proportional to the power setting. A consideration not included was mixture control. The mixture setting determines the fuel to air ratio that enters the engine. For a given power setting, a full range of mixture settings will allow the engine to keep running. To obtain optimum range (or endurance) the fuel consumption must be reduced. This means that we want a "lean" mixture providing a low fuel to air ratio. The proper leaning procedure should be included in the engine manual or the flight manual for the aircraft.

Wind – The effect of wind on range was discussed in Chapter 8. Basically, a headwind will decrease range and a tailwind will increase range. We also saw that some corrections can be made to reduce the effect of a headwind or increase the effect of a tailwind. As well, even though climbing is not normally in our best interest, climbing to get more favorable winds may be worthwhile if it's an option.

Turbulence – Turbulence causes airspeed fluctuations. These variations will prevent the aircraft from maintaining the most efficient airspeed and will therefore decrease the range of the aircraft. As with the wind, a change in altitude may be worthwhile to find more favorable conditions.

Performance Charts

Most aircraft will have range performance charts in the POH or Flight Manual. When using these charts, it is important that we understand what they are telling us. First of all, few aircraft have charts published for *maximum* range conditions. Instead, performance charts normally refer to specific power settings vs density altitudes. These charts will be based on the fuel consumption and the TAS for the given power setting as well as the fuel quantity available. Fuel consumption for a given power setting will not change with altitude, but the TAS will become higher. This means that for a given quantity of fuel, the range will increase with altitude.

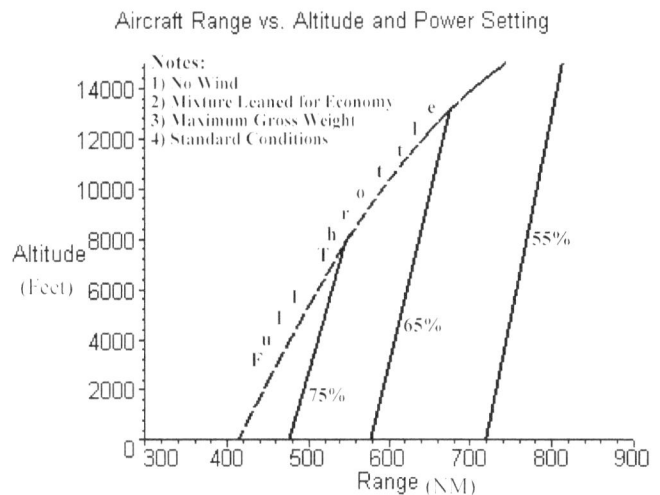

Fig 17-2. - *Range Performance Graph* - This particular graph does not account for fuel burned in the climb, hence the straight lines. Note the condition of "Standard conditions". This implies that the altitudes referenced are actually *Density Altitudes.*

"For a given quantity of fuel" is the operative phrase here. As we climb, large amounts of fuel are burned at a low airspeed, clearly this will reduce the range of the aircraft. Some performance charts ignore the climb and give the range of the aircraft assuming that the aircraft was magically transported to it's cruising altitude. When this is the case, the increase in range with altitude will be approximately linear (Fig 17-2). If on the other hand, the chart accounts for fuel burned in the climb, the graph will follow a curve (Fig 17-3). As a rule, for most single engine general aviation aircraft, normal cruise power settings (55% to 75% power) will give optimum range for the power setting at approximately 6000 ft.

Note with these performance graphs that *optimum* range does not change with an increase in altitude. Recall from Chapter 8 that at higher

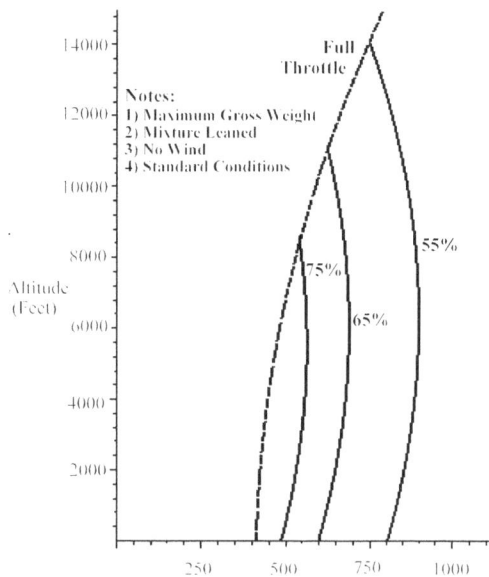

Fig 17-3. - *Range Performance Graph* - This graph accounts for the fuel burned in the climb, hence the curved lines. Range at a fixed power setting will increase with altitude up to a point, then the fuel burned in the climb outweighs the gain in range with altitude.

altitudes we require more power but we also have a higher TAS. Range will only increase for a *given power setting*, this is because as we climb, this power setting becomes closer to the optimum. When establishing the aircraft in flight for maximum range, lower altitudes are better (within reason—picking twigs out of the landing gear usually indicates that we were too low) since we then don't burn excessive amounts of fuel in the climb. This having been said, if the decision to fly for range is made in-flight, and the climb has already been completed, remaining at altitude doesn't cost anything and may be the best bet.

Experimental Procedure

Since most manufacturers don't publish optimum range data, the tangent on the power curve must be found by experiment. The procedure which we will use is based on the shape of the power curve and requires a successive series of airspeed changes. These airspeed changes are intended to find the speed at which the most speed is available for a given amount of power.

Referring to the power curve in Fig 17-4, we can see that in the "normal" cruise range, power changes of a fixed amount will result in approximately equal changes in speed. However, nearing the bottom of the curve these speed

changes become larger due to the changing slope of the curve. This increase in the speed changes is what we are looking for in the airplane. We can see that maximum range is on the high-speed side of the largest change in speed for the selected power change.

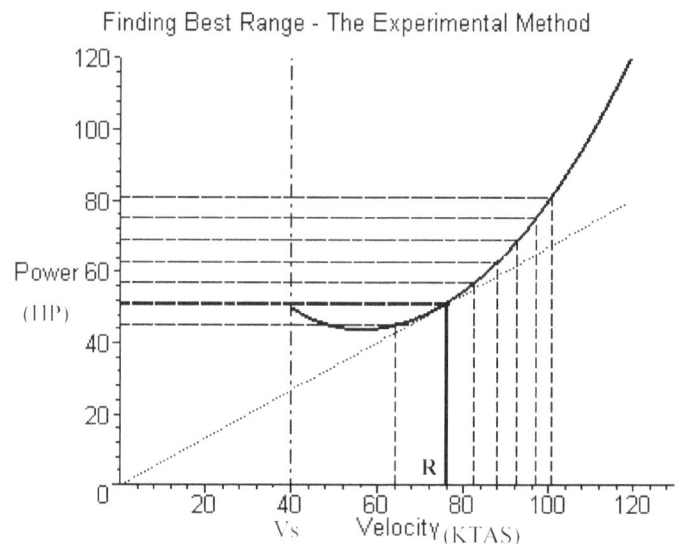

Fig 17-4. - *Power Changes in Equal Increments will have Different Effects at Different Speeds* - This fact allows us to find the speed for best range experimentally by looking for the largest change in airspeed. Maximum Range is on the high-speed side of this change.

Knowing this about the power curve and knowing the factors that affect range, we can develop an experimental procedure for determining the best range speed.

1) Insure the aircraft is in the cleanest possible configuration.
   – Normally this will already be the case in cruise flight.

2) Select an appropriate altitude.
   – Taking into consideration wind, turbulence, and fuel burned in the climb, the best altitude to use will vary from day to day.

3) Shift weight rearward as *practical.*
   – An aft C of G allows for better range performance. This having been said, the C of G should *always* be kept within certified limits and the pilot should be able to reach the controls (hence the word "practical").

4) Change power and note speed changes.
– Select a power change to use (perhaps 100 RPM) and reduce the power in equal increments. Note the airspeed which occurs at each power setting and look for a larger than normal change in speed. At this large change, return the power to it's previous setting to establish the aircraft in flight for best range. Care must be taken to hold altitude as closely as possible. Deviations from the intended altitude will also lead to changes in airspeed which will be misleading.

5) Compensate for wind.
– We saw in Chapter 8 that the wind not only changes the range of the aircraft, but it changes the speed at which best range occurs. The new speed can be determined by shifting the origin of the power curve and redrawing the tangent line (Fig 17-5). Of course, we can't do this in the airplane, and most manufacturers don't publish power curves to work with, so a rule of thumb is useful here.

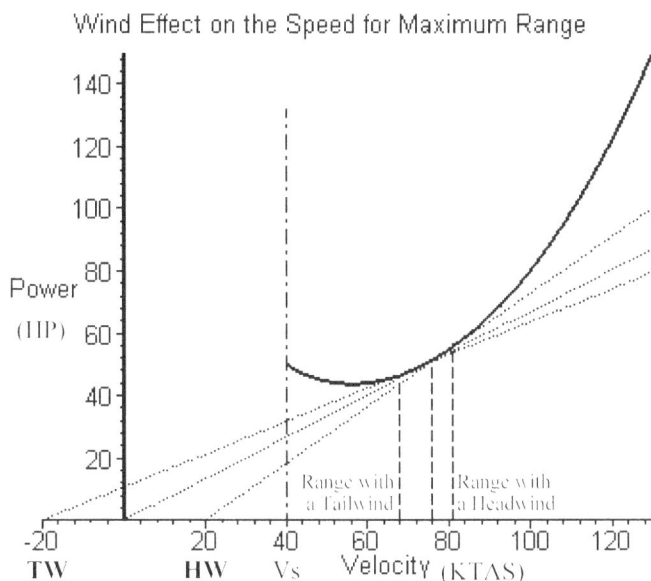

Fig 17-5. - *Wind Effect on the Best Range Speed* - Instead of using the origin to draw the tangent line, we use a point that gives us a *groundspeed* of zero. Drawing the tangent from there gives us a new best range speed.

To correct for the wind we can increase the airspeed by half of a headwind or decrease the airspeed by half of a tailwind. When decreasing the airspeed, however, we should never go below the endurance airspeed—from Fig 17-5 we can see that no matter how strong the tailwind, the tangent will

never be slower than endurance speed. Also from Fig 17-5 we can see that this rule is a reasonable approximation of the actual speed changes necessary.

6) Lean the mixture.
Since the focus of this book is on aerodynamics, I will leave the details of the leaning procedure(s) to engine manuals and other text books.

This procedure is fairly straightforward, and with practice it is easy to use. However, another option does present itself. Most single engine aircraft have a published best glide speed. This speed is generally used to obtain the best glide range in the event of an engine failure. We've seen in Chapter 8 that this speed also occurs at the tangent point on the power curve. Therefore, the best range speed is usually published by the manufacturer—indirectly. This speed can be used to obtain the best cruise range as well, but we must remember to correct for weight using equation 8.7:

$$\text{New } V_{Range} = V_{Range}\sqrt{(GW/MGW)} \quad (8.7).$$

When using this procedure instead of the airspeed change method, it is still necessary to select an appropriate altitude, compensate for wind, lean the mixture, etc.

Endurance

Flight for maximum endurance allows the aircraft to achieve the maximum time aloft. Distance is no longer an issue while flying for endurance, so reducing the fuel flow is the primary concern and speed is unimportant. Endurance could be used when waiting for a runway to be cleared, holding due to traffic, or for other time related objectives.

Factors Affecting Endurance

In Chapter 8, we discovered that flight for maximum endurance occurs at the lowest point on the power curve. We saw that this happens because fuel flow is proportional to the power setting, thus we burn less fuel at lower power. We also saw that numerous factors can influence where the lowest point on the power curve is —in terms of both speed and power required. These factors are similar to the ones for range, but there are some distinct differences. To review and sum up:

Altitude – Unlike range, endurance is directly affected by the selected altitude. Recall that power required is proportional to the TAS and the drag. However, the drag is determined by the CAS. We also know from Chapter 2 that at higher altitudes, a given CAS results in a higher TAS. The effect of this is to increase the minimum power required at higher altitudes. The higher power required will reduce the endurance, therefore lower altitudes are better for achieving maximum endurance.

Weight and C of G – The effects of weight and C of G are similar to the effects they have on range. Added weight will decrease endurance since with more lift, the aircraft will produce more induced drag—this increases the power required for a given airspeed. As for the C of G, a forward C of G increases the required tail down force and thus increases the lift required to maintain equilibrium. This will also increase the induced drag and therefore decrease endurance.

Drag – Increased drag will increase the power required. This will have the effect of decreasing the endurance. Once again, the drag will be directly affected by aircraft configuration (flaps, speed brakes, spoilers, etc.) and condition (new paint and wax vs. a dirty airplane).

Mixture – The effect of the mixture will be the same as it's affect on range. A lean mixture is preferred for maximum endurance.

Wind – Unlike range, wind will have no effect on endurance. Distance travelled is not a consideration for maximum endurance, so the effect that wind has on groundspeed is not important.

Turbulence – Turbulence causes airspeed fluctuations. These variations will prevent the aircraft from maintaining the most efficient airspeed and will therefore decrease the endurance of the aircraft. It may be a good idea to climb out of turbulence (if possible) to improve the endurance. Increased altitude may be unfavorable, but in some cases it may be worth getting out of the turbulence.

Performance Charts

Some aircraft manufacturers will include endurance performance charts in the POH or Flight Manual for an aircraft. These charts will generally relate the endurance of the aircraft to certain power settings—not necessarily the setting for optimum endurance. Invariably, these charts will show an improvement in endurance as the power is decreased.

Another method for obtaining endurance information from the POH/Flight Manual is to check the fuel flow rates vs. the power settings. This information is sometimes included independently, and sometimes is included as part of other performance charts.

Experimental Procedure

Like range, it is usually necessary to determine the speed for optimum endurance experimentally. The objective here is to find the speed at which the lowest possible power setting will allow the aircraft to maintain altitude. Similar to range, the procedure will involve a gradual and systematic reduction in power to find the low point on the power curve (Fig 17-6).

As with range, we can consider the factors affecting endurance, as well as the shape of the power curve, to establish the aircraft in flight for

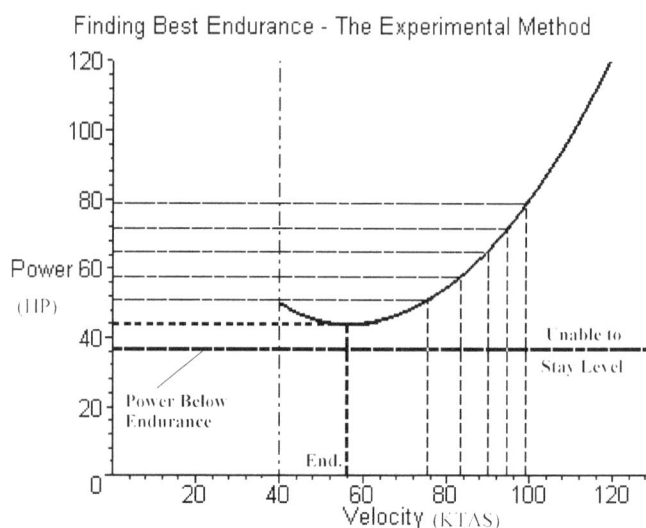

Fig 17-6. - *Finding Maximum Endurance* - A systematic reduction in power until altitude can just be maintained will bring the aircraft to flight for maximum endurance.

best endurance:

1) Insure the aircraft is in the cleanest possible configuration.

2) Select an appropriate altitude.
– In smooth air, the lower the better (within reason). In turbulence, it may be worthwhile to climb to smooth air aloft.

3) Shift weight rearward as practical.
– Like range, an aft C of G is most favorable. Once again, however, the C of G must be kept within certified limits and the pilot should always be able to reach the controls.

4) Reduce power in small increments.
– A gradual reduction in the power setting (perhaps 100 RPM at a time) will allow the aircraft to reach equilibrium at each power setting. When a power setting is reached which does not allow the aircraft to maintain altitude, endurance has been passed.

5) Establish endurance.
– After passing endurance, the aircraft must be returned to the appropriate airspeed. Notice from the shape of the power curve that simply returning the power to the endurance setting will not be sufficient. Having slowed below endurance, the endurance power setting will now place the aircraft below the power required curve (Fig 17-7). This means we should first add more power than necessary, accelerating the aircraft to a speed slightly above endurance. From here, we can reset the endurance power setting and allow the aircraft to settle on the endurance airspeed.

6) Compensate for turbulence.
– We know that turbulence will cause airspeed fluctuations. We can also see from Fig 17-7 that if the airspeed drops below endurance the power setting will not be sufficient to maintain altitude and airspeed. This speed range is called Slow Flight and is detailed in the next chapter. For now, suffice it to say that we want to avoid this range of airspeeds. Because of this, if setting up for endurance in turbulence cannot be avoided, increasing the airspeed slightly above the actual endurance speed is favorable. This reduces endurance somewhat due to the higher power

setting, but allowing the airspeed to drop into the slow flight range will also create the need for a higher power setting—either to maintain equilibrium or to recover into the normal flight range.

7) Lean the mixture.
– As with range, we want to reduce the fuel flow as much as possible. Once again, however, I will leave the details of the leaning procedure to engine manuals and other texts.

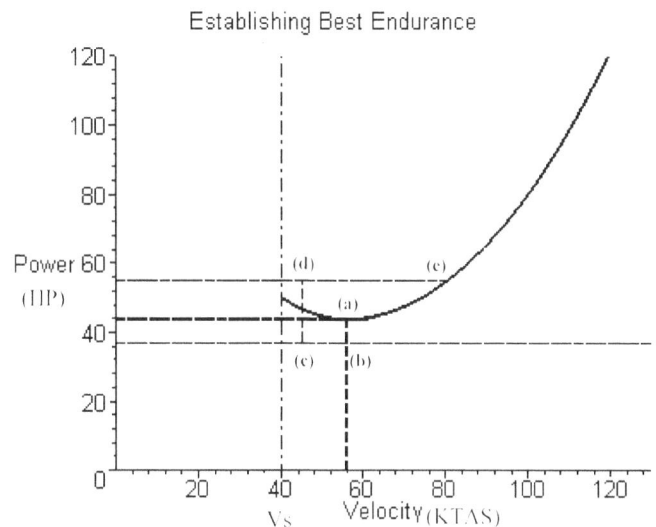

Fig 17-7. - *After Finding Endurance* - Being slower than endurance, extra power must be added to reestablish the endurance airspeed. This is because of the increase in power required at speeds below maximum endurance.
    Consider the following process: following a systematic power reduction, we find ourselves at point (a), but don't know yet that that is maximum endurance. So we continue the power reduction to point (b) and the aircraft decelerates to point (c). We recognize that the aircraft can't maintain altitude in this condition, so we increase power to point (d) (notice that increasing power back to the endurance level is insufficient at his point). From (d), we accelerate to point (e) and then reduce power to the endurance setting that we've now determined. The aircraft settles in at point (a) again and we are established in flight for maximum endurance.

Chapter Summary

The range of an aircraft is the distance that it can fly under a given set of conditions. Maximum range is achieved by setting the aircraft up so that it will fly the greatest possible distance per unit of fuel. On the power curve, the speed appropriate to maximum range is the speed determined by a line drawn from the origin that is tangent to the curve. Numerous factors will affect both the speed for maximum range and the actual distance which can be flown. These include the aircraft weight and C of G, aircraft condition and configuration, the wind and turbulence, the altitude (or more correctly, the fuel burned in the climb to altitude), and the mixture setting.

Range information can often be obtained from performance charts in the POH or Flight Manual for the aircraft. Flight for maximum range can also be determined experimentally. The experimental method involves reducing the power setting in equal increments and observing the changes in airspeed. The high speed side of the largest airspeed change is approximately best range.

The endurance of an aircraft is the amount of time that it can remain airborne under a given set of conditions. Flight for maximum endurance is accomplished by setting the aircraft up to remain airborne for the greatest period of time per unit of fuel. Since the rate of fuel consumption is proportional to the power setting, the lowest possible power setting is needed for maximum endurance. A number of factors will influence the endurance of an aircraft as well as the speed at which maximum endurance occurs. These factors include the altitude, the weight and C of G, the aircraft condition and configuration, turbulence, and the mixture setting.

Like range information, endurance information can often be obtained from the POH or Flight manual. Also like range, flight for maximum endurance can be established experimentally. During the experimental method, the power is reduced in increments. The lowest power setting that allows the aircraft to maintain level flight in equilibrium is the power setting for maximum endurance.

Questions and Problems

1) Ignoring fuel burned in the climb, how does altitude affect range? Why?

2) How do we correct for a strong headwind when flying for maximum range? Does this correction *completely* compensate for the headwind effect?

3) Where does maximum endurance occur on the power curve? Why does it occur at this point?

4) How does wind affect endurance? Altitude? Why?

5) Why is turbulence a problem while flying for maximum endurance?

6) For a given weight, how does the C of G affect maximum endurance? Range? Why?

7) For the following power/airspeed combinations for a given altitude and configuration, where is flight for maximum range located? Maximum endurance?

Power Setting (RPM)		Airspeed (KIAS)
2600	→	121
2500	→	116
2400	→	111
2300	→	105
2200	→	98
2100	→	91
2000	→	84
1900	→	72
1800	→	Unable to maintain altitude (i.e. – stall or descent)

8) If the A/C in Q7 has a 20 kt headwind, where would we find maximum range? Maximum Endurance? What if the wind was a tailwind?

Chapter 18
Slow Flight

At a few points in this book, we have alluded to the idea of "Slow Flight". We have not, however, discussed any of the details of this speed range. Slow flight—also known as the Region of Reversed Command, or being "Behind the Power Curve"—is defined as the range of speeds below maximum endurance and above the stall. This is singled out as a special speed range for two reasons:

1) Contrary to intuition, the power required will actually begin to increase as the speed is decreased. We've seen in previous chapters why this happens—the induced power required becomes prevalent at low airspeeds. We will also see later in this chapter that this trend leads to an unstable speed which can be difficult to control effectively—a slight decrease in the airspeed can lead to a continuing deceleration.

2) Recognition of the slow flight speed range is one of the symptoms used to recognize the approach of an unwanted stall. As well, due to the unstable speed that is characteristic to slow flight, inadvertent entry into slow flight can ultimately lead to a stall.

Slow Flight is a speed range that we generally avoid. However, we will see in Chapters 22 and 23 that there are exceptions to this rule for specialty takeoffs and landings.

As was stated above, slow flight is the range of speeds between the stall and the speed for maximum endurance. This range is illustrated in Fig 18-1. Notice here that slow flight is not a specific airspeed, but a *range* of airspeeds. We can see from the diagram that once the aircraft enters the slow flight range, the power required will increase with a decrease in airspeed—that is, at lower airspeeds the power will have to be increased to maintain a constant altitude/vertical speed.

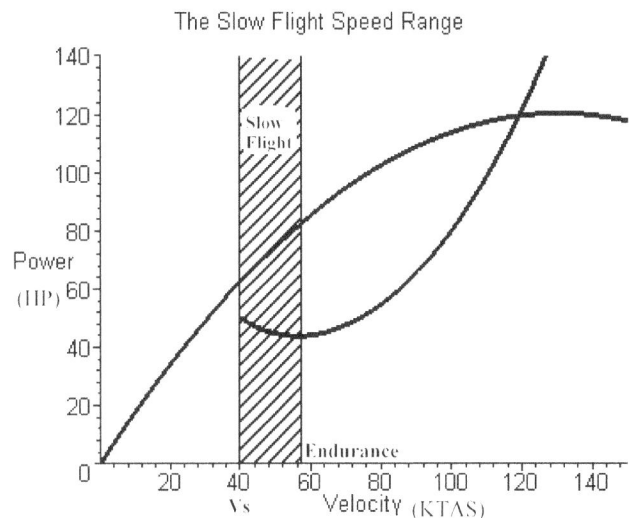

Fig 18-1. - The Slow Flight Speed Range.

Entry

The slow flight entries we will look at here are the *intentional* entries used for flight training purposes. Inadvertent entries into slow flight will be considered later in the chapter in the section on 'Critical Areas'. There are two entry procedures to slow flight that are used commonly. The first involves establishing the aircraft in flight for endurance and then slowing down further. The second method is to set the power below endurance and allow the airspeed to decrease into the slow flight range.

The first method is the most appropriate for pilots who are still in initial flight training and for pilots receiving a checkout on a new type of aircraft. For this technique, the aircraft is set up in flight for endurance using the power reduction method discussed in the previous chapter. Once the aircraft is established at endurance, any reduction in airspeed will place it in the slow flight range. This reduction in airspeed can be accomplished simply by pitching up slightly. The pitch up will result in a *slight* climb initially (if the pitch attitude is adjusted slowly enough, this climb will be negligible). However, any altitude gained can be lost again due to the fact that the aircraft now has a power deficiency (Fig 18-2) and will therefore descend. Once the aircraft is reestablished at the original altitude (or immediately if the climb was negligible) some power must be added to maintain the altitude—otherwise the descent will continue. At this point, the aircraft is established in straight and level slow flight.

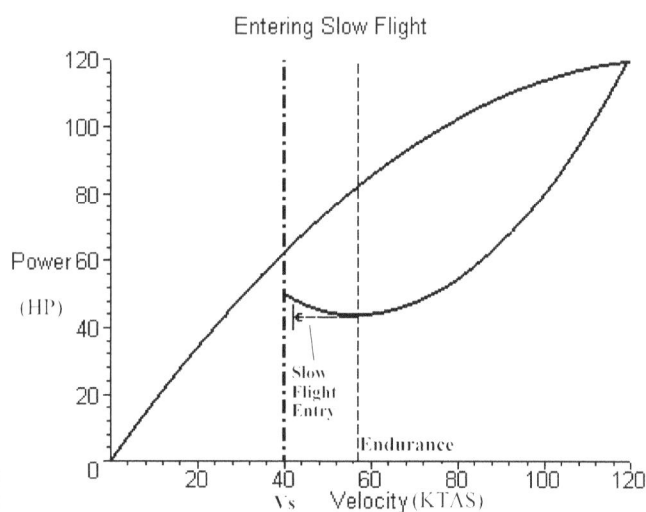

Fig 18-2. - Entering Slow Flight From Endurance - A power deficiency results from the decrease in airspeed.

For pilots familiar with the aircraft type they are flying, the second method works equally well. This method requires that the pilot have a good knowledge of the power settings and corresponding airspeeds for the aircraft type—especially around endurance. Simply reducing the power to some level below the endurance setting will allow the aircraft to decelerate to a speed below endurance. Since the endurance power setting varies with weight and altitude, a setting *well* below endurance must be selected to ensure the aircraft enters the slow flight speed range. Once the airspeed is below endurance (endurance speed also varies with weight, so a speed *well* below endurance must be selected), the power can be adjusted as necessary to maintain altitude. This new power setting will be higher than the endurance setting and the aircraft will now be in slow flight.

Recognition

As was stated earlier, slow flight is used to recognize the approach to a stall and slow flight can actually lead us into an inadvertent stall. For this reason, the ability to recognize slow flight is critical in the operation of an aircraft. This recognition will be based on several symptoms which are characteristic of the slow flight speed range.

First and foremost, slow flight can be recognized by being aware of where the aircraft is on the power curve. In order to do this, we need to be aware of various airspeeds appropriate to the aircraft type (in particular, the stall speed and the endurance speed). As well, we need to be aware of aircraft behavior trends including the airspeed and the power required and/or vertical speed. As we know, the power required increases with a decrease in airspeed once we are in the slow flight range. This, however, doesn't necessarily mean that the power *setting* will be higher since the power setting is controlled by the throttle (and by the prop control in aircraft with a variable pitch prop(s)). Instead, the rate of descent might be higher—if the power setting isn't changed and the power required is increased, the descent rate will increase due to a higher power deficiency. Having said this, if altitude is being maintained, the power setting will have to be increased for a lower airspeed.

Associated with the aircraft's position on the power curve, the airspeed will become unstable—meaning that a slight deviation in speed will lead to larger deviations. This is contrary to flight in the "normal" speed range, which tends to be self correcting after a deviation. The reason for this is illustrated in Fig 18-3. Assuming straight and level flight, consider an aircraft in the "normal" flight speed range (point A). If something were to occur (e.g. – a wind gust or a temporary misadjustment of pitch) that increased the airspeed of the aircraft, we would find ourselves at point B. Since the power setting hasn't been adjusted, we now have a power deficiency. This power deficiency, as we know, will either cause the aircraft to descend or to decelerate. If we maintain a constant altitude, the airspeed will decrease back to it's original value (point A again). Likewise, if the airspeed were to be decreased, we would find ourselves at point C. Without adjusting the power, an excess of power is now being produced and the aircraft will accelerate back to point A.

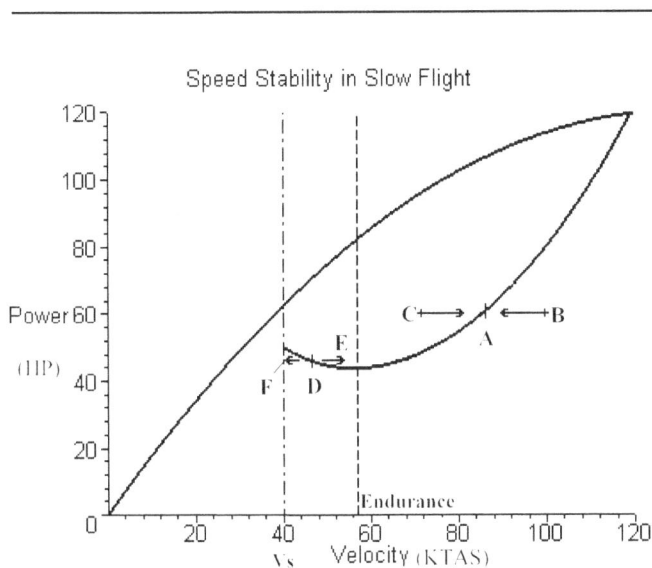

Fig 18-3. - *Speed Stability in Slow Flight* - Changing airspeed in Slow Flight leads to a power condition that tends to aggravate the speed change. In the normal flight speed range, speed changes tend to be self correcting.

Now consider an aircraft in slow flight (point D). If the airspeed was to be increased, we will end up at point E. As we can see from the diagram, this speed and power combination will result in an excess of power—leading to further acceleration (or a climb). As well, a decrease in speed will bring us to point F which results in a power deficiency. This, of course will lead to a further deceleration (or a descent). So as we can see, controlling the airspeed in slow flight will take a very fine touch.

Another flight characteristic that goes hand-in-hand with slow flight is an increase in adverse yaw. Even aircraft that have differential or frise ailerons and thus don't normally have adverse yaw will experience the need for a rudder input with an aileron input. This is a result of the low airspeed. Recall from Chapter 5 that adverse yaw results from an increase in induced drag on the up going wing with a corresponding decrease in induced drag on the down going wing. At lower airspeeds, induced drag is more predominant, therefore this effect becomes more pronounced. As well, parasite drag becomes less predominant at lower airspeeds. Differential and frise ailerons are based on the idea of balancing the induced drag of the up going wing with the parasite drag of the down going wing, so even aircraft with these corrections will experience adverse yaw at low airspeeds.

Along with the increase in adverse yaw, aileron control will be "sloppy" as compared to

aileron control at higher airspeeds. This is a result of the reduced airflow over the control surfaces causing a reduced effectiveness. Elevator and rudder control won't usually be affected to a noticeable degree due to the induced airflow from the propeller.

Last on our list of symptoms (and probably least as well) is the pitch attitude of the aircraft. A nose high attitude *can* be indicative of slow flight—or at least the possibility of slow flight. Be careful, however, not to rely on the pitch attitude as the ultimate indicator of slow flight. The aircraft can be in a nose high attitude and not be in slow flight (we've seen this already in climbs), or the aircraft can be in a nose *low* attitude and be in slow flight. The pitch attitude is best considered to be an indicator of the "odds" of being in slow flight. That is, in a nose high attitude slow flight is *more likely*, while in a nose low attitude slow flight is *less likely*.

So to sum up the recognition of slow flight, we have five things to consider. In the order of importance (importance being determined by the relevance to the slow flight definition and the ease of recognition):

1) Airspeed vs. Power/Vertical Speed

2) Speed Stability

3) Decrease in Control Effectiveness

4) Increase in Adverse Yaw

5) Pitch Attitude.

Maneuvering

There are some distinct differences between controlling the aircraft in slow flight and controlling the aircraft in "normal" flight. These differences can be subtle in some aircraft, but they are very important to prevent an eventual loss of control. The changes result from phenomenon already discussed, namely: airspeed instability, high induced power required, and increased adverse yaw.

Straight and Level – As with flight in the "normal" speed range, straight and level in slow flight means maintaining a constant altitude and direction. The direction is maintained with the rudder while coordination is maintained with aileron (nothing new here yet). Airspeed and altitude are controlled with the pitch and power combination.

The first thing to consider in slow flight is the increase in pressure on the right rudder pedal. Recall from Chapter 4 that the propeller tends to yaw the aircraft to the left. This tendency is increased at lower airspeeds and higher power settings. For this reason, a greater correction (right rudder) will be necessary in slow flight. So in regards to directional control, nothing has really changed in slow flight, however the corrections do become larger.

As for the control of airspeed and altitude, things change somewhat. In the "normal" flight speed range, using pitch to maintain altitude and power to control the airspeed works quite well. This is because when pitch is used to correct small altitude deviations, the resulting speed fluctuations are self correcting (as we discussed earlier in this chapter). In slow flight, however, the airspeed is unstable—thus correcting altitude deviations with pitch will lead to unacceptable speed fluctuations. Because of this, it is easier in slow flight to choose the airspeed as the "critical constant" to be controlled by pitch and use power to control the altitude/vertical speed.

Climbs and Descents – Controlling a descent in slow flight is no different from a descent in the normal speed range. As well, for an aircraft established in a climb nothing changes. There is a difference, however, for the transitions into and out of climbs. This results from the airspeed instability discussed above as well as the fact that the aircraft is close to the stall.

Previously, the procedure APT—Attitude, Power, Trim—was used to establish and recover from a climb. In slow flight this method is inappropriate since it will lead to excessive speed fluctuations. On a climb entry, pitching the aircraft up will cause a decrease in airspeed. If this is done in slow flight, a stall may result. If instead power is added while the pitch attitude is used to maintain the airspeed, the aircraft will climb without the risk of a stall. For the level off from the climb, pitching down will cause a rapid increase in airspeed. During inadvertent slow flight this is exactly what we want, however, in the training environment it is often intended to remain in slow flight while leveling off from a climb. To do this, the power must be reduced to stop the climb while pitch is used once again to control the airspeed. So for climbs in slow

flight, we will use PAT—Power, Attitude, Trim.

Turns – Turning in slow flight requires more coordination than turning in the normal speed range. The first problem is the increase in adverse yaw. As we discussed earlier, this is caused by the low airspeed and the associated increase in induced drag with a decrease in parasite drag. Because of the adverse yaw, it will be necessary to use more rudder than usual while rolling the aircraft. As long as the ailerons are deflected to roll the aircraft in one direction, the aileron drag will cause a yaw in the opposite direction. To compensate, rudder in the direction of the roll will be necessary while rolling into and out of a turn. For example, when rolling into a left turn we will need left aileron to roll into the turn and left rudder to compensate for the aileron drag. When rolling out of a left turn, we will need right aileron to roll the wings level, and right rudder to compensate for the aileron drag.

While established in the turn, more rudder than usual will also be required. This rudder application will be in the direction of the turn and is caused by the difference in speed (and thus the drag) between the two wings. This phenomenon was discussed in Chapter 16, and it was mentioned that it is more pronounced in slower aircraft with longer wingspans. Of course, the wingspan won't change, but in slow flight the airspeed is reduced. This will increase the presence of the wing drag difference.

In addition to the increased need for rudder, an increase in power will be needed even in gentle and medium turns. This is something we've seen before in steep turns. In slow flight the requirement for more power appears at even the shallower angles of bank because of the increased effect of induced drag. The actual amount of extra power required will vary—it will increase as the angle of bank increases.

So for a turn in slow flight, more rudder will be needed than usual and an addition of power will be required. Coordinated rudder will always need to be included with an aileron deflection and rudder will usually be needed to maintain coordination in the turn. Keep in mind, however, that these effects can be complicated by the presence of the propeller. The left turning tendencies may tend to aggravate or mask the need for rudder in slow flight. This is especially true since turns in slow flight usually include changes in power setting.

Recovery

Since slow flight is such a potentially dangerous speed range—particularly when entered inadvertently—recognition and recovery are extremely important. The recognition has already been discussed in detail, so this section of the chapter will concentrate on the recovery. The recovery procedures detailed herein are intended to recover the aircraft from slow flight with *minimum loss of altitude*. This is important since the critical areas discussed in the next section of the chapter are all close to the ground. There are in fact several ways to recover from slow flight—anything that allows the airspeed to increase will work—buy they sometimes involve a significant loss of altitude which is unacceptable.

By referring to Fig 18-4a, we can see that even in the slowest range of slow flight, there is more power available than power required. Keep in mind that power available is the power available with the engine producing full power—it is not necessarily the power being *produced* at any given moment. Fig 18-4a is a reasonable approximation for the power trends in most aircraft at lower altitudes. Since this is the case, it is possible to recover the aircraft from slow flight simply by adding power and accelerating back into the normal flight speed range.

In fact, applying full power is the first step in a recovery from slow flight. Once the power is applied, the altitude should be held by pitching forward—but care should be taken not to descend, it is unnecessary with the excess power. Once established at a safe airspeed, the aircraft should be cleaned up if necessary (retract the flaps and landing gear, etc.). This will allow the aircraft to accelerate better and then to climb. Climbing after a slow flight recovery is generally considered to be the best option (as opposed to leveling off or continuing a descent which may have been a poorly flown approach). This, once again, is because of the low altitude associated with inadvertent slow flight entries.

There are times when Fig 18-4a will not be an accurate representation of the power curve for an aircraft. Instead, the curve illustrated in Fig 18-4b would be more appropriate. Notice that in the slow flight range the power required can be higher than the power available. This is more likely at higher altitudes since the engines ability to produce power decreases as the air density decreases. As

well, an aircraft in a dirty configuration (e.g. – gear and flaps extended) will be more prone to this condition. However, in some aircraft, this can even occur at lower altitudes and in a clean configuration.

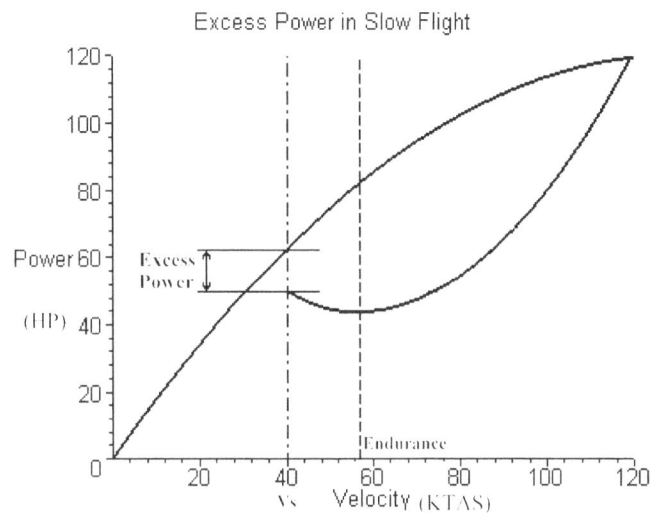

(a) Most aircraft are capable of producing excess power at the slowest of Slow Flight speeds. This means that a recovery can be completed without any altitude loss.

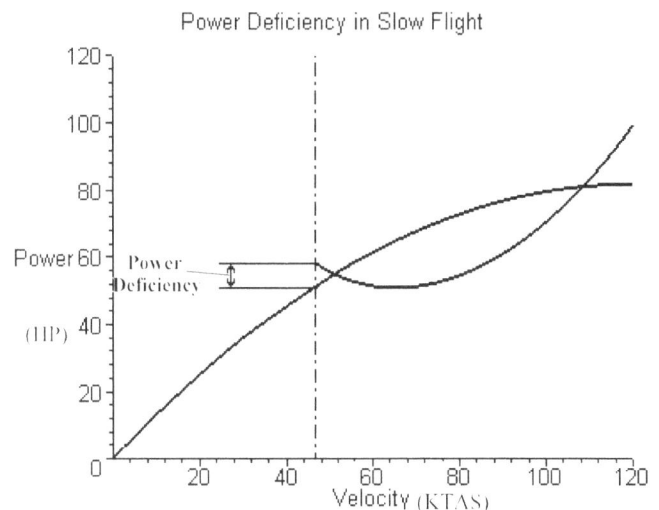

(b) Some aircraft cannot produce excess power at the lower end of Slow Flight. This is more likely at higher density altitudes and means that a recovery cannot be completed without a loss of altitude.

Fig 18-4. - *The Power Curve for Aircraft in Slow Flight.*

When the aircraft is not capable of producing the power required, a descent is unavoidable for the recovery. As with the previous recovery, full power should still be applied to minimize the loss of altitude. The aircraft should then be pitched forward to get the airspeed above point A on the curve—this allows the power available to exceed the power required. The descent should be stopped with pitch at this point and the acceleration should be continued with the excess power available. As before, the aircraft should be cleaned up as soon as safely able and a climb should be initiated.

Critical Areas

At the beginning of the chapter, it was stated that slow flight is a speed range which we generally avoid. For this reason, we need to know what the "Critical Areas" are for this speed range. In other words, what situations commonly lead to inadvertent slow flight entry?

Approach and Landing – The most likely place to see inadvertent slow flight is during the approach and landing. The reason will usually relate to improper use of pitch and power. If the aircraft becomes high or low during an approach, power should be adjusted to regain the optimum approach path (this has already been mentioned in Chapter 14 and will be discussed again in Chapter 23). Unfortunately, it is the tendency of some pilots to use pitch to regain the approach path. Consider what will happen to a pilot who gets low during an approach and uses this technique to make a correction.

To reduce the descent using pitch, one would pitch the aircraft up. This unfortunately would also reduce the airspeed. During a normal approach, the aircraft will often be established somewhere near the best glide speed to begin with, so this reduction in airspeed will bring us back towards endurance. With the speed being reduced below best glide, the glidepath will eventually stabilize at a steeper angle than we began with (Fig 18-5). This means that the aircraft will end up lower than desired even after the pitch correction, so to correct again the aircraft will be pitched up further. We can see where this is going—the speed will continually be reduced until we end up in the slow flight speed range.

Some would argue that this scenario won't happen because while our unfortunate pilot is pitching up, he/she will also add power to maintain airspeed. In theory, this argument works well, but in practice it falls apart quite easily. If we begin adding variables such as heavy traffic, less than ideal weather, a busy radio frequency, *etc.*, the distractions can cause big problems. When the stress level begins to increase, it is easy to fixate. This means using pitch to try regaining the proper

glidepath could very well be the only concern of the pilot. Another consideration is the possibility that the aircraft could decelerate into the slow flight range as the power is being applied. In this case, we still end up in slow flight, but with a higher power setting.

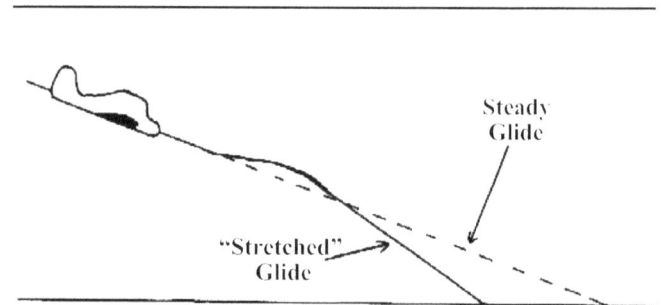

Fig 18-5. - *Controlling the Glide Path with Pitch* - If an approach is too low and a correction is made with pitch, the result will be a steeper approach and further correction. This can lead to a Slow Flight entry.

Takeoff and Climb – The takeoff and climb can also cause problems due to improper pitch control. The first problem area is the initial rotation to lift-off. If this is done too early in the takeoff roll, the aircraft will come off the ground in slow flight. As well, if the pitch adjustment is too abrupt or large, the speed may drop as the aircraft climbs out of ground effect. This could also lead to slow flight.

As for the initial climb, two problems can occur. The first occurs as the aircraft climbs out of ground effect. Recall from Chapter 5 that the tail surface becomes more effective outside ground effect due to the increased downwash. This must be anticipated in the initial climb since if no correction is made, the aircraft will pitch up more than intended. This could lead to the aircraft entering slow flight. Aside from this, a problem can occur during obstacle clearance climbs. If V_x is not adhered to, the speed could drop too low (especially since the most likely problem with speed control is pitching up—a "natural" tendency when trying to clear obstacles). It is possible for V_x itself to be in the slow flight speed range, however, it is undesirable to go any further into the range than necessary—especially since slow flight can lead to a stall.

Overshoot – An overshoot (sometimes referred to as a balked approach or a missed approach) is basically a transition from an approach descent to a departure climb without leveling off or touching down. We can see quite easily how this can cause problems, we are combining the problems associated with the approach and the problems associated with the initial climb into a single maneuver. The key to avoiding inadvertent slow flight in an overshoot is to apply power before pitching up to the climb attitude—thus avoiding the problems caused by a pitch induced speed reduction—and preventing the pitch attitude from becoming too extreme—thus avoiding the problems caused by an excessively steep climb attitude.

Low Level Operations – Any time an aircraft is operated in the vicinity of the ground, slow flight is a potential hazard. Low level operations include aerial photography, forced approaches, precautionary approaches, low level navigation, or any other flight scenario that includes being close to the ground. During these operations, airspeed control is extremely important. In high winds, be conscious of the visual illusions caused by the wind drift. These illusions are not detailed in this text, but they can lead to poor airspeed and coordination control and the associated dangers.

Chapter Summary

Slow flight is the range of airspeeds between best endurance and the stall. This speed range is singled out as important due to the increase in power required as speed decreases and due to the fact that it can be used to recognize the approach to a stall. Intentional slow flight entries are usually done by establishing flight for maximum endurance and then pitching up.

A number of symptoms can be used in conjunction with one another to recognize slow flight. They include a comparison between airspeed and power/vertical speed, speed stability, control effectiveness, increased adverse yaw, and pitch attitude.

Maneuvers in slow flight can be somewhat different from regular maneuvers. This is due primarily to the increased induced drag which is characteristic of slow flight.

The recovery from slow flight is accomplished with the use of full power. Once the power is applied, the pitch attitude should be adjusted to maintain a constant attitude. This will allow the aircraft to accelerate out of the slow flight range and reestablish normal cruise prior to entering a climb. At higher altitudes where engine power is significantly reduced, it may be necessary to descend in order to recover from slow flight.

The critical areas for slow flight are the types of operation which may lead to inadvertent entry into the slow flight speed range. They include takeoff and initial climb, approach and landing, overshoots, and low level operations. During these maneuvers, pilots should be wary of unplanned entry into the slow flight speed range.

Questions

1) What is slow flight? Why is it considered to be an important phenomenon?

2) When slow flight is conducted intentionally, how is the entry normally completed?

3) What are the differences between turns in normal cruise and turns in slow flight? Why do these differences occur?

4) How can slow flight be recognized?

5) In normal cruise, pitch is usually used to maintain altitude. Why isn't this the case in slow flight?

6) How do we recover from slow flight? Why is it important to lose as little altitude as possible in the recovery?

7) What are the critical areas for inadvertent entry into slow flight?

8) Are there times when altitude must be given up to recover from slow flight? If so, why does this occur?

9) Entering and recovering from climbs in slow flight is done differently than in the normal flight speed range. Why is this the case?

10) Does the speed range for slow flight remain the same from day to day and from flight to flight? Why/Why not?

Chapter 19
Stalls and Spins

Stalls and spins are both undesirable phenomenon. They have no practical application whatsoever outside airshow flying. The reason they are practiced in flight training is strictly for reasons of recognition, recovery, and avoidance. We've already seen the theory behind stalls and spins in Chapter 6, now we are going to look at stall and spin training in the aircraft.

In this chapter, we will look at entries, recognition, and recoveries from stalls and spins. It is important for the reader to note that recognition and recovery can vary significantly from one aircraft type to another—and this is especially true for spins. The information in this chapter is based on the "standard" stall and spin information and it is generally correct for most general aviation aircraft. However, care should be taken to obtain information relevant to the aircraft type from the POH or Flight Manual.

Stalls

As we've seen from Chapter 6, when an aircraft stalls it will lose lift as the AOA increases. This loss of lift will cause a corresponding loss of altitude which cannot be stopped without recovering from the stall. This is a condition of flight which we wish to avoid because of the uncontrollable descent. If we stall inadvertently, a recovery must be initiated as soon as possible to avoid excessive altitude loss. Stalls are especially problematic since they usually occur at low altitudes—as we will see in the section on Critical Areas.

Cause and Factors

Stalls are often attributed to various causes—low airspeed, high load factor, high weight, turbulence, etc.—however there is really only one cause of a stall. This cause is exceeding the critical Angle of Attack. All other causes are simply contributing factors to the excessive AOA. The AOA alone determines whether or not the aircraft is stalled.

In Chapter 6 we saw why the stall occurs. As the AOA is increased, the airflow over the top of the wing begins to separate (Fig 19-1) and stops contributing to lift. Eventually we reach an AOA at which the separation is so severe that lift will actually decrease with a further increase in AOA. At this point the aircraft is stalled and the various stall symptoms for the aircraft type will begin to appear. The most critical of these symptoms is a loss of altitude which will just get worse if the aircraft is pitched up further.

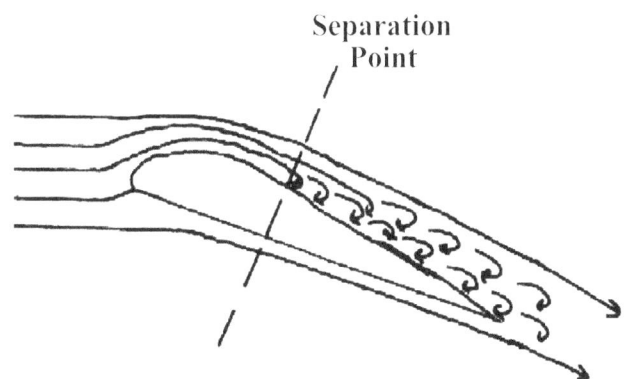

Fig 19-1. - *Airflow Separation* - This separation is caused by the adverse pressure gradient and leads to a decrease in lift.

In most general aviation aircraft, the stall condition is published as a speed instead of an AOA. One must remember, however, that this speed is based on a specified set of conditions. When these conditions are varied, the speed will vary as well. This is because the stalling AOA will remain fixed regardless of the conditions. The details of the speed variation with these conditions has been discussed in detail in Chapter 6. However, to sum up:

1) _Weight_ – The published stall speeds are often defined for the maximum gross weight of the aircraft. A decrease in weight will result in a decrease in stall speed. The reason for this is that the lift required has been decreased, so the critical AOA will occur at a lower airspeed. To quantify the change in the stall speed, we can use equation 6.1,

$$\text{Stall Speed} = V_s \sqrt{(GW/MGW)} \qquad (6.1).$$

2) _C of G_ – A shift in C of G will also change the stall speed because of a change in the lift required. As the C of G moves forward, the tail down force must increase to maintain the aircraft's equilibrium. This tail down force will add to the lift required to be produced by the wings. So the total lift is actually a combination of the weight of the aircraft and the tail down force. Unfortunately, the exact change in stall speed for a given shift in the C of G will vary from one aircraft type to another. Unless the stall speeds are published for different C of G positions, the exact change is unknown. One thing that is certain, however is that as the C of G moves forward the stall speed will be increased and as the C of G moves aft the stall speed will be decreased.

3) _Load Factor_ – As with weight, an increased load factor will increase the lift required and therefore increase the stall speed. Once again, this happens because at a given speed more lift means that a higher AOA is required. So the critical AOA is reached at a higher airspeed with a higher load factor. The change in stall speed with load factor can be quantified in a manner similar to the weight change. Using equation 6.2,

$$\text{Stall Speed} = V_s \sqrt{(n)} \qquad (6.2).$$

where n is the load factor in g's.

Load factors greater than 1 occur in turns, as we discussed in Chapter 8, as well as in pull-ups and turbulence. In turns, load factor increases as the angle of bank increases. Using equation 8.10,

$$n = 1/CosAOB \qquad (8.10).$$

4) _Configuration_ – Changing the configuration of the wing will have a direct impact on the stall speed because it will change the lifting characteristics of the wing. For example, flaps will increase the lifting capacity of the wings because they increase the camber. With an increase in lifting capacity without an increase in lift required (i.e. weight or load factor), the stall speed will be reduced.

5) _Power_ – When the engine is producing power, the thrust will have a component perpendicular to the relative airflow at high AOA (Fig 19-2). This component of thrust contributes to lift and therefore decreases the lift required from the wings. As well, the slipstream from the prop creates an induced airflow over the wing root which decreases the AOA and increases the effective airspeed at the root. The net effect of added power is to reduce the stall speed.

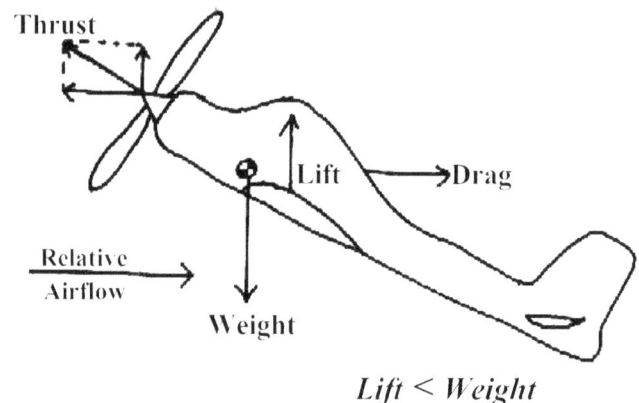

Fig 19-2. - Thrust at a High AOA - A component of thrust acts perpendicular to the relative airflow. This component of thrust contributed to lift.

6) _Ground Effect_ – When an aircraft flies in ground effect, the downwash is reduced and the lift created by the wings acts straight up instead of on a rearward incline (Fig 19-3). As a result of this, more lift is available when flying within a spanlength of the ground. With more lift available and the lift required unchanged, the stall speed of the aircraft will be reduced. This is an important consideration while in the takeoff or landing phase of flight.

7) _Surface Contamination_ – Surface contamination includes such things as mud, slush, and ice. These substances will adhere to various parts of the aircraft—of particular importance here is the wings—and disrupt the airflow patterns that the aircraft was designed to have. Clearly, this will

cause significant problems with drag, but it will also reduce lift due to premature airflow separation. The stall speed of the aircraft will almost invariably be increased by surface contamination. The speed is changed because a) the stall will now occur at a lower AOA, and b) for a given AOA, the lift coefficient will be lower.

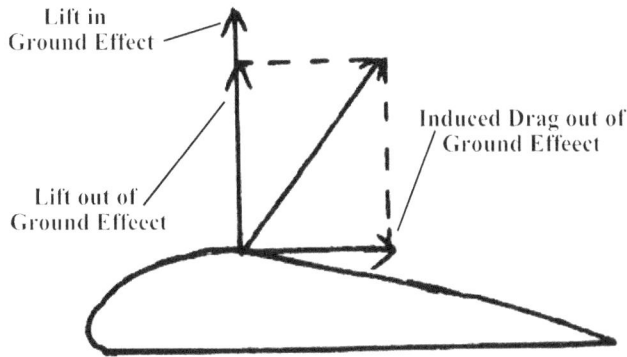

Fig 19-3. - *Ground Effect* - The reduction in downwash increases the lift available, thereby decreasing the stall speed.

Entries and Recognition

There are many different intentional stall entries as well as inadvertent stall entries. In flight training, instructors usually strive to make the intentional entries simulate (as closely as possible) some typical inadvertent entries. Unfortunately, the only stall factors that we can vary once airborne are the power, configuration, and load factor. For this reason, intentional stall entries are usually based on these variables. Generally, stalls are divided up into two main groups, approach stalls and departure stalls. This division is based on the configuration used, whether it is a takeoff or a landing configuration.

The simplest stall entry—and normally the first one introduced to new students—is a stall in clean configuration with the wings level and the power set to idle. The entry into this type of stall is straightforward. Simply reduce the power to idle and maintain the altitude with pitch. This means that as the aircraft slows down, the pitch attitude must be continuously increased until the stall is reached. We must be sure to adapt the control input techniques to the proper slow flight inputs as the aircraft decelerates through the slow flight range. This stall procedure is a reasonable approximation of an approach stall—since a stall on approach generally occurs with the power set at idle or some other low setting.

Other stall entries are simply variations on this basic one. Departure stalls are done with either takeoff or climb power (this often means full power in training aircraft) to simulate conditions during a departure, while approach stalls are done with partial power or no power to simulate the conditions during an approach. These stalls can also be completed in a variety of configurations to simulate real life conditions even more closely. Climbs and descents can be introduced to vary the stalling attitude (in reality, most inadvertent stalls occur during a climb or a descent) and turns can be introduced to vary the load factor.

The stall characteristics of a given aircraft can vary significantly in different configurations. A loss of altitude will always accompany a stall. However, there are other symptoms which become apparent before the loss of altitude (at low altitudes the loss of altitude does become more apparent due to an increase in nearby visual cues). These symptoms include a buffeting of the airframe due to the separated airflow and a nose drop due to changing moments on the wing. The recognition of a full stall is usually based on these symptoms, while the recognition to an approaching stall is based on the symptoms of slow flight as well as the initial stall buffeting. Most aircraft are equipped with some sort of stall warning which will give a visual and/or an audible warning of an impending stall. These devices are an asset, however they can fail and should not be relied on as the sole indicator of a stall.

Some aircraft are less stable than others in a stall. These aircraft will be more prone to rolling, or "dropping a wing" in a stall. As well, a given aircraft may be more prone to instability in different configurations. The use of power is destabilizing in a stall and can often lead to a "wing drop". As well, climbing and descending turns tend to lead to a wing drop—the reason for this was discussed in Chapter 6. The initial roll and/or yaw caused by this instability can potentially lead to a spin.

Recovery

Once a stall has been recognized, recovery is imperative. This recovery can be initiated from the approach to a stall or from the actual stall. Either way, the recovery works in the same way.

The main difference is in the altitude lost—the sooner a stall is recognized, the less altitude will be lost in the stall and subsequent recovery.

The key to a stall recovery is lowering the AOA. This goes against all intuition for new pilots since one would think "pull up to stop a descent". In a stall, pulling up will simply increase the AOA and thereby deepen the stall and increase the descent rate.

Step one in a stall recovery is always to pitch down to lower the AOA and break the stall. The stall itself, as well as pitching down, will lead to an unavoidable loss of altitude. This altitude loss can be minimized by adding full power as the AOA is reduced. Keep in mind that the power is *not* what is breaking the stall—the reduction in AOA does this. Power is simply used to minimize the loss of height, and thus minimize the chance of an unpleasant encounter with the ground.

When the aircraft is pitched down, the pitch input should only be enough to break the stall (the exact amount will vary from aircraft to aircraft and even from stall to stall). Any more than this will increase the altitude lost. In some aircraft (aircraft with lots of power to spare) lowering the nose may not be necessary—full power while maintaining a constant pitch attitude may be sufficient. This does not, however, mean that the power is breaking the stall (not directly, anyway).

The power may be sufficient to change the flight path of the aircraft. This means that the relative airflow changes and therefore the AOA changes, possibly enough to break the stall. This procedure is more common in multi-engine aircraft since the slipstream from the prop will actually create induced airflow over a substantial portion of the wings, thereby reducing the AOA in these areas and assisting recovery.

While recovering from the stall, we also need to be careful not to aggravate the condition by deflecting the ailerons. We saw in Chapter 6 that using ailerons in a stall can initiate a spin by imbalancing the stall of the two wings. If the aircraft begins to roll in a stall, the rudder should be used to stop the roll. The tendency to stop the roll with ailerons must be anticipated and prevented—the only safe and effective way to stop a roll and prevent a spin is to use the secondary effect of yaw. This means that if the aircraft rolls in one direction, opposite rudder must be used to stop the roll and prevent a spin.

Secondary Stalls

In the recovery from a stall, the airspeed is going to be low. Because of this, high load factors are unacceptable in the post-stall dive recovery. A common problem is to try to pull out of the dive too abruptly. This will lead to a secondary stall, which is a stall during a stall recovery.

Secondary stalls will increase the altitude loss during a stall recovery, so they should be avoided at all times. In the dive recovery after a stall, be sure to *ease* out of the dive. This is extremely important since close to the ground it will be a natural tendency to try arresting the descent by pitching up further.

If a secondary stall is encountered, the recovery is exactly the same as the initial stall. The AOA must be reduced in order to recover and continue the pullout of the dive.

Critical Areas

The critical areas for stalls are the same as those for slow flight, for the same reasons. They are the takeoff and climb, approach and landing, overshoot, and low level operations. Recall from Chapter 18 that slow flight is often looked upon as a warning sign that a stall is approaching. In order to stall the aircraft, we must pass through the slow flight range first. As with slow flight, stalls during low level maneuvers (all of the critical areas) are especially dangerous due to the proximity to the ground—the best course of action is to avoid stalls altogether. Having inadvertently entered a stall, however, we must recognize the stall and recover immediately—with an absolute *minimum* of altitude loss.

Spins

A spin is a possible consequence of a stall. If one wing becomes more stalled than the other, the imbalance will induce a yaw/roll combination that is unstable. This means that the yaw and roll will aggravate the situation and make the imbalance even greater—leading to more yaw and roll. During a spin, the aircraft will be rotating and descending rapidly, and the airspeed will be low and steady. Clearly, an inadvertent spin is something we wish to avoid—especially close to the ground.

Causes and Factors

Anything that will cause one wing to stall more than the other can potentially lead to a spin. Since a stall is a result of exceeding the critical AOA, anything that causes the two wings to be at different AOA's can lead to a spin. This can include the AOA difference caused by climbing and descending turns (see Chapter 6). As well, a roll induced by rudder (via the secondary effect of yaw) or aileron can lead to an imbalanced stall as we discussed in Chapter 6 (also keep in mind the possibility of the ailerons reversing due to a wingtip stall). Turbulence in the atmosphere can cause a gust to strike one wing. During controlled flight, this just induces a roll. However, during a stall a spin can result.

The factors affecting the spin characteristics of an aircraft have been discussed in detail in Chapter 6. However, to sum up, numerous factors can influence spins, including:

1) *Weight* – A heavier aircraft will have more inertia and will therefore be more difficult to recover from the spin.

2) *C of G* – A forward C of G allows for a longer rudder arm (Fig 19-4). Therefore the same rudder deflection will create a greater yawing moment and the aircraft will be easier to recover from the spin.

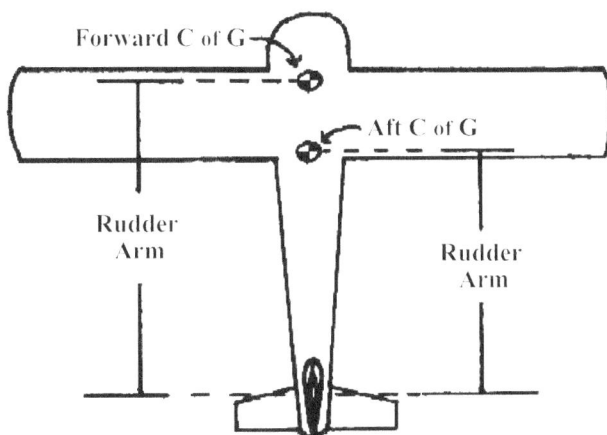

Fig 19-4. - C of G Effect on the Rudder Arm - The rudder controls movement about the vertical axis (yaw). The vertical axis passes through the C of G. So the further the rudder is from the C of G, the longer it's arm will be and the more effective it will be. Since the rudder is fixed in place, movement of the C of G determines the length of the rudder arm.

3) *Weight Distribution* – An aircraft with the weight concentrated near the C of G will have less rotational inertia, whereas an aircraft with the weight distributed further from the C of G will have a greater amount of rotational inertia. More rotational inertia will make the aircraft more difficult to recover from the spin.

4) *Power* – Power in a spin can have varying effects depending on other aircraft design considerations. Generally, power will increase the rate of rotation in the spin and flatten the spin attitude. Recovery from the dive after the spin has been broken will also be more difficult due to an extremely rapid buildup in airspeed.

5) *Configuration* – Like power, the configuration can have varying effects depending on other aircraft design considerations. Generally, turbulent air from the flaps will dampen the effectiveness of the rudder and make it more difficult to recover from the spin.

6) *Rudder Effectiveness* – Rudder effectiveness will have a direct impact on the probability of recovering from a spin since, as we will see shortly, the rudder is normally used to stop the rotation.

Critical Areas

In order for an aircraft to spin, it must stall first. From this statement, we can reason that the critical areas for spins are the same as those for slow flight and stalls. This is indeed the case. Inadvertent spins are most likely during the takeoff and climb, approach and landing, overshoot, and low level operations.

Consider an aircraft on an approach to land. If this aircraft were to begin flying through the final approach while turning from base leg to final, the turn radius would have to be reduced in order to roll out of the turn established on the approach. Normally, the turn radius is reduced by increasing the angle of bank (as it should be in this case as well). However, a common error is to "lead" the turn with rudder, that is to try reducing the turn radius by using rudder into the turn. This causes two problems. First, the secondary effect of yaw causes the aircraft to roll further in to the turn. Second, in a banked attitude, yaw into the turn will cause the aircraft to pitch down as well (Fig 19-5).

To prevent the roll, aileron to the outside of the turn must be used—otherwise the aircraft will simply continue to roll. Now the aircraft is in uncoordinated flight with a "cross-control" (ailerons and rudder applied opposite to one another). As well, to keep the aircraft from pitching down, the elevator must be used to pitch up. This will increase the load factor which will in turn increase the stall speed (which was already increased due to the initial turn).

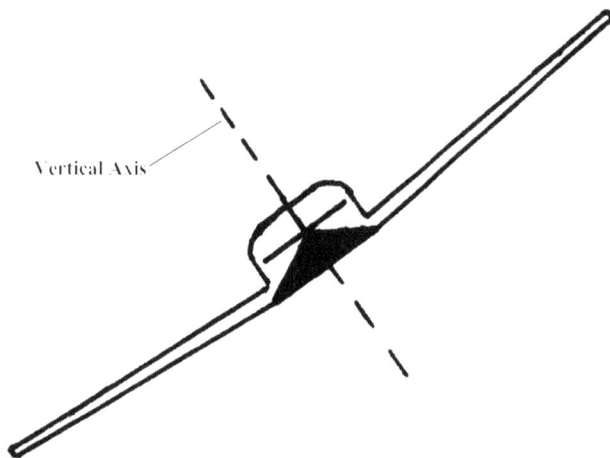

Fig 19-5. - *Yaw in a Banked Attitude* - Since the vertical axis is at an incline, yawing will cause the aircraft to "pitch" up or down.

If the turn was begun at a low airspeed, we are now in danger of stalling. We've already seen the effect of an aileron deflection in a stall. In this case, if we were turning left, the ailerons would be deflected to the right. This could lead to the left wing dropping in the stall. This situation is aggravated by the fact that the rudder is also deflected—but to the left, which will increase the tendency for the left wing to drop. As well, most turns from base to final are descending turns, this means that the low wing (left wing) is at a higher AOA to begin with and will also tend to drop for this reason.

In this scenario, several incorrect control inputs were applied. However, it remains one of the more common scenarios for stall/spin accidents. That is, this situation leads to many of the inadvertent low altitude spins that lead to accidents.

There are several other common scenarios that can lead to inadvertent stall/spin entries. Invariably, a series of incorrect control inputs will ultimately lead to the stall/spin. If we can avoid these erroneous control movements and be extra cautious in the critical areas, stall/spin accidents are easily avoided.

Intentional Spin Entries

In the training environment, spins are often done intentionally to familiarize pilots with the approach to and causes of spins, as well as recognition and recovery. Intentional spin entry procedures can vary significantly from one aircraft type to the next. In this section we will look at the "typical" spin procedure. The reader should be sure to consult the appropriate manufacturers information for the correct procedure for a given aircraft type.

As stated in Critical Areas, an aircraft must stall in order to spin. This is an important point that must be understood—an aircraft can seem to be in a spin, but if it is not stalled it is not spinning (more about "apparent spins" in Chapter 20).

Clearly, the first step to spinning is entering a stalled condition. From there, we need to create an imbalanced stall—which means that we need to get the two wings at different AOA. This is normally done by using rudder to induce a roll (however if the stall is deep enough and the wing tips are stalled, aileron will also do the trick). As the roll develops, the two wings are travelling in different directions (Fig 19-6), so they will have different relative airflows. Different relative airflows mean different AOA, therefore the two wings will be stalled by different amounts. The downgoing wing will have a higher AOA and will be stalled more. As we saw in Chapter 6, this aggravates the roll and causes more yaw in the same direction. As the roll and yaw increase, the AOA difference increases as well. The spin will continue to develop until it reaches a point where the rotation becomes cyclic.

It typically takes two to three rotations for the spin to stabilize completely into the fully developed stage. Normally, training spins are not taken to this stage, instead recovery is initiated at some point in the first rotation. As with the stalls, this is to simulate as much as possible a real encounter with inadvertent spins—since the critical areas are at low altitude, more than a full rotation will usually mean reaching the ground.

Spins (as well as stalls) are practiced at an altitude that allows recovery by at least 2000 ft

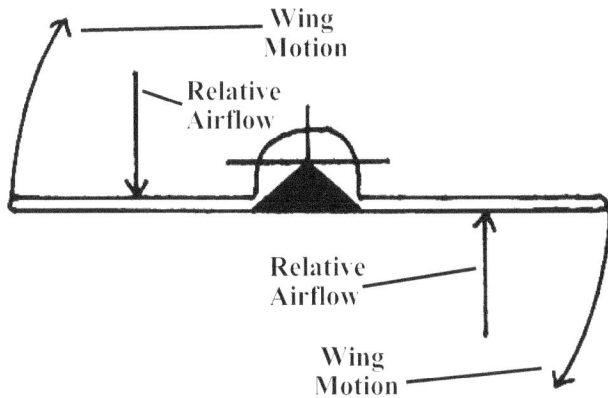

Fig 19-6. - *Roll in a Stall* - The two wings have different relative airflows. This means that they have different AOA as well.

above ground level. This takes away some of the realism of stall and spin training, but it clearly is a necessary compromise to provide for safe training.

Recovery

In order to recover from a spin, the balance of rotating moments acting on the aircraft must be upset in such a way as to stop the rotation. Once the rotation stops, the aircraft is merely stalled and a stall recovery can be executed. As with the entry, the recovery procedure can vary significantly from one aircraft type to the next. Be sure to consult the manufacturers documentation for accurate information.

First, we need to minimize the aircrafts tendency to remain in the spin. This means reducing the power to idle to reduce the spin rate and retracting flaps to improve the effectiveness of the rudder. In most aircraft certified for intentional spins, the spins are done with the power already set at idle and the flaps already retracted. However, these two steps can be critical in an inadvertent spin entry.

Typically, the rotation is stopped with opposite rudder. Ailerons are not used for reasons discussed previously in Chapter 6. The rudder induces a yaw in the direction opposite the spin direction. This upsets the balance of moments and slows or completely stops the rotation of the spin. Recovery from the ensuing stall is the same as previous stalls with one possible difference. The AOA must be reduced to break the stall, but the use of power may be inadvisable. This is because

many aircraft recover from a spin in such a nose low pitch attitude that airspeed builds rapidly even without power. Adding power at this stage may mean running the risk of exceeding airspeed limits of the airframe or RPM limits on the engine. As well, even if we remain within all operating limits, pulling out of a dive at high airspeed requires much more altitude than at a low airspeed (Fig 19-7).

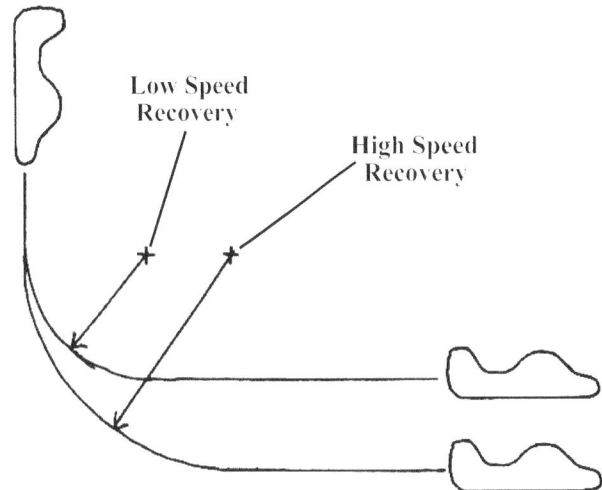

Fig 19-7. - *Dive Recoveries* - At higher airspeeds, more altitude is required to pull out of a dive. This is because the flight path radius increases with the square of the airspeed (recall uniform circular motion in Chapter 1).

Spin Certification

It is important to note that not all aircraft are certified for intentional spins. There are in fact airworthy aircraft that will not recover from a spin once it is developed. For this reason, it is vital that pilots refer to manufacturers information prior to conducting spins in any aircraft. Even aircraft that are certified for intentional spins are usually subject to additional restrictions in regard to weight, C of G location, and configuration.

The requirements that an aircraft must meet are complex, but we will sum up the important points here. First of all, the following requirements must be demonstrated through actual in-flight spin trials. Calculations done by design engineers are not enough to satisfy the certification requirements.

The crux of spin certification is in the number of rotations required. In order to be used for intentional spin training, an aircraft must be capable of conducting a spin of six rotations and recovering within two rotations after anti-spin control inputs have been applied. This requirement applies to the configuration for which spin certification is intended (normally the clean configuration). An allowance is made for aircraft that will not remain in a spinning condition. If the aircraft recovers from the spin on its own with pro-spin control inputs prior to the sixth rotation, the recovery requirement is deemed to be met.

In addition to this, the aircraft must be able to do a one rotation spin in *any* configuration. Recovery in this case must also be within two rotations of the application of anti-spin controls. Note that these other configurations are not permitted to be used for *intentional* spins, but the requirement is in place to allow for some margin of error during training.

Last, but certainly not least, it must be impossible to put the aircraft into an unrecoverable spin no matter what control inputs are made during the spin. This requirement covers the use (and misuse) of all three primary controls as well as the power.

Chapter Summary

Stalls and spins are both potentially hazardous maneuvers which must be avoided near the ground. Neither the stall or the spin has any real practical application. They are practiced strictly for the purpose of recognition, recovery, and avoidance. Stalls and spins are especially dangerous since all of the critical areas leading to inadvertent entry are close to the ground.

Stalls result from an increase in AOA beyond the critical angle. Airflow separation over the top of the wing leads to a decrease in lift and an increase in drag. This leads to an unavoidable loss of altitude which cannot be stopped without first recovering from the stall. A spin will occur if the stall is aggravated by roll and/or yaw. If the two wings are meeting the relative airflow at different angles, each wing will be stalled to a different degree. It is even possible for one wing to be stalled while the other remains flying. This imbalance is unstable and can lead to a repetitive autorotation around an axis which is vertical to the ground.

Numerous factors can influence the stall and spin characteristics of an aircraft. These factors include the weight and C of G, load factor, configuration, ground effect, surface contamination, power, and weight distribution.

There are several different ways to enter a stall. Which one used depends on the stall configuration being used. In all cases, the AOA must be increased beyond the stalling angle. Recognition of the stall during the approach and entry is based on the symptoms of slow flight, the airframe buffeting, and the nose drop resulting from the stall. As well, many aircraft are equipped with visual or aural stall warnings which will alert the pilot of the high AOA. Unfortunately, these warnings are mechanical devices which are not foolproof. They should be used, but not relied on.

Recovery from the stall involves decreasing the AOA by pitching down sufficiently. Full power is also used in order to minimize the altitude loss during the stall. If a roll should begin during the stall, it should be corrected with the rudder despite all instincts to use ailerons. The effect of ailerons in a stall can be unpredictable and should be avoided.

In most aircraft, intentional spins are entered by applying full rudder during a stall. The resulting yaw initiates autorotation and leads into a spin. Recoveries are normally initiated very early in the spin cycle by applying opposite rudder to halt the rotation and a forward pitch input to reduce the AOA and recover from the stall.

In order for an aircraft to be used for intentional spins, it must be certified for such activity. This certification requires that the aircraft meet minimum requirements in terms of how many rotations can be completed safely, how many rotations are required for recovery, and how forgiving the aircraft is to incorrect control inputs during the spin.

Questions and Problems

1) Why do stalls occur? How can they be avoided?

2) An aircraft's normal stalling speed is 58 knots. During a 40° bank turn from base leg to final approach, what is the stall speed?

3) If the aircraft in Q2 has a normal MGW of 2750 lbs., what speed would it stall at loaded to 2200 lbs.?

4) How do we recognize a stall? Are there symptoms that warn of the *approach* to a stall?

5) After a takeoff over an obstacle, an aircraft is placed in a climbing left turn and then inadvertently stalled. What would be the most probable sequence of events during the stall and recovery (assuming that the recovery can be completed prior to contacting the ground)?

6) What is the danger associated with the use of ailerons during a stall? If ailerons cannot be used, how do we correct for an undesirable roll in a stall? If no correction is made, where can this roll lead?

7) What is a secondary stall? How can they be avoided?

8) What are the critical areas for inadvertent stalls and spins? Why are these areas more dangerous than others? How do we avoid stalls and spins?

9) When is power added during a stall recovery? During a spin recovery? Why is there a difference?

10) In aircraft certified for intentional spins, why is there normally an aft C of G limit which is more restrictive than the weight and balance limits for normal operations?

11) Assuming that an aircraft certified for intentional spins will not recover of it's own accord, what is the minimum number of rotations that it must be able to complete safely? What is the maximum number of rotations that are allowed in the recovery?

Chapter 20
Unusual Attitudes

Unusual attitudes are, for all intents and purposes, temporary loss of control of the aircraft. During an unusual attitude the aircraft will be in an attitude that was not intended by the pilot, and the pitch and/or bank attitude will generally be extreme. Airspeeds and load factors are usually outside acceptable limits or quickly approaching these limits. Recovery from this situation is critical and must be initiated immediately.

The most common cause for an unusual attitude is spatial disorientation due to flying without visual reference. However, unusual attitudes can also result from meteorological turbulence or an encounter with the wake turbulence of a larger aircraft. As well, an unusual attitude can result simply from a prolonged distraction of the pilot (recall spiral divergence from Chapter 7) or from a poorly executed aerobatic maneuver.

Stalls and spins both qualify as unusual attitudes. However, since they have previously been discussed in detail they will only be mentioned in passing in this chapter.

Unusual attitudes are generally subdivided into two main types—nose high and nose low. Nose high unusual attitudes can often lead to stalls and spins when left unchecked. Nose low unusual attitudes are often spiral dives or they result in spiral dives. Both cases require immediate and correct recovery procedures, however they also require distinctly different recoveries. For this reason the ability to recognize and differentiate between nose high and nose low unusual attitudes is very important.

Nose High

Nose high unusual attitudes are exactly what the name implies. The pitch attitude is abnormally high and as a result the airspeed is bleeding off rapidly—even with a high power setting. As we've seen, this can lead to a stall and possibly a spin. Clearly, a recovery prior to a complete loss of control due to a stall is called for.

Recognition

The primary source of attitude information is always the horizon. In the case of an instrument recovery without visual reference, this attitude information comes strictly from the flight instruments. We will assume for our purposes here that these recoveries are being completed under visual conditions. A nose high attitude, by definition, can be recognized by the fact that the horizon is below the cruise position (Fig 20-1a). In the case of a nose high *unusual* attitude, the

(a) Nose High Attitude

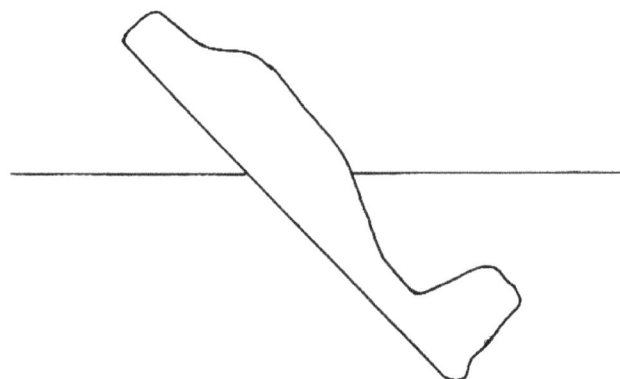

(b) Example of a Nose High Unusual Attitude

Fig 20-1. - Attitudes and Unusual Attitudes.

horizon will be *well* below the cruise position (Fig 20-1b). As well, most unusual attitudes include bank in one direction or the other. This can be recognized from the incline of the horizon across the windscreen.

Corresponding to the nose high attitude, there will be an increase in altitude (if we are not already stalled) and a decrease in airspeed. A decreasing trend in the airspeed is an extra piece of information that can assist us if disorientation occurs.

Recovery

Once the nose high unusual attitude has been recognized, recovery should be commenced immediately. The recovery procedure is based on the assumption that a stall is imminent or has already occurred.

First of all, the AOA must be reduced. This means moving the control column forward to lower the nose. This reduced AOA will often mean a partial load factor (less than 1g) and may cause some temporary discomfort. However, partial loads are better than remaining in or inducing a stall. As the AOA is being reduced, full power should be applied to minimize the altitude lost and the time taken to recover to a safe airspeed. As well, the wings should be leveled with rudder (remember the problems that ailerons could possibly cause).

Notice the similarity between this recovery and the recoveries from slow flight and stalls. There is essentially no difference, with one possible exception being the loss of altitude in a stall recovery. A nose high unusual attitude is simply another place where inadvertent slow flight, stalls, and spins can occur. The recovery technique reflects this possibility.

Nose Low

Nose low unusual attitudes are the counterpart to nose high unusual attitudes. The causes are much the same, however the recognition and recovery are distinctly different.

Recognition

In a nose low attitude, the horizon will be above the cruise position (Fig 20-2a). In a nose low *unusual* attitude, the horizon will be *well* above the cruise position (Fig 20-2b). As with nose high, nose

low unusual attitudes are often accompanied by a banked attitude. In extreme nose low attitudes, the horizon may be so high that it is necessary to look up to see it. If this is the case, it is important to look up and determine which way the aircraft is banked or if it is banked at all.

Corresponding to the nose low attitude, the airspeed will be high and increasing. As well, if the aircraft is banked the rate of turn can be high along with the load factor. The vertical speed will also be increasing rapidly. This condition is usually referred to as a spiral dive.

(a) Nose Low Attitude

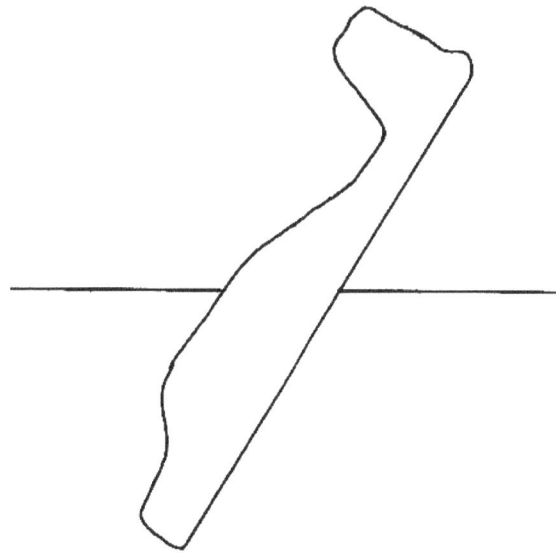

(b) Example of a Nose Low Unusual Attitude

Fig 20-2. - Attitudes and Unusual Attitudes.

A common problem with nose low unusual attitudes is distinguishing between a spiral dive and a spin. The difference is important since the recoveries are different. The key lies in the airspeed. In a spin, the airspeed will be low (near the stall speed) and steady. In a spiral, the airspeed will be increasing. So when the outside attitude indicates a spin or a spiral, we need to cross check the airspeed to know for sure which one is occurring.

Recovery

During a spiral dive, maximum speed and load limits for the aircraft are approached rapidly and are easily exceeded. For this reason, a timely recovery from a spiral dive is important. The tendency once a spiral is recognized is to pull up to reduce the airspeed. This, however, is exactly what we do *not* want to do. The result of a straight pull up from a spiral dive will be to overstress the wings and possibly worsen the spiral. The wings *must* be level prior to any pitching movements.

There are three essential steps to a spiral dive recovery:

1) Reduce the power to idle.
2) Level the wings.
3) Ease out of the dive.

Reducing the power to idle obviously is to help prevent the airspeed from building too rapidly. Not only will the engine stop contributing to the aircraft's motion, the prop will actually act as a brake at high airspeeds and low RPM. This is because of the change in AOA of the prop blade with airspeed and RPM (Fig 20-3). The propeller blade will actually have a negative AOA and will produce thrust in the reverse direction.

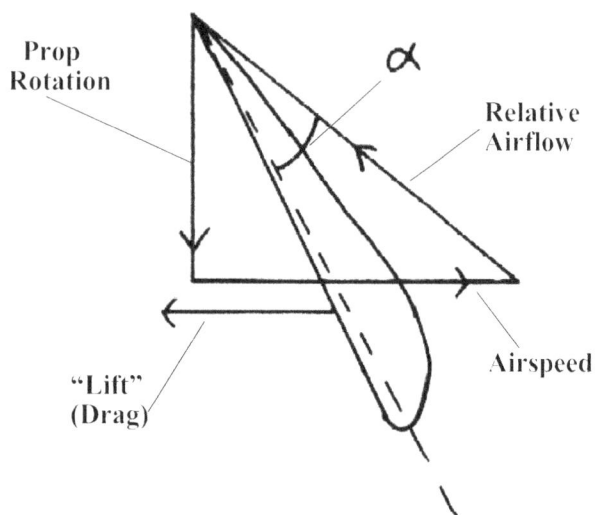

Fig 20-3. *- Prop Blade AOA at High Airspeed/Low RPM -* The high airspeed combined with the low rotational speed of the prop cause the blade to have a negative AOA and create thrust to the aft – i.e. drag.

Leveling the wings is critical if we want to get out of the dive without overstressing the airframe. This step can be combined with reducing

the power—in fact, the two should be done simultaneously—but it should not be combined with the pitching movement. Pitching and rolling at the same time will cause additional bending and twisting stresses on the inside wing (Fig 20-4) which may end up being overstressed. In rolling the wings level, both aileron and rudder should be used. Aileron of course will produce a rolling movement. Rudder will also contribute to the roll via the secondary effect of yaw. Being in coordinated flight at this point is not critical since the airspeed is high, therefore with low load factors a spin is not a worry, the drag produced by a slip is actually desirable since we are trying to slow down, and the roll caused by the yaw is exactly what we are trying to produce. Thus the problems associated with uncoordinated flight are not problems at all in this case. Keep in mind that at the high airspeeds typical of spiral dives full deflection of the ailerons and rudder should be avoided due to being above V_a. Recall from Chapter 9 that $1/3$ control deflection can be used right up to V_{ne} and linear interpolation is acceptable for intermediate speeds.

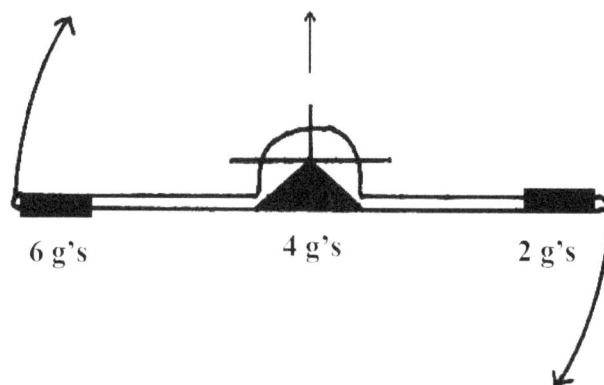

Fig 20-4. *- Simultaneous Pitch and Roll -* "Rolling g" and the twisting moment created by the ailerons combine to stress the inside wing more than if the two movements were separate. This can potentially lead to an overstress condition.

In the third step, ease out of the dive, "ease" is the operative word. Pulling abruptly on the control column could overstress the aircraft or it could lead to an accelerated (high g) stall. The aircraft should be pitched up to a climb attitude and held there until the airspeed is in a reasonable range. At that point, power can be applied to commence a climb back to altitude.

Aircraft Limits

During spiral dives—inadvertent spirals as well as intentional spirals done for training—speed and load limits are critical considerations. Preventing these limits from being exceeded is a priority since once they have been passed the aircraft is no longer structurally reliable.

Chapter Summary

Unusual attitudes are unplanned, and often extreme, flight attitudes. They can be grouped as either nose high or nose low. Nose high unusual attitudes result in a rapidly decaying airspeed and the associated risk of a stall/spin. Recovery involves immediately reducing the AOA while adding power and leveling the wings. Nose low unusual attitudes result in a rapidly building airspeed and load factor and the associated risk of overspeeding/overstressing the airframe. To recover, the power is immediately reduced to idle, the wings are leveled, and the aircraft is pitched up out of the ensuing dive. It is important to keep the pitch and roll of a nose low recovery as two separate movements. Otherwise, the inside wing may be overstressed due to the added load and twisting moments induced by the aileron deflection.

Questions

1) During an unusual attitude, how would we distinguish between a nose high and nose low attitude?

2) During a nose low unusual attitude, how do we distinguish between a spin and a spiral? What are the recovery procedures for each?

3) Why shouldn't pitching and rolling movements be combined during a spiral dive recovery?

4) During a spiral dive recovery, how can we approximate the maximum safe control deflection?

Chapter 21
Slips

Up to this point, we have always assumed that the aircraft has been in coordinated flight and that as pilots, we will always endeavor to maintain coordinated flight. Under most circumstances, this is true, however there are times when it is actually desirable to set the aircraft up in *un*coordinated flight. For this reason, slips are used.

During a slip, the aircraft is intentionally placed in uncoordinated flight. The controls will be "crossed"—meaning that the ailerons and the rudder will be deflected in opposite directions—and the aircraft will be in a banked attitude. The aircraft's tendency to turn in the banked attitude will either be reduced (in a slipping turn) or eliminated altogether (in a forward slip or a side slip) by the use of opposite rudder. Slips are used for two reasons—either to lose altitude or to land in a crosswind.

Forward Slips

During a forward slip, the aircraft is placed in uncoordinated flight in order to create drag and increase the descent rate without a corresponding increase in airspeed. On older aircraft, prior to the common use of flaps, forward slips were the primary means of controlling descent rate (in addition to power). In a forward slip, the aircraft is essentially flying sideways through the air. This slip presents more of the airframe to the relative airflow and thereby increases the parasite drag of the aircraft significantly.

In order to maintain the slip without the aircraft reacting with either lateral or directional stability, the control surfaces must be deflected in a manner which encourages uncoordinated flight. This means that the controls must be crossed—the ailerons are deflected in one direction while the rudder is deflected in the other. Otherwise, the slip

will induce a roll or a yaw that will eliminate the slip.

To enter the forward slip, rudder is first applied in the desired direction of the yaw. As the aircraft yaws, the rolling tendency must be eliminated by the use of opposite aileron. As well, the aircraft will need to be rolled slightly in the direction opposite the yaw in order to prevent a "flat" turn. The aircraft will now be flying in a slip towards the low wing.

Rudder is applied first in order to yaw the aircraft away from it's flight path. This must be done to maintain a constant flight path along the ground. A slip could also be established by rolling to a banked attitude first and then stopping the turn with opposite rudder. However, this technique would result in a flight path that is different from the original one.

The forward slip is normally used during an approach to land. However, if the slip is continued right to the ground the aircraft will land with side loads on the landing gear. These side loads could very well be sufficient to shear off the landing gear and result in a very short (and noisy) stop. For this reason, the aircraft must be returned to coordinated flight prior to touchdown. To do this, we can simply neutralize the rudder and ailerons to allow the aircraft to coordinate itself.

Another consideration during forward slips is the direction of choice for the slip. Aerodynamically, one direction is as good as the other. However, it is often preferable to slip to the left (right rudder and left aileron) to provide for better visibility. For a pilot sitting on the left side of the aircraft, which is the generally accepted norm, a slip to the right will move the engine cowling into the field of view while a forward slip to the left will leave the pilots vision unobstructed. In the case of a flight instructor, a first officer, or a pilot who simply prefers the right seat, a slip to the right would have the same effect.

Side Slips

Aerodynamically speaking, a side slip is exactly the same thing as a forward slip. The difference is in the application. A side slip is used to maintain a straight path over the ground with the aircraft pointed along that path. The usefulness of this maneuver rests in crosswind landings.

During a crosswind approach, the aircraft must be flying into the wind in order to maintain a track along the extended runway centerline. This means that the longitudinal axis of the aircraft will not be lined up parallel to the centerline even though the flight path will be (Fig 21-1). Landing in this condition is unacceptable since it will impose excessive side loads on the landing gear. During the landing, the aircraft must be maneuvered so that these side loads are removed—the aircraft must be pointed in the direction in which it is travelling.

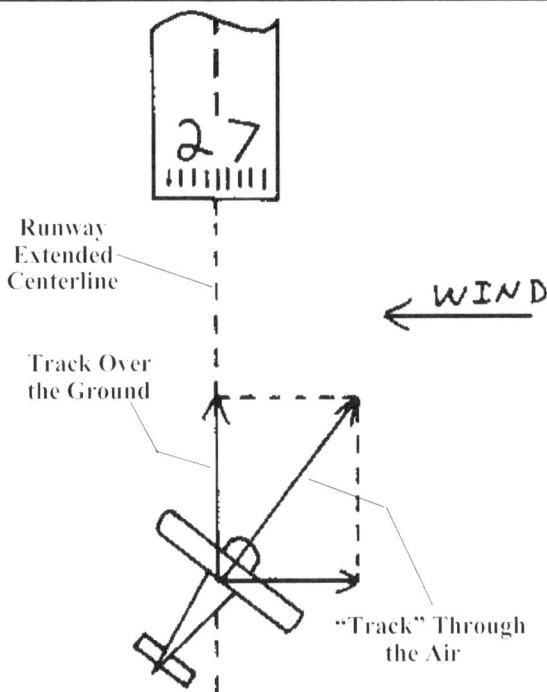

Fig 21-1. - *Approaching in a Crosswind* - The aircraft must be pointed into the wind so that the sidewards component of the aircraft's velocity cancels out the crosswind component. Unfortunately, landing like this can damage the aircraft or lead to a loss of control due to side loading on the gear.

To understand what we are doing in a side slip, first consider an aircraft flying an approach with no winds. If the heading of the aircraft while in coordinated flight is not parallel to the centerline, neither will the flight path (Fig 21-2a). Now consider placing this aircraft in a forward slip with

the rudder applied toward the runway (Fig 21-2b). The flight path in this case hasn't changed, but the direction in which the aircraft is pointed has—the aircraft is now pointed in a direction parallel to the centerline.

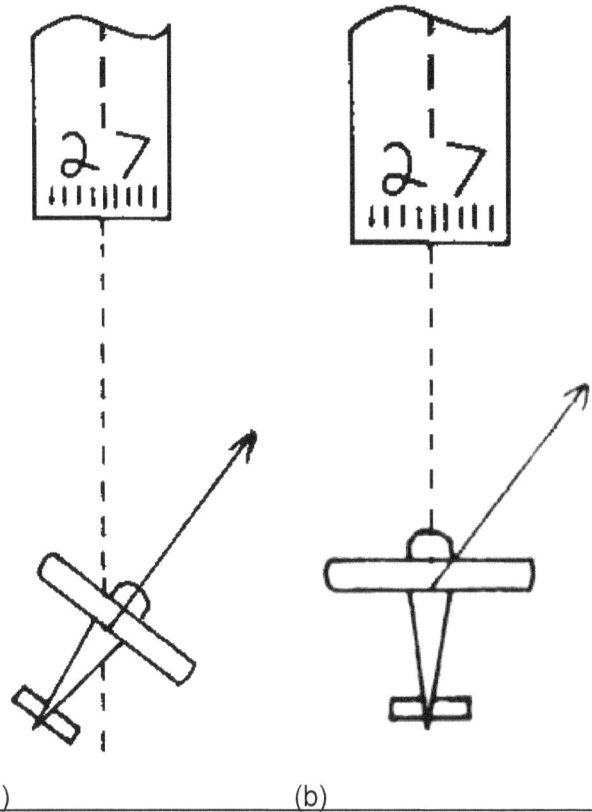

Fig 21-2. - *Crosswind Approach Without a Crosswind* - (a) In coordinated flight with a crab angle, the flight path crosses the correct approach path, and (b) entering a forward slip so that the aircraft is pointed parallel to the runway maintains the same flight path as before.

To complete the picture, we can now consider these two scenarios with a crosswind. When the aircraft is pointed into the wind, it's velocity can be divided into two components—one along the runway centerline and one perpendicular to the centerline. To maintain a flight path along the straight approach, the sidewards component of the velocity must be equal to the crosswind component. For this reason larger crosswinds require larger heading corrections. This change in heading to maintain a track along the ground is called "crabbing", and the difference between the heading and the flight path or track is called the "crab angle".

We can begin now with an aircraft crabbing to maintain the runway centerline (Fig 21-3a). Landing in this condition would be too stressful on the landing gear, so we need to align the gear with the track. If we were to establish this aircraft in a "forward" slip by applying rudder towards the runway and opposite aileron, the flight path will not change, but the aircraft will be pointed along the runway (Fig 21-3b). This aircraft appears to be in coordinated flight to the outside observer, but in order to maintain the track along the ground parallel to the heading a slip is necessary. This application of the slip is referred to as a side slip.

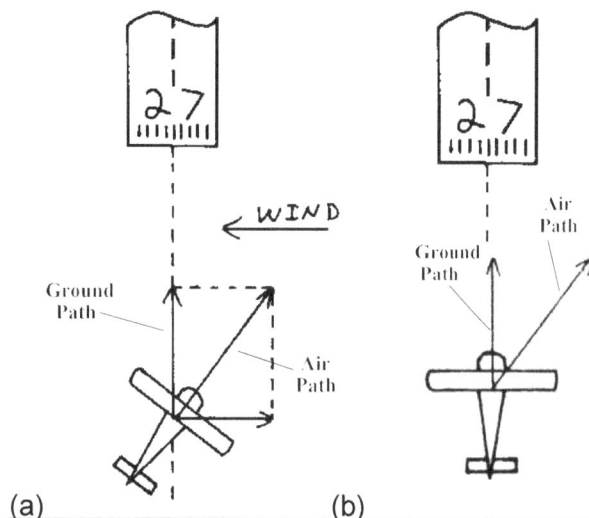

Fig 21-3. - *Crosswind Approach With a Crosswind* - The control inputs and flying techniques are the same as those used in Fig 21-2, but now a crosswind is present so that the component of aircraft motion across the runway is canceled out. This results in a straight approach and a straight touchdown with the aircraft oriented in the correct direction.

Unlike the forward slip, the sideslip is continued right to the ground. This difference is due to the fact that the forward slip will produce side loads on the landing gear, but the side slip is intended to eliminate those same side loads.

Slipping Turns

A slipping turn is simply a combination of a forward slip and a turn. Once established in the slipping turn, the aircraft will be in a banked attitude in one direction, with rudder applied in the other direction to reduce the rate of turn and produce a slip.

There are two methods to use in entering a slipping turn. The first is to establish a turn first, and then apply rudder to the outside of the turn (e.g. – a right slipping turn will require left rudder). The other method is to establish a forward slip, and then increase the angle of bank so that the aircraft turns. Both methods will provide for the same end result.

Chapter Summary

A slip is occurring when the relative airflow is at an angle to the longitudinal axis as seen from above. This situation is uncoordinated flight and is generally avoided for reason discussed throughout this text. However, under some circumstances the characteristics of a sideslip can be used to our advantage. The circumstances include needing to lose altitude at an increased rate and landing in a crosswind.

The forward slip is used to lose altitude in a hurry (relatively speaking). To enter this maneuver, rudder is applied in the desired direction and opposite aileron is used to roll the aircraft away from the direction of the yaw. The slip thus produced increases drag and necessitates an increase in descent rate in order to maintain a constant airspeed. Recovery from the slip is required prior to ground contact in order to avoid excessive side loading on the landing gear.

Sideslips are the exact same thing as forward slips from an aerodynamic perspective. However, in application, they are somewhat different. During a sideslip, the uncoordinated flight condition creates a lateral component of motion which cancels drift due to a crosswind. Sideslips are generally initiated very late in the approach so as to avoid the altitude loss associated with a forward slip. The exception to this rule would be when a loss of altitude is needed along with a crosswind correction.

Slipping turns are a combination of a forward slip and a turn. They are generally used when a slip is initiated prior to turning onto the final approach.

Questions

1) Describe the mechanics of a forward slip. When would a forward slip be used?

2) How does a sideslip eliminate drift during a crosswind landing? Why is it important to eliminate this drift?

3) What is a slipping turn? When would one be used?

4) In general, what direction should a forward slip be conducted in? Under what circumstances would you deliberately conduct a slip in the opposite direction (Hint: the answer to this one is not in the text—you'll have to put some thought into it)?

Chapter 22
Takeoff

The takeoff is the portion of flight during which the aircraft departs contact with the ground. The commonly accepted definition of the takeoff includes the ground roll, the lift-off, and the initial climbout. During the takeoff, the aircraft is accelerated down the runway to an appropriate lift-off airspeed. At this point, the aircraft is "rotated", or pitched up to a climb attitude, and the climb to altitude commences. Under most circumstances, the centerline of the runway is the desired path along the ground for the roll, and the extended centerline is the desired flight path for the initial portion of the climb.

In this chapter, we will consider the techniques normally used for takeoff. There are several types of takeoffs to consider, each of which have slightly different techniques to apply. These types include the "normal" takeoff, the short field takeoff, and the soft field takeoff. As well, each of these methods can be conducted with obstacle clearance in mind or with a crosswind.

Generally, takeoffs are conducted as into the wind as possible. However, there are times when a crosswind is unavoidable. Sometimes even a tailwind is favored due to practical considerations such as runway gradients or obstacles on the departure of the into-wind runway. The ability to deal with these conditions is critical. Just as important is the ability to recognize when the conditions of the day are beyond the safe capability of the aircraft or the pilot.

Normal

The normal takeoff is, as the name implies, the most common type of takeoff used. During a normal takeoff, takeoff distance available is not a limiting factor, nor are obstacles, winds, or the runway surface condition. The normal takeoff procedure is usually the simplest type to perform.

For that reason, it is generally the first to be taught to students. In most aircraft, the trim is set to approximately neutral (roughly halfway between full forward and full aft) prior to takeoff.

The Takeoff Roll

During the takeoff roll, the aircraft is transitioning from taxiing to flying. It is very important to understand the necessary control inputs during this transition. Early in the roll, the inputs are the same as those used in taxiing. While late in the roll, the inputs become more and more like those used in the air.

Under most circumstances, the takeoff is completed as into the wind as possible. Full power (or takeoff power in some aircraft) is applied once the aircraft is established in position on the centerline of the runway (or the selected takeoff path). As power is applied and the aircraft accelerates, the left turning tendencies of the propeller will begin to take effect—this must be anticipated and corrected with right rudder. Directional control on the runway is maintained with the rudder and the centerline is held by using the rudder pedals. As the airspeed increases and the wings begin to produce lift, the ailerons must be used to hold the wings level.

Rotation and Climb

At the beginning of the takeoff roll, the elevator is usually held at approximately neutral. When the airspeed nears the desired lift-off speed, some aft pressure can be applied on the control column to "rotate" the aircraft to the takeoff attitude.

A specific target airspeed is not often used during a normal takeoff. Instead, an airspeed range is selected (e.g. – between V_s x 1.2 and V_s x 1.3).

As the aircraft enters this range, the elevator should be deflected to pitch the aircraft up. However, lift-off should not be forced at this point. As the airspeed continues to build, the elevator will become more effective and the aircraft will pitch up of it's own accord. As well, since the airspeed is increasing and the AOA is increasing, the lift being produced will increase. Eventually, the aircraft will lift off and smoothly climb out.

During the early climb out, ground effect must be taken into consideration. We already know that the stall speed increases as we leave ground effect. For a properly executed normal takeoff, this is not a concern since the lift-off airspeed is well above the stall to begin with. However, the change in downwash and the subsequent change in elevator effectiveness can be a problem, especially in high-wing aircraft. As the aircraft leaves ground effect, the elevator becomes more effective due to it's increased AOA which is caused by increased downwash. If the elevator is held in a constant position, the aircraft will pitch up as this happens. Care must be taken to move the control column forward slightly as ground effect becomes less and less pronounced—otherwise, an undesired pitch up will occur.

Once ground effect is cleared, the climb can proceed just like any other climb. A target airspeed should be selected depending on what type of performance is desired, and that airspeed should be held by using the pitch attitude of the aircraft.

Considerations

During a normal takeoff, there are many things to consider. They primarily include runway selection and centerline selection (distance available, obstacles, and surface condition come into consideration for other types of takeoffs). Other considerations such as takeoff clearance/radio call and traffic separation are subjects of other texts.

The runway is normally selected based on the wind. Whichever runway out of those available is the closest to being into the wind is selected. This allows for a lower lift-off speed (groundspeed) and therefore a shorter takeoff run. As well, this will minimize the need for crosswind corrections.

Under most circumstances, the actual centerline of the runway is used as the takeoff centerline. However, under some conditions it may be preferable to use an offset centerline. For

example, during winter operations the surface to the sides of the centerline may be dry pavement while the centerline itself may be icy. As well, in rain, the paint that is on the centerline will tend to promote hydroplaning since it is non-porous and the water pools on it. However, the unpainted pavement to the sides has small grooves which will reduce pooling and help prevent hydroplaning.

Short Field

The short field technique is used to minimize the takeoff distance required. This procedure differs from the normal takeoff in a number of ways, including the method of power application, rotation speed, and possibly configuration.

Takeoff Roll

In Chapter 8, we discussed the pros and cons of using flaps to minimize the takeoff distance. It was determined that the optimum configuration will vary from one aircraft type to another, and the best source of accurate information relevant to a particular type would be the POH or the Flight Manual. Some aircraft require the use of a clean configuration (i.e. – no flaps or other high lift devices), while other aircraft will require the use of partial flaps, or even full flaps in some cases. The configuration called for by the manufacturer should be established prior to beginning the takeoff roll.

For the short field takeoff, runway length available is a primary consideration, therefore no runway should be left unused. This means taxiing to the extreme beginning of the surface prior to beginning the roll. As well, we don't want to use up any distance at less than the maximum possible acceleration since at lower accelerations the takeoff run becomes longer. This means that takeoff power should be applied prior to releasing the brakes.

Once the brakes have been released and the aircraft begins to accelerate down the runway, the flight controls are used in the same manner as in the normal take off. The rudder pedals are used to maintain directional control while the ailerons are used to keep the wings level. The use of the elevator, however, is slightly different from a normal takeoff as will be detailed in the next section.

Rotation and Climb

Unlike the normal takeoff where the rotation is gradual, the short field has a distinct rotation at a specific airspeed. This airspeed is selected based on the stall speed plus some reasonable safety factor (e.g. – V_s x 1.1) and can be adjusted for weight. If the airspeed is too low, we run the risk of a stall, while if the airspeed is too high, the takeoff run becomes much longer (recall from Chapter 8 that the takeoff distance will increase by the *square* of the lift-off speed).

When the required speed is reached, the rotation maneuver is accomplished by moving the control column aft to pitch the aircraft up. It is important that this movement be distinctive without being abrupt. An abrupt rotation at such a low airspeed can impose a load factor which can lead to a stall. As the aircraft lifts off and climbs out of ground effect, elevator effectiveness must be taken into account. As with the normal takeoff, most aircraft will have a tendency to pitch up further as ground effect is cleared. At the low airspeeds associated with a short field takeoff, this could result in a stall if it is not anticipated and corrected for.

Early in the climb, the aircraft should be reconfigured into the clean configuration if another one was used for the initial takeoff (e.g. – 20° of flaps). The most important point about retracting high lift devices close to the ground is to do so gradually at a safe altitude and in a safe climb condition. The term "safe altitude" needs some clarification. Many instructors will teach students to begin retracting flaps above 500 ft AGL. This rule is fine for beginning students, it provides for a good safety margin. However, it must be remembered that flaps and other high lift devices will degrade the climb performance of the aircraft. For this reason, once proficiency is achieved, another method is more appropriate.

This method is to begin the configuration change once a sustained positive rate of climb is established. *Sustained* is the key word. The aircraft may begin to climb, however due to an error in flying technique or perhaps due to a vertical or horizontal wind gust the aircraft may again begin to settle. Once a positive rate of climb is established *and* maintained, high lift devices can be retracted gradually. In most cases, gradually means in stages. For example, if flaps are extended to 20° for the takeoff, we may wish to retract them 10° at a time.

The reason that high lift devices need to be retracted slowly is twofold. First of all, as the aircraft returns to a clean configuration, the stall speed increases. At the low airspeeds associated with a short field takeoff and initial climb, this increase in stall speed could be disastrous. Secondly, the pitch changes and lift changes that occur on the wing during surface retraction tends to lead to some sink or reduction in climb rate. As a result of this, some ground clearance as well as some climb is necessary to avoid descending back to the ground. As an added precaution, the surfaces can be retracted slowly so that the aircraft can regain the climb before further retraction occurs.

Obstacles

The need to clear obstacles after the takeoff may change the techniques used somewhat. After a short field takeoff without an obstacle, the climb speed normally used is V_y (best rate of climb). This speed allows us to get away from the ground as quickly as possible. However, when clearing an obstacle, the *angle* of climb is much more important. This means that we will use V_x (best angle of climb) instead of V_y.

As we discussed in Chapter 8, some aircraft manuals will specify a different speed for an obstacle clearance takeoff. This speed, if different from V_x, will always be lower. This is because a takeoff is an accelerating maneuver while V_x is based on an aircraft in equilibrium. In clearing an obstacle, power that would otherwise go into accelerating the aircraft needs to be used to climb.

As well as a different speed, some aircraft will require a different configuration for an obstacle clearance takeoff. The reason for this was discussed in Chapter 8. Essentially, high lift devices may shorten the takeoff roll, but they will also degrade climb performance once clear of the ground. Depending on the height of the obstacle, it may be advantageous to lengthen the takeoff roll somewhat to gain climb performance.

For an obstacle takeoff, the optimum climb speed and configuration given by the manufacturer will be based on a 50 ft obstacle (unless stated otherwise). This fact needs to be taken into account when taking off over higher obstacles. If the configuration used for a 50 ft obstacle isn't the clean configuration (besides the landing gear) and/or the climb speed is not V_x, the optimum procedure will change as the obstacle gets higher. The reasons for this were discussed in Chapter 8. The exact obstacle height at which to change the takeoff procedure will vary from one aircraft type to the next. If the information is not published, familiarity with the aircraft type and the expected performance is the only way to know the best procedure. No hard numbers or even rules of thumb can be presented here since they will vary so significantly between aircraft types.

Performance

When conducting a short takeoff or a short takeoff with an obstacle, the performance capabilities of the aircraft should be determined prior to departure. Discovering at 30 ft that the aircraft can't outclimb a 50 ft obstacle could be unpleasant. Most manufacturers will include operational performance data in the POH or Flight Manual. This information can be referenced prior to flight whenever there is some doubt as to the aircrafts ability under a given set of conditions.

Fig 22-1a is a typical takeoff performance graph for a light aircraft. Fig 22-1b is the same type of information in chart form. Manufacturers can vary in their method of presenting performance data, but these two methods are the most common. Notice that the numbers presented are based on conditions which are included on the charts. If these conditions are not included, one can assume that the data is based on a best case scenario (e.g. – paved level dry runway, proper configuration, perfect piloting technique, etc.).

Often, manufacturers will include correction factors or separate charts/graphs for determining performance when conditions are different from those specified. The variables most commonly include weight, wind, density altitude, surface condition, and runway gradients (sloping runways). The charts included in Fig 22-1 include some correction factors. However, if this information is not provided by the manufacturer, it is sometimes useful to apply rules of thumb to the variables.

These rules of thumb will relate back to equation (8.2), which gave us the takeoff distance:

$$d = \frac{Wv_{LO}^2}{2g(T-D)} \quad (8.2).$$

Remember that this equation is very simplified and wouldn't be very accurate in determining actual

takeoff distances. However, we can use it to determine a good approximation of takeoff distance proportions when an actual distance is known.

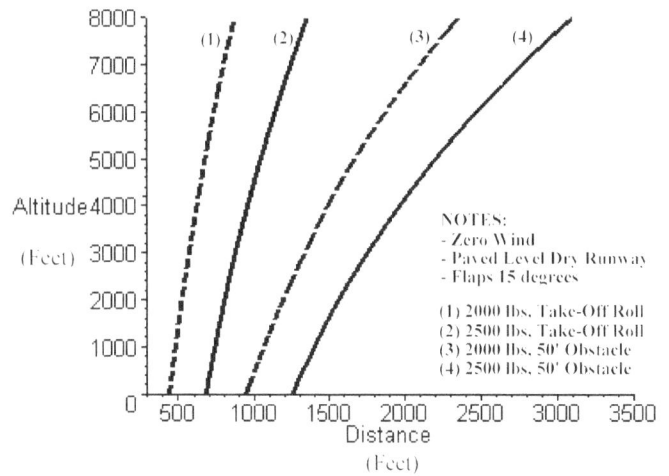

(a) Takeoff Performance Graph.

(b) Takeoff Performance Chart.

Fig 22-1. - *Manufacturer's Takeoff Data.*

Weight – The first variable we will consider here is weight. Intuitively, it can be reasoned that an increase in weight will increase the takeoff distance and vise versa, a decrease in weight will decrease the takeoff distance. We saw in Chapter 8 that from equation (8.2) this change in distance will be proportional to the square of the weight change if the lift-off speed is adjusted for weight:

$$d = D(GW/MGW)^2 \quad (22.1a),$$

where (d) is the new takeoff distance and (D) is the original takeoff distance. If the speed is not adjusted for the weight change, we simply don't square the weight ratio,

$$d = D(GW/MGW) \qquad (22.1b).$$

If an obstacle is to be cleared, the effects of weight must also be considered for the initial climb. Recall from Chapter 8 that the sine of the climb angle is a function of the excess thrust and the weight. Higher excess thrust and lower weight lead to a greater climb angle. At lower weights, more excess thrust will be available due to the change in induced drag. However, considering this effect is not practical, so we will not consider it. As well, ignoring this effect will lead to an error on the side of caution—that is, the distance we calculate will be slightly longer than the actual distance required.

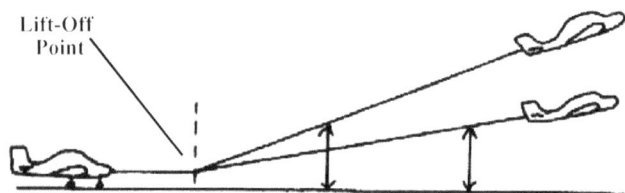

Fig 22-2. - *Distance to Clear an Obstacle At various Climb Angles.* - Notice that doubling the climb angle approximately halves the distance required to clear a given obstacle height.

Recall from equation (8.4) that as the weight decreases, the sine of the climb angle increases:

$$Sin a = \frac{ET}{W} \qquad (8.4).$$

At small angles (general aviation aircraft rarely can sustain a climb steeper than 10°) it is an acceptable approximation to say that the angle will change by the same proportion as the sine of the angle. In turn, it is a fair approximation (once again, for small angles) to say that the distance covered to get to a specific height will change by the proportion of the angle change (Fig 22-2). For this reason, we come up with the following rule of thumb:

$$d_o = D_o(GW/MGW) \qquad (22.2),$$

where (d_o) is the new distance required to clear the obstacle and (D_o) is the distance required at maximum gross weight. The distances in this rule of thumb are the distances required *after* the takeoff roll, that is they include only the climb portion of the takeoff.

Wind – Wind will also have a direct effect on the takeoff distance. Clearly, a headwind will decrease the roll while a tailwind will increase the roll. The lift-off velocity in equation (8.2) is the groundspeed, so the takeoff distance will change according to the square of the change in lift-off groundspeed. Considering this, we can determine the following rule of thumb:

$$d = D\frac{(v_{LO} - w)^2}{v_{LO}^2} \qquad (22.3),$$

where v_{LO} in this equation in the lift-off TAS and (w) is the headwind component (for a tailwind, the value of (w) would be negative, leading to an increase in takeoff distance).

The wind will also have a direct impact on the distance required to clear an obstacle once airborne. With a headwind, the groundspeed is reduced but the rate of climb is unaffected. This will clearly reduce the distance needed to climb over an obstacle.

Consider an aircraft with a fixed rate of climb. At this rate of climb it will take a certain amount of time to reach the specified obstacle height. If the forward speed over the ground is reduced, the distance covered by the aircraft in this fixed climbing time will be reduced proportionally. From this, we get the following rule of thumb:

$$d_o = D_o\frac{TAS-W}{TAS} \qquad (22.4).$$

As with equation (22.2), this rule applies only to the climb portion of the takeoff. The ground roll must be calculated separately.

Density Altitude – As the density altitude increases, we run into two problems. First of all, the power available from the engine decreases as the altitude increases. We saw in Chapter 4 why this happens. The air density decreases as we get higher, and since power available is proportional to the air density, the power available will decrease with altitude. The second problem is the increase in the TAS for a given EAS at higher altitudes. This phenomenon was discussed in Chapter 3. Since the lift-off speed is based on EAS, the TAS lift-off speed is higher at higher altitudes. From equation (8.2) this will increase the takeoff distance.

In the lower few thousand feet of the standard atmosphere, air density decreases by approximately 3% per 1000 ft. From this, the power available will decrease approximately 3% per 1000 ft. Ignoring the effects of drag, this loss of power will lead to an increase in takeoff distance of approximately 3% for every thousand feet above standard sea level. However, ignoring the effects of drag is not reasonable, it leads to a large error. Referring back to equation (8.2), we can see that the net thrust is thrust minus the drag (T-D),

$$d = \frac{Wv_{LO}^2}{2g(T-D)} \quad (8.2).$$

For a given speed, thrust is proportional to power, so a 3% decrease in power will also mean a 3% decrease in thrust. However, since the *net* thrust is less than the thrust from the engine, this decrease will be greater than 3%. It is fair to say that for most light aircraft, the average drag on takeoff is equal to approximately 20% of the average thrust. This brings the decrease in net thrust up to about 4% per 1000 feet of density altitude.

As well, the increase in lift-off speed will have an effect. For a given EAS, the corresponding TAS increases approximately 2% per 1000 ft in the lower levels of the atmosphere. Since the takeoff distance is affected by the square of the lift-off velocity, the distance will increase by about 4% per 1000 ft of altitude as a result of a higher lift-off speed. The total effect of power loss and the increase in lift-off speed is to increase the takeoff roll by approximately 8% of the sea level distance per thousand feet of density altitude.

As for the obstacle clearance distance, the 3% decrease in thrust leads to more than a 3% decrease in *excess* thrust. Generally, the distance to clear an obstacle (after lift-off) will increase by about 12% to 15% per 1000 feet of density altitude.

Surface Condition – The surface condition of the runway will have a direct impact on the takeoff distance since rolling friction will be affected. Most takeoff performance data is based on a paved level dry runway. If the surface condition varies from this standard, the takeoff distance will change. The effect of the rolling friction will vary depending on

the weight and the power of the aircraft—light aircraft with high power will be less affected by a rough or soft surface.

As the rolling friction is increased, the takeoff distance will increase. This means that for a grass strip, the roll will be longer than on pavement. As well, long grass will extend the roll more than short grass, etc. For typical single engine trainers, increasing the pavement roll by the following amounts (in percentages of the total distance required to clear a 50' obstacle) are fair approximations:

Gravel (no potholes)	– 10%
Short Grass (< 3")	– 15%
Long Grass (< 6")	– 30%
Snow/Slush (< 3")	– 50%.

These numbers are intended for use at sea level. At higher density altitudes, where engine power is reduced, the effects of rolling friction become more significant.

The surface condition will have no effect on the climb performance of the aircraft, so the distance required to clear an obstacle *after* lift-off will remain unchanged.

Runway Gradients – If the runway is sloped, a component of the weight will act in favor of or against the acceleration of the aircraft. Taking off on a downgrade will shorten the roll since the aircraft will accelerate faster. Taking off on an upgrade will lengthen the roll since the weight will now be acting against the acceleration.

Inclines are normally measured in degrees. However, runway gradients are published in the Canada Flight Supplement in percentages. The percentage incline of a runway is determined simply by comparing the difference in height of the two ends of the runway to the length of the runway. For a 1% upgrade, the takeoff distance should be increased by about 10%, and for a 2% upgrade, the distance should be increased by about 20%. With a 1% downgrade, the takeoff distance should be decreased by about 5%, while for a 2% downgrade, the distance can be decreased by about 10%. These effects are for sea level and will become more pronounced at higher altitudes where takeoff thrust is reduced.

As with surface condition, the runway gradient will not affect the climb performance of the aircraft. So the distance from lift-off to obstacle clearance will not be changed by a gradient.

Considerations

Centerline – As with the normal takeoff, short field takeoffs are normally conducted on the actual centerline of the runway. However, under some circumstances it may be advisable to use an offset centerline.

Wind – As well, takeoffs are normally conducted as into the wind as possible to minimize the takeoff roll. This rule also applies to short fields, however there are some notable exceptions. When considering obstacles and/or gradients, it may be advisable to accept a tailwind departure. For example, taking off with a 10 knot tail wind on a slight downgrade may be favored over a 10 knot headwind with a slight upgrade and a 50 ft obstacle.

Turbulence – Atmospheric turbulence will inevitably cause airspeed fluctuations and undesirable load factors. These fluctuations and loads create two problems during short field takeoffs. First, proximity to the stall during a short field procedure makes speed and load fluctuations dangerous. For this reason, the rotation speed should be increased slightly during turbulent conditions (half of the gust factor is the generally accepted rule—this increase must also be considered during performance calculations). Second, when clearing an obstacle, maintaining V_x accurately is difficult at best.

Another type of turbulence to consider is the wake turbulence left behind large aircraft (Fig 22-3). This turbulence is to be avoided at all times since the roll rates at a vortex core often can't be countered sufficiently with aileron and the descent rates between the two vortices often can't be outclimbed. Wake turbulence avoidance is usually achieved by waiting for it to dissipate after a large aircraft departs. As well, wake turbulence invariably travels downward after it forms, so remaining above it will also suffice. Since the turbulence is invisible (all it is is air in motion) a performance takeoff is a good idea. Even after waiting for dissipation, a short field takeoff with obstacle clearance will help insure avoidance.

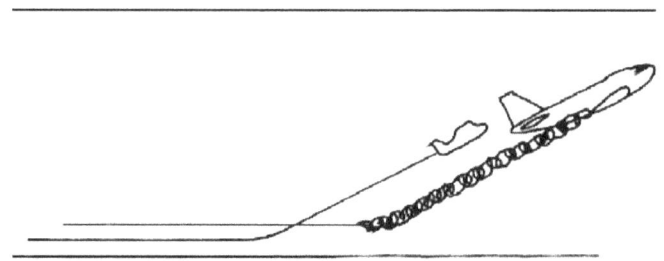

(a) Taking off after a heavy aircraft - Since Wake Turbulence settles, remaining above the leading aircraft's flight path will ensure separation.

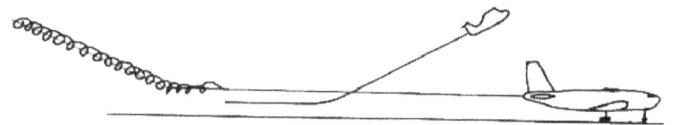

(b) Taking off after a heavy aircraft lands - To provide wake turbulence clearance in this case, make sure that the rotation point is well after the point where all of the leading aircraft's wheels are on the ground.

Fig 22-3. - *Wake Turbulence Avoidance* - Wake turbulence travels downwards, so always stay above the flight path of a heavier aircraft.

Aborting Takeoff – When takeoff distance is critical (e.g. – the water off the end of the runway is just a tad too cold to swim in), 100% reliance on performance calculations is ill advised. If an error is made during the calculations or if the aircraft is just not performing as advertised (there could be an unknown engine problem), it would be good to know before it's too late. For this reason, a decision point should be determined prior to beginning the roll. This decision point may be the intended lift-off point if extra runway is available to stop on, or it may be a point prior to the intended lift off where a specific speed should be reached.

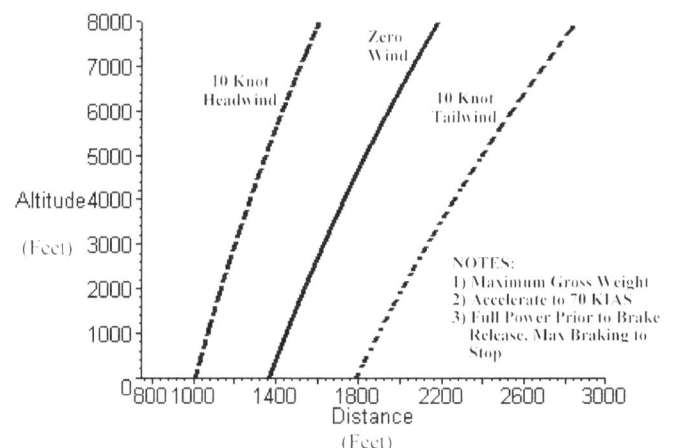

Fig 22-4. - *Accelerate-Stop Distance Graph.*

Ideally, working with an accelerate-stop distance performance chart (Fig 22-4) would give the ideal decision point. Unfortunately, these charts are not usually available for single engine aircraft. In multi-engine aircraft they are used quite frequently (but for a different reason associated with an engine failure on takeoff). In lew of these charts, adding the takeoff and the landing distances together is a reasonable approximation. If this distance is less than the runway available, the decision point can safely be the intended lift-off point. At this point, if the aircraft cannot safely remain airborne, the engine should be brought back to idle and the aircraft should be stopped.

When the accelerate-stop distance (or the takeoff + landing distance) is greater than the distance available, the decision to discontinue the takeoff must be made prior to the point of intended lift-off. In determining a new decision point, we can refer back to equation (8.2),

$$d = \frac{Wv_{LO}^2}{2g(T-D)} \qquad (8.2).$$

Consider for a moment what would happen if we were to cut the lift-off speed in half—the takeoff distance would be quartered. Of course, we can't simply cut the lift off speed in half, but we can use this to determine a new decision point. By the time one quarter of the intended distance has been used, half of the lift-off speed should have been achieved. Using similar reasoning, by the time one half of the intended distance has been used, seven tenths of the lift-off speed should have been achieved.

Keep in mind that the lift-off speed which we use in this rule of thumb is the lift-off groundspeed. So winds should be taken into consideration. For example, an aircraft at sea level is taking off with a rotation speed of 75 knots. If the winds are 15 knots straight down the runway and the calculated takeoff distance is 900 ft, what should the decision point be?

To answer this question, we first need to know the lift-off groundspeed,

75 - 15 = 60 knots.

Half of the lift-off speed (30 knots) should be achieved one quarter of the way down the runway (225 ft). We can now convert the groundspeed of

30 knots back into an airspeed,

30 + 15 = 45 knots.

So if the aircraft hasn't reached 45 knots of airspeed at 225 ft down the runway, the takeoff should be aborted.

Using this same example without correcting for winds, if the aircraft has reached 38 knots (75/2 rounded off) by the 225 ft mark, the takeoff would be continued. However, according to the above (correct) calculation, this would be unacceptable.

Procedure – A short field procedure which occasionally presents itself in the flight training environment is to add flaps just prior to rotation. The argument for this procedure is that the rotation speed can still be reduced without the disadvantage of aerodynamic drag during the roll. There are, however, some problems with this theory.

The flaps may well add *aerodynamic* drag during the takeoff roll, but the increased lift also reduces the rolling friction. On some aircraft, the two effects negate one another. However, even on aircraft where aerodynamic drag is higher than the reduction in rolling friction, the advantage gained by not using flaps during the roll is minute. There is, however, a significant disadvantage to this procedure. At the point of rotation the pilots attention is divided between flying the aircraft and extending the flaps. Chances are, the effect will be either to over rotate and risk stalling (not to mention degrading climb performance even if a stall does not occur) or to delay rotation until a slightly higher airspeed which will increase the ground roll and negate the slight advantage gained by delaying the use of flaps.

Soft or Rough Field

The soft field takeoff is used to takeoff from runways which are unprepared or poorly maintained. As well, the soft field technique can be used on runways which are well prepared but are not paved (e.g. – grass or gravel strips) or are temporarily unmaintained (e.g. – snow covered). This technique centers around minimizing the stress on the landing gear during the roll as well as minimizing the risk of "digging in" the nose gear and therefore losing control.

Takeoff Roll

During a soft field takeoff roll, most aircraft require some degree of flaps or other high lift devices. This allows the aircraft to lift-off at a lower airspeed and therefore minimize the stress on the aircraft during the roll. As well, using flaps allows the partial removal of weight from the landing gear at lower speeds.

On a soft field, the takeoff roll is often continued directly from the taxi roll, since stopping may allow the aircraft to sink into the ground and prevent further rolling. On fields that are rough but not necessarily soft, the takeoff roll can also be started much like a short field technique (we can begin the roll from a stop and add takeoff power prior to brake release).

Just prior to beginning the roll, or just as the roll begins, full aft control column should be applied. This removes weight from the nose gear and in short order will get the nose gear off the ground altogether. As the aircraft pitches up and the nose gear leaves the ground, back pressure should be relaxed enough to prevent a tail strike. However, a nose-high attitude should be maintained for the remainder of the roll. The high AOA provided for by the nose high attitude allows the weight of the aircraft to transfer from the wheels to the wings sooner and allows the aircraft to lift off at a lower airspeed (just barely above the "in ground effect" stall).

As with the normal and short takeoff, directional control can be maintained with the rudder and the wings can be kept level with the ailerons. The need for a significant amount of right rudder should be anticipated since the nose wheel leaves the ground very early in the roll.

Rotation And Climb

A "rotation" as such is not required during a soft field takeoff due to the nose high attitude used throughout the roll. However, as the aircraft lifts off, the nose should be lowered (pitched forward) immediately to prevent climbing out of ground effect. Recall from Chapter 3 that as we climb out of ground effect induced drag begins to build. This induced drag is also associated with a reduction in lift which can in turn lead to a stall. For this reason, accelerating to a safe climb speed prior to leaving ground effect is of the utmost importance.

Once a safe airspeed is established, the climb can be commenced. The procedure to use for the climb is much the same as with a short field takeoff. If obstacles are not a concern, the climb can be commenced at V_y and flaps can be retracted gradually. If an obstacle is a concern the climb can be executed at V_x (or the speed specified by the POH or Flight Manual) and flaps should be left extended (if they were used) until the obstacle is cleared. As with the short field procedure, it may be more favorable not to use flaps at all when clearing an obstacle. To find out, check the POH or Flight Manual.

Performance

A soft field takeoff will require more distance than a short field takeoff. It would initially appear that since similar (if not the same) configurations are used, the lower airspeed used for a soft field would result in a shorter roll. In fact, this is not normally the case for three reasons. First, the nose high attitude used during a soft takeoff creates drag which will lengthen the roll. Second, the soft or rough ground itself will create drag which lengthens the roll. Third, accelerating in ground effect requires distance. The distance used during this acceleration may not be a part of the ground roll per se, but climbing is not an option, so it should be considered part of the takeoff distance.

If we are using the soft-field technique on one of the previously mentioned surfaces, we can first determine the distance required using a short field technique on the given surface. Once this distance is known, we can increase it by another 50% to account for the different takeoff technique that is used.

If we are operating from a surface that doesn't give us the option of using a short field technique (i.e. – it is too soft or rough and a soft/rough field technique *must* be used), we can use the following percentages to *estimate* the distance required:

Mud (< 3")	–	60%
Mud (≥ 3")	–	100% to ∞
Snow/Slush (≥ 3")	–	100% to ∞
Wet Grass	–	60%
Long Grass (≥ 6")	–	100% to ∞
Rough surface	–	40%.

The rough surface quoted here may include cracked up pavement, a grass strip with frequent ruts in it (don't forget to count the grass as well), or a gravel strip with frequent ruts in it (don't forget to count the gravel), etc. As with the first set of percentages, these numbers pertain to sea level and will increase at higher density altitudes.

Considerations

Centerline – On some soft fields, it is advantageous to chose a takeoff path other than the physical centerline of the runway. If the surface to the left or right of the actual center is in better condition, using it will reduce stress on the aircraft and improve takeoff performance.

Loose Particles – The term soft field often includes strips which have loose particles on the surface (e.g. – gravel strips). When this is the case—and takeoff distance is not critical—it may be advisable to add power gradually as the aircraft accelerates. The motion of the aircraft once full power is applied will reduce the amount of debris that gets pulled through the prop. This will improve the service life of the prop.

Another consideration when operating off a gravel strip is the use of flaps. If the use of flaps is not necessary, leaving them retracted will minimize damage to the airframe structure as pebbles and rocks get flicked back by the prop.

Crosswind

Unfortunately, it is not always possible to select a runway that is directly into the wind. When this is the case, a crosswind takeoff will be necessary. A crosswind on takeoff will create the tendency for the aircraft to roll away from the wind and yaw into the wind. Effectively, taking off in a crosswind means taking off in a sideslip—the airflow over the aircraft is at an angle to the longitudinal axis (Fig 22-5). For this reason, control inputs change slightly during a crosswind takeoff (or, more correctly, rudder and aileron inputs become more pronounced and elevator inputs change slightly). These changes become more pronounced (and therefore more important) as the crosswind component on the given runway increases toward the aircrafts crosswind limitation.

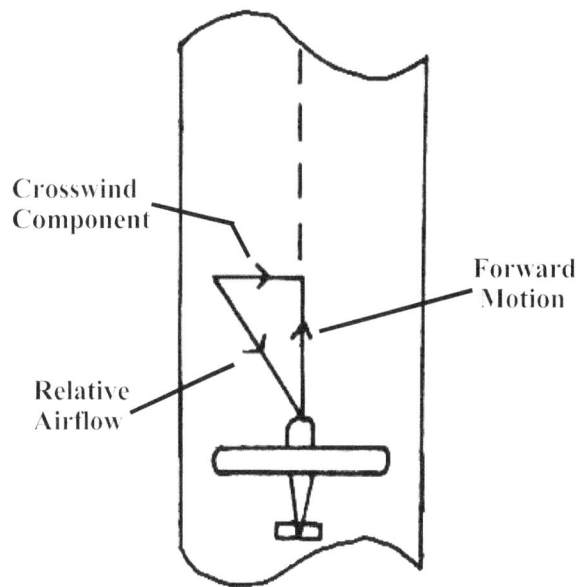

Fig 22-5. - *Takeoff in a Crosswind* - The aircraft is effectively in a slip during the takeoff roll.

Takeoff Roll

Normal and Short – As with the normal and short takeoffs without a crosswind, ailerons are used to maintain wings level while the rudder is used to maintain directional control. The difference between this and a crosswind takeoff is merely one of degree.

At the beginning of the roll, the ailerons should be deflected fully into the wind. This prevents the aircraft from rolling away from the wind as speed increases. As well, increased airspeed means increased control effectiveness, so as the aircraft accelerates toward lift-off, the ailerons should be gradually returned to neutral to prevent rolling *into* the wind. Ideally, this control shift should be timed so that the ailerons reach neutral just as the aircraft becomes airborne.

As for the rudder, some input will be necessary to prevent the aircraft from yawing into the crosswind (this is commonly referred to as "weathervaning"). If the aircraft yaws into the wind, the possibility of running off the runway exists.

An extra consideration for short field procedures is the use of flaps. Flaps tend to increase the aircraft's weatervaning tendency. As well—particularly on high wing aircraft—turbulent airflow off the flaps may dampen the effectiveness of the rudder. This will clearly reduce the ability of the aircraft to correct for a crosswind.

Soft/Rough – A soft field procedure presents us with an added problem in crosswinds. Early in the takeoff roll, the rudder is ineffective due to the low airspeed. This means that directional control must be maintained by using the nose wheel to steer. As a result, the nose wheel must remain on the ground for a longer period during a crosswind takeoff than is normally desirable. This means modifying the procedure by waiting until some acceptable airspeed is built prior to raising the nose. Unfortunately, this increases the stress on the nose gear and may not be acceptable on some fields.

In some aircraft, directional control must be maintained at low airspeeds with differential braking since the nose gear simply swivels and is not controllable. When this is the case, performance will be noticeably degraded and if the field is soft enough lift-off may not be possible.

Rotation And Climb

The actual rotation and lift-off method used in a crosswind is also somewhat different from a non-crosswind situation. As soon as the aircraft is airborne, drift will occur and the aircraft will travel sideways along the ground. If contact is reestablished with the ground at this point, the side loads may damage the airframe or cause a loss of control.

For this reason, the lift-off in a crosswind must be positive and must insure that ground contact will not be reestablished. This means two things. First of all, the rotation speed needs to be higher than usual (this added airspeed must be considered in performance calculations). Second of all, the added airspeed needs to be used to make the rotation somewhat more abrupt than usual so that some positive separation from the ground can be established.

Once the ground is cleared, the intended flight path is normally along the centerline or the extended centerline of the runway (Fig 22-6a). However, immediately upon becoming airborne, the aircraft will begin drifting with the wind. To eliminate this drift, we simply need to coordinate the aircraft. Remember that during the takeoff roll the aircraft is in a sideslip due to the crosswind. At the point of lift-off, rudder can be used to yaw the aircraft into the wind—thus coordinating the flight. The amount of rudder required will probably be slight since the aircraft has a tendency to yaw into the wind anyway.

As the sideslip is eliminated, the directional change will be sufficient to correct for drift and maintain a flight path that is straight along the runway centerline (Fig 22-6b).

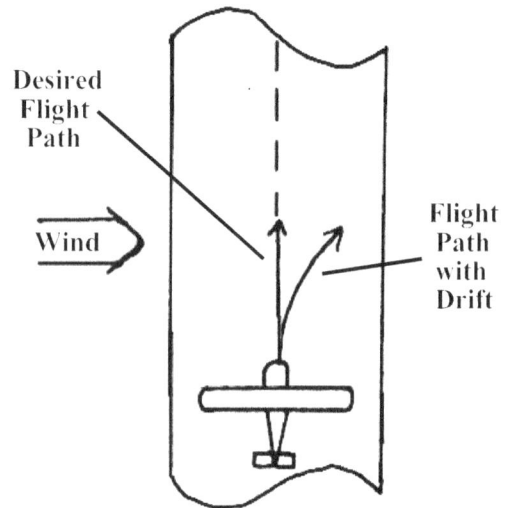

(a) Desired Flight Path and Flight Path With no Correction.

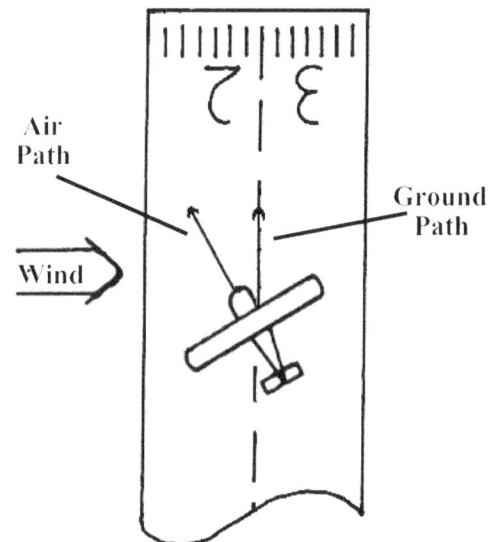

(b) Eliminating the Slip Provides the Correct Amount of "Crab".

Fig 22-6. - *Crosswind Takeoff* - After rotation/lift-off, immediately coordinating the aircraft will eliminate the slip and provide enough "crab" to maintain the desired takeoff flight path.

Considerations

Aircraft Limitations – Due to the need for control inputs to correct for a crosswind, different aircraft will be capable of dealing with different amounts of crosswind. Many manufacturers will publish crosswind limits for an aircraft in the form of a "demonstrated maximum crosswind component". This demonstrated crosswind is the crosswind component successfully compensated for by test pilots during the air trials of the aircraft. The aircraft may or may not be able to deal with larger crosswinds, but the manufacturer does not assure us of any abilities beyond the demonstrated limit.

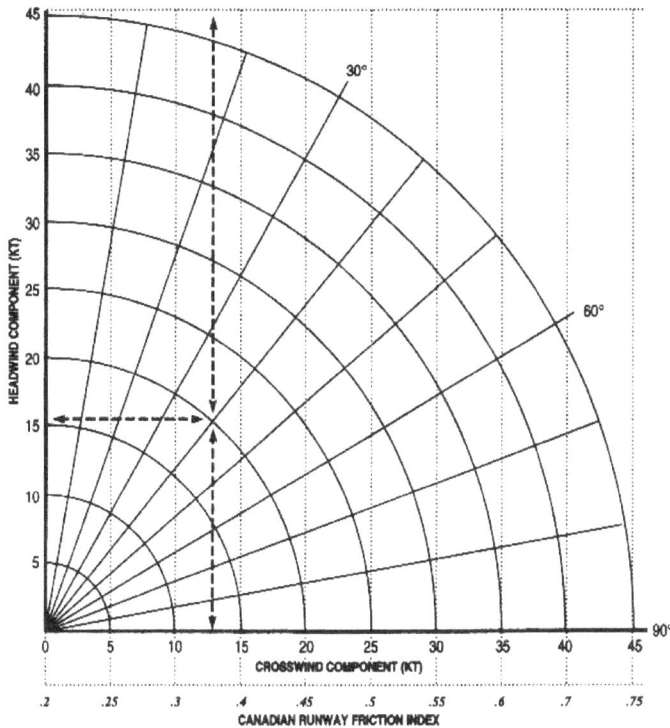

(a) Headwind/Crosswind Component Chart

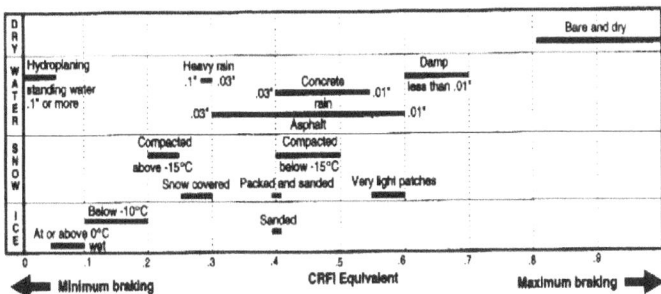

(b) Runway Surface Condition to CRFI Conversion Chart

Fig 22-7. - *Crosswind Charts.* - These charts can be used the determine whether or not a crosswind is acceptable from both an aerodynamic standpoint *and* a runway friction standpoint. Both charts are adapted from the A.I.M. Canada. You can find these charts in their original form by referring to the AIR section of the A.I.M. or the General section of the Canada Flight Supplement.

When a crosswind limit is not published, a rule of thumb can be used. In order for an aircraft to receive a type certificate, it must be able to takeoff and land with a crosswind component equal to 20% of it's stall speed. This means that even without manufacturers information, a minimum crosswind limit can be established. As with the demonstrated crosswind component, the aircraft may or may not be able to correct for more than this minimum requirement, but we always know that at least this minimum is available.

In determining the crosswind component on a runway, the wind strength and direction must both be considered. The crosswind is the component of wind that is perpendicular to the runway (that is to say that there is no such thing as a "60° crosswind" or a "45° crosswind" etc., all crosswinds are at 90°). Fig 22-7 is a chart which can be used to determine the headwind and crosswind components. Notice at the bottom of the chart, there is a row of numbers labeled as 'Canadian Runway Friction Index' (known in most other countries as the 'James Brake Index'). These numbers will be discussed in the next section.

Charts similar to those in Fig 22-7 can be found in the A.I.M. and the Canada Flight Supplement. As well, the wind side of flight computers can be used to split winds into components.

Canadian Runway Friction Index – Even when an aircraft is aerodynamically capable of taking off in a crosswind. The Canadian Runway Friction Index (CRFI) must be considered. The CRFI is an expression of the braking action on a runway. For dry pavement, the CRFI is approximately 1. However, for other surfaces (grass, snow, ice, etc.) the CRFI will be lower. At low airspeeds, the friction between the tires and the runway prevent the aircraft from going off the edge of the runway. If the CRFI is too low, an aircraft that would otherwise be capable of correcting for a crosswind may be unable to takeoff safely.

At many large airports, CRFI information is available from the control tower. However, there are also many aerodromes where CRFI information is not available. When this is the case, the chart in Fig 22-7b can be used to estimate the CRFI.

Gusts (Unsteady Wind) – A gusty crosswind is more difficult to deal with than a steady crosswind. In execution, a takeoff in a gusty crosswind condition will require constant readjustment of the control inputs in order to compensate. There isn't much that can be said here beyond that one point, practice and exposure to the conditions is required.

As for crosswind calculations, always base the crosswind on the peak gust that is occurring at the time. Using the steady wind value or an average can lead to a loss of control due to an underestimated crosswind.

Centerline – As is the case for other types of takeoffs, crosswinds normally use the actual centerline of the runway. However, an offset centerline may be preferred in some cases due to the surface condition or CRFI considerations.

When taking off in strong crosswinds where the aircraft limits are uncertain (unpublished by the manufacturer), using the upwind side of the runway is advisable. This allows for more "drifting room" if things don't go well and the takeoff must be aborted.

Wind Shear

Normally, wind shear is a problem related to IFR flying. However, under some circumstances, it can occur in VFR weather—particularly in mountainous terrain and during high winds. Wind shear is a sudden (almost instantaneous) change in wind speed and/or direction. This can cause problems for aircraft due to inertia. The *groundspeed* of an aircraft will momentarily remain constant as the aircraft passes through a shear surface. However, since the wind has shifted, we will experience a corresponding shift in airspeed.

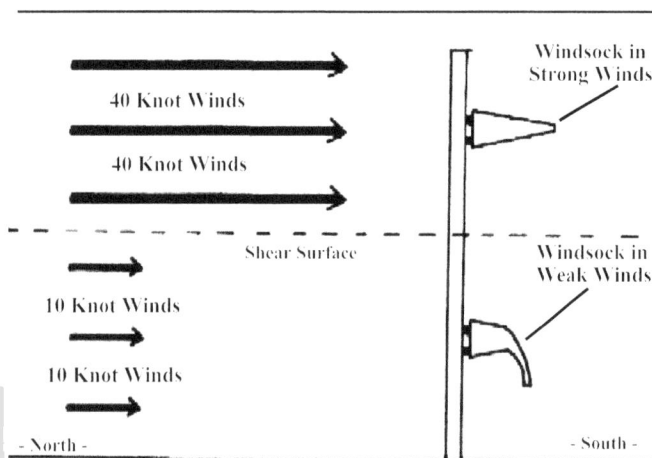

Fig 22-8. - *Wind Shear.* - A sudden shift in wind speed and/or direction can cause an unexpected change in airspeed due to aircraft inertia.

To see what's happening during wind shear, consider a pole that extends up through a shear

surface that is parallel to the ground (Fig 22-8). The measured winds above the surface will be different from the measured winds below the surface. For example, the winds above the surface may be out of the north at 40 knots, while the winds below the surface may be out of the north at 10 knots. An aircraft climbing through the shear surface while flying to the south would experience a sudden 30 knot drop in airspeed. In the meantime, another aircraft climbing to the north would experience an airspeed increase of 30 knots.

Increased Performance

Wind shear that causes a sudden gain in airspeed is referred to as *increased performance wind shear*. This type of wind shear doesn't constitute much of a hazard, just an inconvenience. Simply maintaining the proper climb attitude will allow the airspeed to return to it's proper value after a few seconds.

Decreased Performance

A loss of airspeed is caused by decreased performance wind shear. This type of wind shear is more dangerous. Notice that increased and decreased performance wind shear can both occur at the same shear surface depending on which direction the aircraft is travelling in when it passes through the surface.

If the airspeed of the aircraft is low enough to begin with and the wind shear is strong enough, the result could be a stall or an approach to a stall. If this occurs, an immediate stall recovery is called for. The aircraft must be pitched forward to reduce the AOA and allow the airspeed to rebuild. Once a safe climb speed is regained, the climb can be continued. The wind shear should be reported to ATC or FSS so that other pilots can be forewarned of it's presence.

Chapter Summary

The takeoff includes the roll down the runway (or other surface), the lift-off and the initial climb. Several types of takeoffs can be performed, they include the normal, short, and soft/rough takeoff. As well, each of these techniques can be conducted with a crosswind and/or an obstacle to clear after lift-off.

The normal takeoff is straightforward. Once the aircraft is aligned in position on the runway, takeoff power is applied and the roll begins. Directional control is maintained with the rudder and lateral control is maintained with the ailerons. As the aircraft accelerates into the rotation speed range, some aft pressure is applied to the control column and a slow pitching movement is begun. Gradually the weight of the aircraft is transferred from the wheels to the wings and the aircraft lifts off.

Short field takeoffs are intended to minimize the runway distance required. For this reason, a more specific and reduced rotation speed is used. As well, some aircraft use high lift devices in order to reduce the stall speed and therefore reduce the rotation speed even further. Factors that will affect the takeoff performance include the weight of the aircraft, winds, density altitude, runway surface condition, and runway gradients.

Soft or rough field takeoffs are intended to minimize the stress on the landing gear and to prevent a loss of control on unprepared or poorly prepared surfaces. Lift-off is accomplished at the lowest possible airspeed. This means that it is necessary to take advantage of ground effect to accelerate prior to climbing. Without this acceleration, a stall may very well be the result. As with the short field takeoffs, many aircraft use high lift devices during soft/rough field operations.

Conducting a takeoff with a crosswind is not fundamentally different from a takeoff without a crosswind. A crosswind simply means that directional and lateral inputs become larger. During the roll, ailerons are deflected into the wind in order to prevent a roll away from the wind. As well, the rudder must be used more assertively in order to maintain directional control. The crosswind limits of the aircraft are often published. When they are not, 20% of the stall speed can be used to determine the minimum allowable crosswind limit of the aircraft.

Wind shear is a rapid shift in wind speed and/or direction. Due to the inertia of the aircraft, this shift will cause a sudden increase or drop in airspeed. An increase in speed is caused by an increased performance shear, and a decrease in speed is caused by a decreased performance shear.

Useful Rules of Thumb

Short Field

Performance

$$d = D(GW/MGW)^2 \qquad (22.1a)$$

$$d = D(GW/MGW) \qquad (22.1b)$$

$$d_o = D_o(GW/MGW) \qquad (22.2)$$

$$d = D\frac{(v_{LO} - w)^2}{v_{LO}^2} \qquad (22.3)$$

$$d_o = D_o\frac{TAS-W}{TAS} \qquad (22.4)$$

– Increase both the takeoff roll by 8% to 10% per 1000 ft of density altitude.

– Increase obstacle clearance distance required (after lift-off) by 12% to 15% per 1000 ft of density altitude.

– Taking off on surfaces other than pavement, increase the pavement distance by approximately the following amounts:

Gravel (no potholes)	– 10%
Short Grass (< 3")	– 15%
Long Grass (< 6")	– 30%
Snow/Slush (< 3")	– 50%.

– For runway gradients, increase the roll by approximately 10% for a 1% upslope. Decrease the roll by approximately 5% for a 1% downslope.

Considerations

– Taking off in gusty or turbulent conditions, increase the rotation speed by ½ of the gust factor.

– To determine the approximate accelerate-stop distance, add the takeoff and landing distances.

– When the runway is not long enough for the accelerate stop distance. An abort point should be chosen based on the fact that half of the lift-off groundspeed should be reached at one quarter of the required lift-off distance.

Soft or Rough Field

Performance

        When taking off from a soft or rough field, increase the takeoff distance by approximately the following amounts:

Mud (< 3")	–	60%
Mud (≥ 3")	–	100% to ∞
Snow/Slush (≥ 3")	–	100% to ∞
Wet Grass	–	60%
Long Grass (≥ 6")	–	100% to ∞
Rough surface	–	40%.

Crosswind

Considerations

        When an aircraft doesn't have a published crosswind limit, use 20% of the stall speed.

Questions and Problems

NOTE: For questions that pertain to performance, assume that no manufacturers data is published beyond the information given.

1) Given a choice of several runways, what are some factors that would be considered in choosing one to use?

2) How does ground effect affect a normal takeoff?

3) What are the differences between the normal and short field rotations?

4) An aircraft has a MGW of 3275 lbs. and a sea level takeoff distance of 1480 ft with a rotation speed of 67 knots. Approximately what will the takeoff distance be for this aircraft loaded to weigh 2800 lbs.?

5) If the aircraft in Q4 needs a total distance of 2150 ft. to clear a 50 ft. obstacle, approximately what distance will it need to clear a 50 ft. obstacle at MGW with a 15 knot headwind? What about a 5 knot tailwind? The aircraft has an obstacle clearance climb speed of 74 knots.

6) What are the pros and cons of using flaps during a short field takeoff?

7) If the aircraft in Q5 were taking off from an airfield at sea level surfaced with short, dry grass, what would be the approximate distance required to clear a 50 ft. obstacle with a 10 knot headwind?

8) Once again referring to Q4, if this aircraft was to takeoff from an airstrip with a density altitude of 3000 ft. while loaded to a weight of 3000 lbs., what would the approximate takeoff distance be without an obstacle?

9) An aircraft has a MGW of 2980 lbs. and a takeoff distance of 1020 ft. at sea level with a rotation speed of 62 knots. If this aircraft was to takeoff with the winds 60° off the runway at 15 knots gusting to 25 knots, what should the new rotation speed be, and approximately what would the take off distance be?

10) If the aircraft in Q9 has a stall speed of 54 knots, is the crosswind within limits according to the rule of thumb presented in this chapter? If not, is it possible that the takeoff could be within safe limits?

11) If the aircraft in Q9 was to be operated out of a grass strip at a 3400 ft. density altitude, what would the most likely take off procedure be after heavy rain the previous night? What would the approximate required takeoff distance be? The total distance required to clear an obstacle at sea level is 1850 ft.

12) What would be the result of passing through a decreased performance wind shear surface during climbout? Are there any hazards associated with this phenomenon?

13) What is the significance of the procedure used to depart from a soft field? Are there any differences between this and departing a very rough field? Are there any hazards associated with this procedure?

Chapter 23
Approach and Landing

It is a common belief that the landing is the most difficult maneuver to execute in an airplane. In actual fact, it is one of the easier things to *do*, however it is probably the most difficult maneuver to *learn*. For this reason, it's important to have a thorough understanding of the control functions and aircraft behavior during the landing.

During the landing phase of flight, the aircraft is returned to the ground. The vertical speed should be as low as possible to prevent damage to the aircraft or a bounce. As well, the forward speed should usually be reduced as much as is safely possible in order to reduce the distance required to land.

In the present chapter, we will look at the techniques to use for the approach and the landing. Like the takeoff, there are several different (but closely related) landing techniques used to match the given conditions. These include the "normal" landing, the short field landing, and the soft/rough field landing. Also like the takeoff, these approach and landing techniques can be conducted with obstacle clearance and/or crosswind considerations.

Normal

The "normal" landing is used at airports where the length and surface condition of the runway are not important factors in determining the safety of the landing. This means that the surface is firm and reasonably smooth, and the runway is several times as long as the aircraft's minimum required landing distance. The normal landing is considered to be the easiest to perform—not to mention the safest for new students, who normally require some room for error—and is therefore the first to be taught. The basic techniques used here will be carried forward into all other types of landings.

Approach

An approach to land is simply an application of the descending techniques discussed in Chapter 14. The speed selected for the approach is generally about 1.3 V_s. This speed allows for a reasonably slow approach while still providing a large margin between the stall and the approach speed. As with any descent, this speed is maintained by adjusting the pitch attitude of the aircraft—pitch up to slow down and pitch down to speed up. The approach path (or glidepath) is controlled with the use of power. Assuming a constant airspeed is maintained, higher power settings result in a shallower approach due to a slower descent rate, and lower power settings will result in a steeper approach due to a higher descent rate (Fig 23-1).

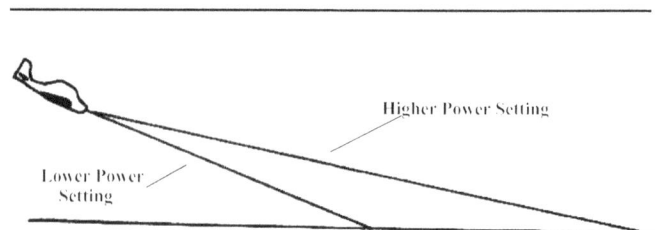

Fig 23-1. - The Effect of Power on the Approach Path. - If a constant airspeed is maintained, higher power settings will result in a more shallow approach path. This will increase the distance to the touchdown point.

In controlling the approach path, power should be used to control the aiming point. We saw in Chapter 14 that the visual indication of the aiming point is a stationary point on the windscreen (Fig 23-2). This is the point at which the aircraft would touch down if no flare was performed. The desired aiming point will vary depending on the intentions of the pilot (e.g. – due to following traffic and/or an

ATC request, an intentional long landing may be made to provide for a more rapid exit of the runway) and the type of aircraft. Generally, however, for light aircraft the 500 ft point on the runway is used (some runways include 500 ft markers, which makes this more convenient).

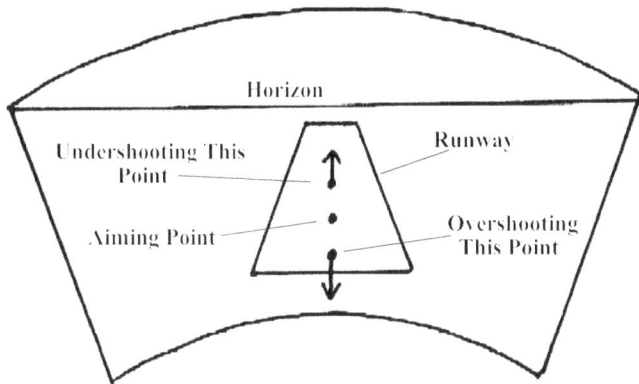

Fig 23-2. - The Stationary Point Indicating the Aiming Point. - Looking out the windscreen during the approach, three points of significance can be found. The point that appears to be stationary is the current aiming point. Any points that are moving upward in the windscreen are being undershot. And any points moving down in the windscreen are being overshot. The current aiming point can be changed by changing the power setting.

If the intended aiming point is stationary in the windscreen once the approach is established, the glidepath does not need to be adjusted—the approach can be continued to the flare. Often, however, adjustments are required. These adjustments are made by changing the power setting. If the desired aiming point is moving up in the windscreen, the approach is too steep and power must be added. As the power is added, the pitch attitude should also be adjusted to maintain the appropriate airspeed. If the aiming point is moving down in the windscreen, the approach is too shallow and power must be reduced. Once again, as the power setting is changed, the pitch attitude needs to be readjusted to maintain the approach airspeed.

Flare

The flare maneuver is intended to reduce the descent rate to a level which is tolerable to the airframe structure and to the crew and passengers. During the approach, the rate of descent is normally much too high to make a safe ground contact—the

result could be structural failure or a bounce (which can lead to more problems). As well, the nose low attitude used for the approach will lead to the nose wheel contacting the ground first. This means that if a failure occurs, the nose gear will likely be the point of failure and a prop strike will follow. Along with a hard impact, a rapid deceleration will result and the occupants of the aircraft could be injured.

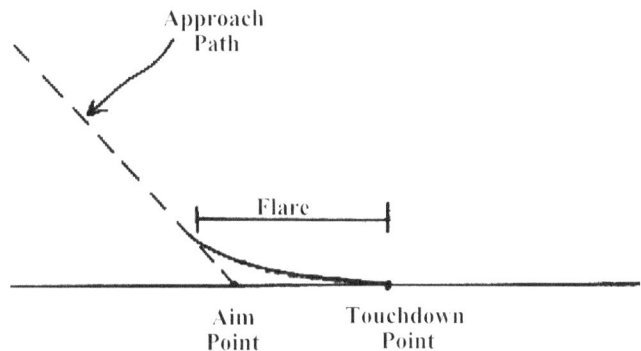

Fig 23-3. - The Flare Effect on the Touchdown Point.

Instead of allowing these things to occur, the flare uses the excess airspeed available during the approach and stops the descent just as the runway is reached. This changes the approach path so that the actual touchdown point is further down the runway than the aiming point (Fig 23-3). Often, the intended touchdown point is around the 1000 ft mark of the runway, this is why the aiming point for the approach is chosen further back.

To execute the flare, a gradual but continuous pitch adjustment is made to reduce the airspeed close to the ground. This adjustment should begin from approximately the height of a wingspan to half of the wingspan and should ideally end at the runway surface. As the airspeed is reduced, the flight path changes so that the descent is more shallow and the descent rate is reduced. As well, the change in pitch attitude allows the main wheels to touch down first. The flare should be judged so that the aircraft touches down just as it runs out of airspeed.

To accomplish this, consider using pitch to control the vertical speed of the aircraft. As we pitch up, the airspeed will be reduced as well as the vertical speed. If no more adjustment is made, the aircraft will recommence the descent at a steeper angle than original (at this point most aircraft are below the best glide speed). However, if the pitching movement is continued, the airspeed will continue to decrease and the vertical speed will remain low. This is precisely what we want to do when flaring to land.

Often, landings are difficult simply because the mind set is incorrect. Instead of trying to land the aircraft, we should be trying to level off at 6 inches above the runway to prevent the aircraft from landing. With this in mind, we can simply recall from Chapter 13 that at lower airspeeds a higher pitch attitude is required. For this reason, as the airspeed bleeds off in the flare, the aircraft must be pitched up. As for judging the flare height, visual reference to the end of the runway is critical.

The landing technique described here is often referred to as the "full stall" landing—meaning that the aircraft stalls just as it touches down. There are other methods to landing an airplane, but this one tends to be the most effective and the easiest to accomplish consistently well.

Directional control is maintained with the rudder. If the aircraft is not pointed straight down the runway, yawing towards the runway is necessary. This means applying rudder pressure towards the runway to make the correction. Any lateral drift away from the selected centerline of the runway also needs to be corrected. This can be done by applying aileron and establishing a banked attitude away from the direction of the drift. During crosswind landings these techniques will result in a sideslip which will correct for the crosswind. This will be discussed in more detail later in this chapter.

Rollout

Once the aircraft is on the ground, the flare is complete—but the aircraft is still "flying". For this reason, the controls can't simply be released and forgotten. The necessary inputs should be held at least until the aircraft has decelerated to taxi speed, and then the necessary taxi inputs can be applied. The elevator position established late in the flare should normally be held at least until the nose wheel lowers to the ground due to the reduced airspeed. Releasing the elevator prior to this point can lead to a sudden drop of the nose which could possibly overstress the gear.

The directional control established with the rudder cannot be simply forgotten once the aircraft is on the ground either. If it is, the aircraft may veer off the runway—especially in crosswind operations.

The same stands for the ailerons. If the ailerons are returned to neutral, the aircraft may begin an uncommanded roll—that is, one wing may lift back off the ground. As with the rudder, this is more of a consideration in crosswind operations.

Considerations

During an approach and landing, there are many variables to contend with and decisions to be made. It is important that these variables receive due consideration and that the decisions made are informed decisions. Items such as traffic separation and ATC clearances are topics for other texts, however some considerations come under the umbrella of aerodynamics and it's applications.

Gusty Winds – When an approach is flown in gusty winds, the airspeed will fluctuate above and below the target airspeed. This reduces the safety margin between the approach speed and the stall speed and may be hazardous when large gusts are present. For this reason, there is a commonly accepted rule of thumb—add half of the gust factor to the normal approach speed. This means that if the wind speed is 20 knots gusting to 30 knots, the gust factor is 10 knots and the approach speed should be increased by 5 knots. This adjustment becomes more important during performance landings when the approach speed is reduced further than normal.

Flaps and Slips – Most, if not all, light aircraft can make a safe landing without the use of flaps. Flaps generally are used to increase the descent angle and thus help to clear an obstacle or to correct for an approach flown inadvertently too high. As well, flaps reduce the stall speed and therefore allow a slower approach leading to a shorter landing.

During a normal approach and landing, obstacles and landing distance available are not critical considerations. However, partial flaps are often used anyway. The reason for this is the increase in visibility provided by the lower pitch attitude (Fig 23-4). If flaps are extended without a change in power setting, pitching down will be required to maintain the same airspeed. This raises the horizon in the windscreen and provides better visibility for the pilot. In addition to this, the stall speed (and therefore touchdown speed) improves the safety of the landing.

Forward slips can also be used for similar reasons. If an approach is inadvertently flown too high, a slip will increase the descent rate (and hence the descent angle) to allow for a correction. When a slip is used only to lose altitude and not to correct for a crosswind, it is generally preferred to slip towards the pilots side to improve visibility.

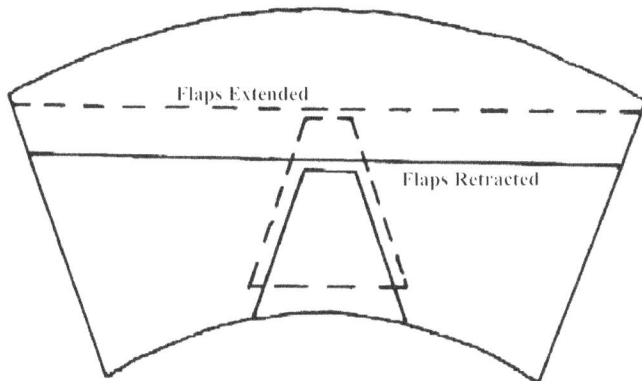

Fig 23-4. - *The Effect of Flaps on the Pilot's Visibility.* - Assuming a fixed airspeed, extending the flaps dictates the use of a lower pitch attitude for the approach. This improves the pilots visibility.

Choice of Centerline – As with the takeoff, the actual centerline of the runway is not always the best choice for the landing path. An offset landing may be preferred due to considerations such as ice patches, snow drifts, or water on the painted portions of the runway.

Bouncing and Ballooning – Bouncing and ballooning are very common problems when learning to land. A bounce occurs when the vertical touchdown speed is too high and the aircraft bounces back into the air. Ballooning results from over-flaring and entering a climb from the flare maneuver. Both the bounce and the balloon are minor problems when handled properly, however they are potentially dangerous when allowed to continue or get aggravated.

Two things can go wrong after a bounce or a balloon. First of all, if no recovery is initiated, the low airspeed and nose high attitude will result in slow flight and possibly a stall. The second possibility is a "porpoise". A porpoise occurs when the pilots correction inputs get out of phase with the aircraft's reaction. The result is an unstable pitch oscillation which eventually leads to a hard impact on the nose gear. The porpoise begins when the pilot pitches the aircraft down to correct the bounce or balloon. This input comes along at just about the same time as the aircraft's dynamic pitch characteristics would have lowered the nose anyway. The combined effects lead to an over correction which is then corrected for by pitching the aircraft up again.

To avoid a porpoise and the possibility of inadvertent entry into slow flight or a stall, a proper recovery from a bounce or balloon must be initiated immediately. For a slight bounce or balloon, the recovery simply involves maintaining a slightly nose high attitude and continuing the flare to a touchdown. After a large bounce or balloon, the addition of power is required to reestablish the aircraft in controlled level flight above the flare height. From there, power can be once again removed and another flare initiated. If the bounce/balloon is very large and/or the airspeed is dangerously low, full power should be applied and after a safe airspeed has been achieved an overshoot can be initiated (overshoot techniques are discussed later in this chapter).

Wake Turbulence – All heavier than air aircraft produce the wing tip vortices and downwash which were discussed in Chapter 3. These vortices persist for some time after the passage of an aircraft and can be a hazard to following aircraft. In the vicinity of an airport—particularly during takeoff and landing—awareness of these vortices (often referred to as wake turbulence) and avoidance procedures can be critical. Roll rates encountered at the core of a vortex can be well in excess of an aircrafts maximum roll rate by aileron. As well, descent rates encountered between the two vortices may be well in excess of an aircrafts maximum climb rate.

Vortices are created during the production of lift. So it is reasonable to say that aircraft producing more lift (i.e. – heavier aircraft) will produce stronger vortices. This is exactly the case and as a result, wake turbulence becomes a concern when lighter aircraft are following heavier aircraft.

The first step in avoiding these vortices is in understanding how they behave. Vortices are associated with the downwash produced by the wings. Because of this, they will always travel downward. So remaining above the flight path of a preceding aircraft will guarantee wake turbulence separation. As well, wake turbulence will gradually dissipate due to the internal friction of the air. Usually the vortices will be gone two to three minutes after they are formed. This means that providing sufficient separation behind a heavier aircraft will allow the vortices to dissipate so that they will not be a danger.

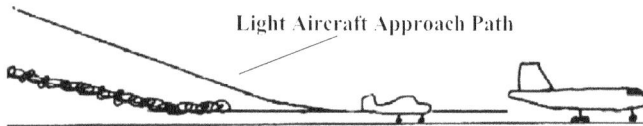

(a) Landing Behind a Heavier Aircraft – Approach above and touchdown beyond the heavier aircraft.

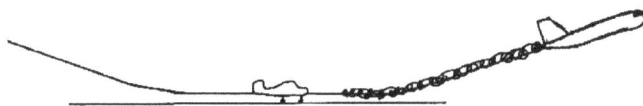

(b) Landing after a Heavier Aircraft's Departure – Aim to be on the ground and stopped prior to the heavier aircraft's lift-off point.

Fig 23-5. - Wake Turbulence Avoidance - Always remain above the lead aircraft's flight path. Since wingtip vortices always travel downwards, this will keep you away from them.

When an approach is flown behind a larger aircraft, wake turbulence avoidance procedures should be used. This simply means remaining above the leading aircraft's flight path and touching down beyond that aircraft's touchdown point (Fig 23-5a). Since wing tip vortices travel downward and break up in ground effect, this will insure avoidance. When approaching after a larger aircraft's departure, it is a good idea to be stopped prior to the rotation point of that aircraft (Fig 23-5b).

Short Field

On some runways, the landing distance available eliminates the option of performing a normal landing. When this is the case, a short field technique is used to safely get the aircraft on the ground. The primary difference between the normal landing and the short field landing for most aircraft is simply in the approach speed and the use of brakes. The approach configuration may also be different, but this is primarily to reduce the stall

speed and therefore reduce the approach speed safely.

Approach

During performance landings, most aircraft require the use of maximum available flaps (or any other high lift devices installed on the aircraft). This allows the stall speed to be reduced and therefore the approach speed will be reduced. High lift devices will also produce extra drag which serves three purposes for us. First off all, with more drag, the approach descent will be steeper (recall from Chapter 8 that the glide ratio is equal to the L/D ratio). This fact comes in very handy when we are clearing obstacles on the approach and don't want to waste runway after the obstacle (Fig 23-6). Second, in the flare more drag will cause a more rapid deceleration. This means that the aiming point and the actual touchdown point will be closer together than in a normal approach and landing. Lastly, early in the rollout aerodynamic drag is a large contributor to the total braking action of the aircraft. More braking action of course means a shorter landing roll.

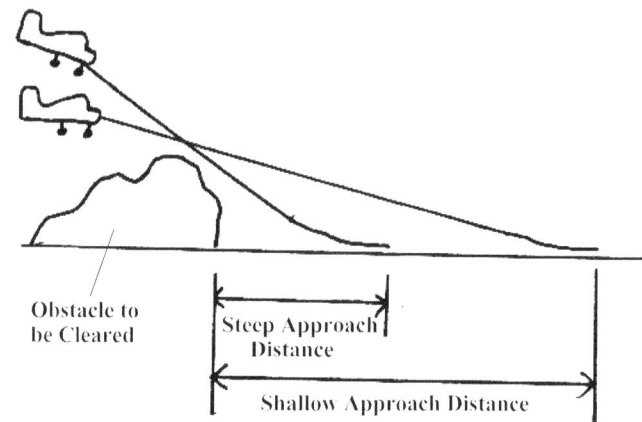

Fig 23-6. - Obstacle Clearance Descent - A steep descent minimizes the runway overflown after the obstacle is cleared.

In order to reduce the distance used up in the flare even further, a lower airspeed is often used for the approach. If this speed is not published by the manufacturer, a commonly accepted rule is to use 1.2 times V_{so}. This speed allows for a very slow approach while still providing an acceptable margin from the stall.

As with the normal approach, the desired airspeed should be maintained with the use of pitch, while the approach path is controlled with power.

Flare

The flare technique for a short field landing is very much the same as for the normal landing. The only difference is that the "hold-off" may not be continued as long since ground contact means the presence of rolling friction which will shorten the landing distance. The sooner the aircraft is firmly established on the ground, the sooner brakes can be applied. This having been said, the flare is still a necessary maneuver to reduce landing loads and establish the nose high touchdown attitude which prevents damage to the nose gear.

As with the normal landing, directional control is maintained with the rudder and lateral drift control is maintained with the ailerons.

Rollout

During a normal landing, the rollout simply involves allowing the aircraft to decelerate to a safe taxi speed. On a short field, however, the rolling distance is to be minimized. This means that we need to determine the optimum method for slowing the aircraft to a stop.

It was stated earlier that high lift devices provide us with aerodynamic drag that will help us slow down early in the roll. There is, however, a disadvantage to using these devices. Even though drag is being produced, so is lift. As long as the wings are producing lift, the full weight of the aircraft will not rest on the landing gear and as a result, the brakes will not be as effective (recall from Chapter 1 that frictional drag is a function of the normal force holding two surfaces together—in this case weight). On many aircraft, the flaps (or other high lift devices) will do more harm than good late in the roll. This means that at some point they should be retracted to maximize braking (unless other issues dictate leaving them extended, e.g. – the flap switch is right next to the landing gear switch and confusing the two in the rush of landing could be disastrous).

Maximum braking is usually applied as early as possible in the landing roll without "locking" the brakes. Besides retracting the high lift devices, braking effectiveness can be improved by holding the control column full aft during the rollout. With the elevator deflected upward, the tail down force being produced lifts the nose wheel and transfers

some weight onto the main wheels which is normally where the brakes are. If spoilers are available, they can be used to increase the amount of weight on the gear as well.

Performance

As with the short field takeoff, aircraft performance is often a major concern during a short field landing. For this reason, when landing at a questionable aerodrome, the capabilities of the aircraft should be determined prior to attempting the landing.

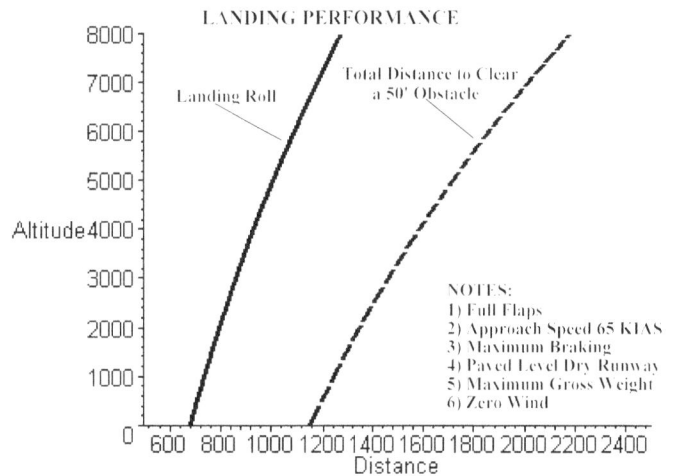

(a) Landing Performance Graph

WT.	WND.	Sea Level		2000'		4000'		6000'	
		Roll	50'	Roll	50'	Roll	50'	Roll	50'
2500 lbs	-5	720	1241	840	1518	979	1862	1142	2290
	0	620	1057	715	1293	834	1586	973	1950
	10	423	730	494	893	576	1095	672	1346
2250 lbs	-5	648	1117	756	1367	882	1677	1029	2061
	0	552	952	644	1164	752	1428	876	1755
	10	381	657	445	804	519	986	605	1212
2000 lbs	-5	576	993	750	1215	784	1490	915	1832
	0	490	846	572	1035	678	1270	779	1560
	10	339	584	396	715	461	876	538	1077

CONDITIONS:
1) Paved Level Dry Runway
2) Weights and Winds as indicated
3) Approach at 65 KIAS
4) Standard Conditions

NOTES:
1) Add 19% to Ground Roll Distance for Short Dry Grass.
2) Add 10% to Ground Roll Distance for a 1% Downslope.

(b) Landing Performance Chart

Fig 23-7. - *Typical Manufacturer's Landing Data.* - Always be sure to note any conditions attached to manufacturers data. This will enable you to recognize other conditions.

Fig 23-7a is a landing performance graph. Fig 23-7b is the same information presented in chart form. These graphs/charts are very similar to those used for takeoff performance, and are used in much the same way. They provide information on what the aircraft is capable of under various atmospheric and runway conditions. The conditions specified on the charts must be adhered to in order for the numbers to be accurate.

Manufacturers will often include correction factors for the variables that will affect the aircraft's performance. However, when this information is not provided by the aircraft manufacturer, rules of thumb can be applied to approximate the expected performance from the aircraft. Each of these rules of thumb relates in some way to equation 8.15:

$$d = \frac{Wv_{td}^2}{2gD} \qquad (8.15).$$

Weight – The effect of weight on the landing distance is somewhat deceiving. It is often assumed that a lighter aircraft will stop shorter since it has less momentum to dissipate during the rollout. In fact, a reduction in weight will also reduce the effectiveness of the brakes, so the reduction in momentum doesn't shorten the roll. We see then that weight has no *direct* effect on the landing distance (we discovered this in Chapter 8).

In fact, a reduction in weight will reduce the landing roll because of the lower approach and touchdown speeds that can be achieved. The approach and touchdown speeds are based on the stall speed, which in turn changes with the square root of the weight change. Since equation 8.15 indicates that the landing distance changes with the *square* of the touchdown speed, the actual landing distance will change proportionally to the weight change:

$$d = D(GW/MGW) \qquad (23.1).$$

Keep in mind that this rule of thumb is only valid if the speeds used for the approach and landing are adjusted for weight.

Wind – Wind will change the touchdown groundspeed, and therefore the landing distance. Obviously, a headwind will reduce the ground roll due to a decreased touchdown speed and a tailwind will increase the ground roll due to an increase in touchdown speed. Since, from equation 8.15, the

landing distance is proportional to the square of the touchdown speed, we get the following:

$$d = D\frac{(v_{td} - w)^2}{v_{td}^2} \qquad (23.2).$$

Wind will also change the distance required to clear an obstacle since the groundspeed will be changed but the vertical speed will not. The new distance to clear the obstacle will be proportional to the groundspeed change:

$$d_o = D_o\frac{TAS-W}{TAS} \qquad (22.3).$$

These are the same rules of thumb that were used for estimating takeoff performance.

Another effect that wind will have is to reduce the approach groundspeed. This means that either the rate of descent will need to be reduced on the approach by carrying extra power, or the approach descent will have to be started later (in order to use eq. 22.3 for obstacle clearance, the approach must be flown steeper and therefore must be started later).

Turbulence – Turbulence will cause airspeed fluctuations during the approach. The short field approach speed doesn't normally allow a safety margin wide enough to allow for these fluctuations. Because of this, it is important to adjust the approach speed according to turbulence. A good rule is to increase the approach speed by half of the gust factor. Be sure to account for this speed change in performance calculations.

Density Altitude – We've seen previously that at higher density altitudes a given EAS will result in a higher TAS. In equation 8.15, the touchdown speed is the *true* touchdown speed, not the indicated, calibrated, or equivalent touchdown speed. Approaches are normally flown at a fixed indicated speed (which for our purposes can mean a fixed equivalent airspeed). This means that at higher altitudes the touchdown speed will be higher even though the same speed is read off the airspeed indicator.

In the lower atmosphere, the TAS increases by approximately 2% per 1000 feet of density altitude increase. Since the distance increases with the square of the speed increase, the landing distance will change by approximately 4% per 1000 feet of density altitude (1.02² = ~1.04). Because the engine is not used for landing, loss of engine performance will not affect the landing distance.

   Surface Condition – The surface condition will have a direct impact on the braking effectiveness of the aircraft during the roll out. Dry pavement is the most optimum surface to land on for braking efficiency, but sometimes performance landings are executed on other types of surfaces. The degraded braking performance must be taken into consideration when determining aircraft performance.

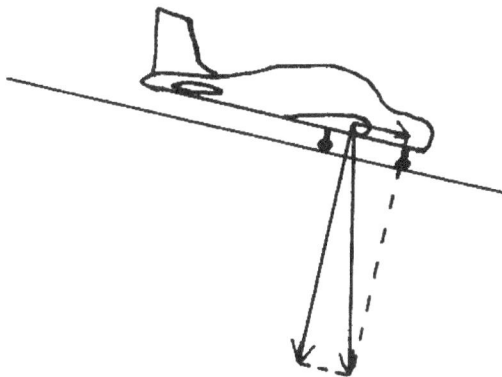

(a) Landing on a Downgrade – Distance is increased

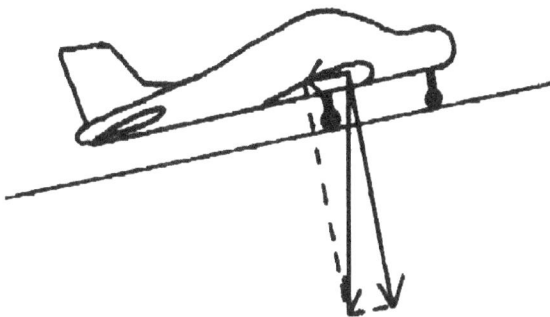

(b) Landing on an Upgrade – Distance is decreased

Fig 23-8. - *Runway Gradients.* - The forward or aft component of weight (as applicable) will influence the distance required to land.

   On soft surfaces, the sinking of the wheels will add to the drag on the aircraft, but the inability to use maximum braking will lead to an overall reduction in drag. Standard landing performance data is usually based on dry pavement, so landing on another surface requires an increase in distance over the anticipated "pavement" value. The

following are some approximate percentages to add when landing on other surfaces:

Wet Pavement	– 20%
Firm Turf	– 20%
Short Grass	– 30%
Long Grass	– 50%
Wet Grass	– 100%
Mud or Snow	– 100%

   Runway Gradients – Landing downhill will cause a component of the aircraft's weight to be acting in the direction of motion (Fig 23-8a). As a result, the landing will be longer. Landing uphill is just the opposite—a component of weight will act to slow the aircraft down (Fig 23-8b), thus shortening the landing roll.

   As a general rule, a 1% upgrade will shorten the roll by about 5% and a 1% downgrade will lengthen the roll by about 10%. Unlike the gradient corrections for takeoff, these numbers don't vary much with altitude since engine power is not used for landing.

Soft or Rough Field

   As with the soft and rough field takeoffs, the main concern with a soft or rough field landing is to avoid excessive gear loads which may be damaging and to avoid "digging in" which may cause a loss of control. This means that touching down at a low airspeed and a low vertical speed is of the utmost importance.

Approach

   The approach to a soft or rough field is carried out almost identically to a short field approach. The difference is that when landing distance is not a big concern, a slightly higher approach speed may be used (1.3 V_{so}). When landing distance is a concern, the short field approach speed can be used. This higher speed allows for better control in the flare. Once again, airspeed is controlled with pitch and approach path is controlled with power.

Flare

   The flare technique used for landing on a soft or rough field is somewhat different than the normal or short fields. Power can be used to help arrest the descent and to improve the elevator effectiveness in the flare. Only a small amount of power should be used—too much will cause an acceleration or a climb. This flare method takes a longer period of time and a longer distance to complete. However, it also provides for a much softer and somewhat slower touchdown—which is the objective during a soft field landing. If the distance available is critical, the power can remain at idle for the flare. However, this method requires a much finer touch to accomplish the same objective and should be left until some experience is gained.

Rollout

Once the main wheels are on the ground the power should be reduced to idle—otherwise, the aircraft will be prone to lifting off again. Application of aft elevator should be continued in order to hold the nose wheel off the ground as long as possible. Even when the nose wheel settles, aft elevator should be held to minimize the stress on the nose gear. Once the nose wheel has settled, the flaps can be retracted. The use of brakes should be kept to a minimum to reduce the chance of the nose wheel digging in and causing an upset.

Performance

The landing distance required for a soft or rough field landing is significantly higher than that required for a short landing. This is primarily due to the power used in the flare and the absence of maximum braking. The actual difference will vary quite a bit from one aircraft type to another and with piloting technique. A good rule for most light aircraft is to multiply the distance required for the ground roll portion of a short landing by 3 or 4. The distance to clear an obstacle if necessary won't change.

Crosswind

Any of the landing techniques detailed so far may have to be executed in a crosswind. When this is the case, the techniques used vary slightly. As with the crosswind takeoffs. The difference is more in the degree of control inputs as opposed to the type of control inputs.

Approach

A crosswind approach is flown in a "crab" to maintain the runway centerline. This crab angle allows the aircraft's velocity to have a sideward component to cancel-out the crosswind component (Fig 23-9). The net effect should be that the flight path is along the runway's extended centerline. The exact crab angle that needs to be used will vary depending on the crosswind component and the approach speed being used. With large crosswinds and low approach speeds, the crab angle will be large.

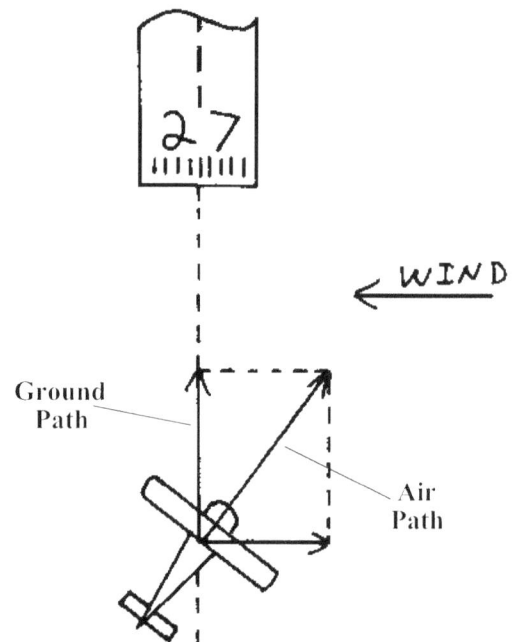

Fig 23-9. - *"Crabbing" on Approach* - The component of airspeed perpendicular to the runway cancels the drifting effect of the crosswind.

As for approach speed and configuration, the choice will be based on the type of landing intended. In a large crosswind, however, the amount of flaps used is often reduced due to their tendency to increase weathervaning and reduce rudder effectiveness. This configuration limit and the resulting increase in approach speed must be accounted for when determining landing performance in a crosswind.

Short Final

If the approach is continued in a "crab", the landing gear will suffer side loads upon touchdown and may fail under the stress. For this reason, the crab approach is abandoned on short final and replaced with a sideslip approach (it is actually favorable to eliminate the crab while flaring, but this technique takes considerably more coordination and is generally left until some experience is gained). The slip is initiated by applying rudder toward the runway. This will cause a roll which must be eliminated with the use of aileron. As well, enough extra aileron must be used to roll the aircraft into the wind enough to eliminate the lateral drift resulting from the crosswind. Once the aircraft is established in the slip, the flight path and the heading will both be parallel to the runway centerline. This is the desired condition for landing in a crosswind without stressing the landing gear.

Flare

Once in the flare, a more precise adjustment of the side slip is possible due to increased visual cues. At this point, it is important to understand the function of each of the controls—it is pretty simple really, all of the controls are used in the same manner as any other landing. This having been said, things will seem quite a bit different to the unaccustomed.

The elevator and the power are used in the exact same manner as any other landing. The difference in control use becomes apparent when considering the rudder and ailerons. Once again, the function of the controls hasn't changed any, but the degree of use will increase. The rudder and ailerons will always be crossed in a properly executed crosswind landing. Using the rudder for directional control, the aircraft can be pointed directly down the runway. As this is being done, the angle of bank can be adjusted with the ailerons to eliminate lateral drift across the runway.

An increase in the angle of bank while maintaining a straight direction with the rudder will increase the sideslip angle and therefore cause the aircraft to move into the wind more. If the initial crosswind correction is insufficient, this adjustment will allow for a straight landing. On the other hand, if the aircraft is drifting into the wind during the flare, decreasing the angle of bank while maintaining straight with rudder will decrease the slip and eliminate the drift.

The sideslip is continued to the touchdown since the intention is to eliminate side loads when contact is made with the ground. Since the aircraft is in a banked attitude, the upwind main wheel will touchdown first, then the downwind wheel, and finally the nose wheel.

Rollout

During the rollout from a crosswind landing, the crosswind should not be forgotten about. Even though the aircraft is on the ground, air is still flowing over the wings and aerodynamic reactions are still taking place. If the controls are released during a crosswind rollout, it is possible that the upwind wing will become airborne again and the aircraft will roll away from the wind. If this is allowed to continue, the downwind wingtip will eventually strike the ground and control will be lost.

To avoid this condition, the ailerons should remain deflected into the wind during the entire rollout from a crosswind landing. The rolling effect of the ailerons will then cancel the rolling effect of the wind.

The use of rudder during a crosswind rollout also deserves some consideration. Initially, as the main wheels touch down, the rudder is deflected opposite the crosswind. If the aircraft has a free swinging nose wheel, this input can be maintained throughout the landing roll. However, if the nosewheel is steerable via the rudder pedals, this input will cause problems when the nose wheel touches down. For this reason, in an aircraft with a steerable nose wheel, the rudder pedals must be neutralized immediately prior to the nose wheel touching down. Once the nose wheel is firmly on the ground, directional control can be maintained via nose wheel steering.

Considerations

Use of Flaps – When conducting short and soft field landings, it is normally favourable to use full flaps. In a crosswind, this may not be the case. Flaps cause two problems in a crosswind. First of all, they increase the aircraft's tendency to weathervane once on the ground. This may contribute to a loss of directional control during the rollout. Secondly, on some aircraft (particularly high winged aircraft) the turbulent airflow coming off the flaps flows over the vertical stabilizer (Fig 23-10) and reduces the effectiveness of the rudder. Since rudder effectiveness is usually the limiting factor

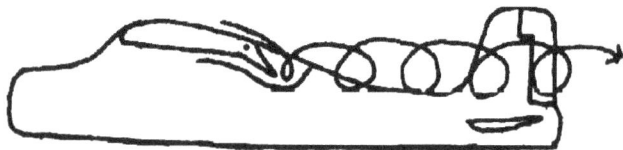

Fig 23-10. - *Effect of Flaps on the Rudder.* - The turbulent airflow coming off of the flaps can blanket the rudder and reduce it's effectiveness.

when landing in a crosswind, this effect can be detrimental.

Crosswind Limits – The crosswind limitations established for takeoff apply equally during the approach and landing. If the manufacturer has not published a crosswind limit, 20% of the stall speed can be used. Again, crosswind and CRFI charts similar to those in Fig 22-7a and 22-7b can be used to determine the crosswind component.

Canadian Runway Friction Index – As with the takeoff, the runway surface condition may restrict the aircraft's crosswind capability more than aerodynamic limitations do. Snow, ice, etc. may reduce the friction between the tires and the runway enough to make the aircraft uncontrollable in a crosswind. Canadian Runway Friction Index (CRFI) information can be found in the Canada Flight Supplement and is applied to landings in the same way that it is applied to takeoffs.

Gusty Winds – As with the takeoff, landing in a gusty crosswind is far more challenging than landing in a steady crosswind. A constant readjustment of the correcting inputs will be required—especially in the flare.

Again, use the peak wind for crosswind calculations. Using the steady wind speed or an average wind speed can lead to an underestimation of the winds that you will have to deal with on the landing.

Overshoot

During some approaches, The decision may be made to discontinue the approach and go around for another try—or to divert to another airport. Once this decision is made, an overshoot (A.K.A. – 'missed approach' or 'go around') is initiated. An overshoot is basically a transition from

an approach descent to a departure climb without leveling off or touching down.

The overshoot procedure essentially combines the level out from a descent with a climb entry. Recall from Chapter 14 that a descent recovery is initiated with power:

P – Power

A – Attitude

T – Trim.

This procedure is also used for the overshoot. Full power (or climb power) is applied and the aircraft is pitched up into a climb attitude. Once the overshoot climb is established, the pitch trim can be adjusted appropriately. If flaps or other high lift devices are extended for the approach, they should be retracted prior to trimming since they will have an effect on the required trim adjustment. As well, these devices will detract from the aircraft's climb performance which in some cases could be critical.

If an approach leads to a dangerous situation such as a very large bounce/balloon or inadvertent entry into slow flight or a stall, the recovery is often followed by an overshoot. This is a better alternative than continuing the approach and landing from a poor position.

Wind Shear

As we discussed in Chapter 22, wind shear is a sudden change in wind speed and/or direction. This shift in wind will cause a fluctuation in airspeed due to the inertia of the aircraft. As with wind shear on takeoff, wind shear during an approach can be divided into two fundamental categories: increased performance and decreased performance.

Increased Performance

As with the takeoff, a sudden increase in headwind or decrease in tailwind will result in an increase in airspeed due to the inertia of the aircraft. Unlike an increased performance wind shear on takeoff, however, increased performance wind shear during an approach is more than an inconvenience—it can cause some serious problems—especially for pilots flying an IFR approach.

Initially, the sudden increase in airspeed will cause the aircraft to rise above the desired approach path as the excess airspeed bleeds off. To get back onto the desired approach path, a power reduction will be necessary. As the approach path is regained, an increase in power will be required to maintain it. This power increase will bring the power setting higher than the original setting due to the increased headwind decreasing the groundspeed and thus dictating a lower rate of descent to reach the desired aiming point. This whole process is illustrated in Fig 23-11.

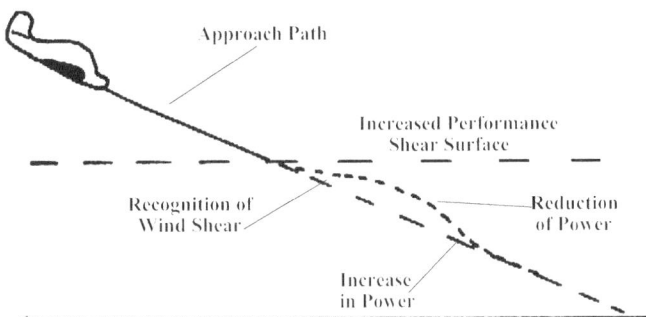

Fig 23-11. - *Increased Performance Flight Profile.*

Decreased Performance

A decreased performance wind shear is caused by a sudden increase in tailwind or a sudden decrease in headwind. The decreased performance shear is much more dangerous than the increased performance shear because not only does it lead to an inaccurate approach, but the drop in airspeed can lead to a stall. Combined with the risk of a stall, the aircraft will inevitably descend below the desired glidepath due to the loss in airspeed.

Upon recognition of a decreased perform-ance wind shear, power should be applied to regain control of the descent. If the possibility of a stall is even remotely suspected, full power should be applied and a stall recovery should be initiated. As the extra power allows the desired approach path to be regained, the power can be reduced to maintain that approach path. The reduction in power will result in a power setting lower than the original setting due to the higher groundspeed dictating a higher descent rate.

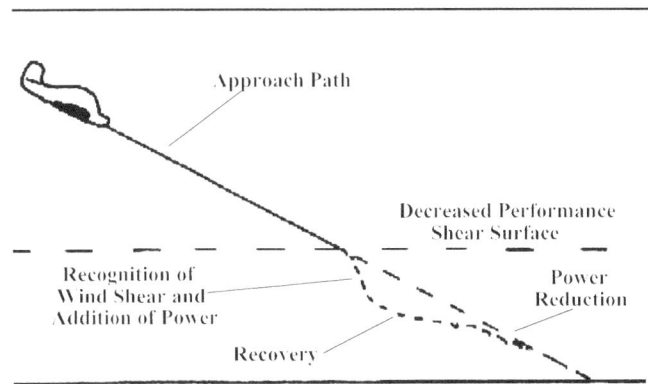

Fig 23-12. - *Decreased Performance Flight Profile.*

So to sum up, an encounter with a decreased performance wind shear required a large increase in power (possibly full power) to prevent a stall and regain the approach path followed by an even larger decrease in power to maintain the approach path (Fig 23-12).

Chapter Summary

The landing maneuver is generally considered to include the approach to land, the flare, and the landing roll. During a normal landing, the approach speed is high enough to provide a large margin between the approach and the stall. In most aircraft, normal approaches can be flown in any configuration. However, many pilots prefer to use partial flaps in order to improve visibility with a nose low attitude. This also provides the benefit of a reduced stall speed.

The flare maneuver is intended to use the excess airspeed available in the approach to arrest the descent and provide for a controlled touchdown. In the flare, the pitch attitude is gradually increased in order to reduce both the airspeed and the descent rate. Rudder is used for directional control and ailerons are used for lateral control (as usual). After the touchdown, control inputs should be maintained since air is still flowing over the wings.

A short field approach is intended to minimize the runway required for the landing. This is done by reducing the approach and touchdown speeds as much as possible, creating as much drag as possible, and using maximum available braking once on the ground. Factors that can affect the short field performance of an aircraft include the weight, the wind, the density altitude, the runway surface condition, runway gradients, and turbulence (due to the adjusted approach speed).

Soft field approaches are very similar to short field approaches. The main difference is in the flare. The soft field flare usually involves the use of some power to help arrest the descent and provide for a softer touchdown. During the rollout, it is even more important than usual to maintain the proper control inputs due to the possibility of a wheel digging in to the surface.

During crosswind landings, the approach should be flown in a "crab" to maintain the extended centerline of the runway. On very short final or in the flare, rudder and aileron inputs should be introduced to create a sideslip. This sideslip allows the aircraft to maintain a straight track over the ground while being aligned with the runway. It is important to realize that the controls don't work any differently in a crosswind landing, but the rudder and aileron inputs will be larger than normal.

An overshoot is a transition from an approach descent to a departure climb without touching down. They are executed by applying climb power, pitching up into a climb attitude, and adjusting the trim.

Wind shear can affect an approach in a manner similar to it's effect on a departure. Increased performance wind shear will cause a sudden increase in airspeed which must be compensated for. Decreased performance wind shear will cause a sudden decrease in airspeed which must be compensated for.

Useful Rules of Thumb

Normal

Approach

Normal approach speed = ~1.3 x V_s

Flare

The best height to begin the flare for most aircraft is between the wingspan and half of the wingspan (this is about the time that ground effect begins to develop).

Considerations

During approaches in turbulence, increase the approach speed by half of the gust factor.

Short Field

Approach

Short field approach speed = ~1.2 x V_{so}

Performance

$$d = D(GW/MGW) \qquad (23.1)$$

$$d = D\frac{(v_{td} - w)^2}{v_{td}^2} \qquad (23.2)$$

$$d_o = D_o \frac{TAS\text{-}W}{TAS} \qquad (22.3)$$

When landing on unprepared surfaces, add the following approximate percentages to the landing roll:

Wet Pavement	– 20%
Firm Turf	– 30%
Short Grass	– 40%
Long Grass	– 70%
Wet Grass	– 100%
Mud or Snow	– 100%

When landing uphill, decrease the landing roll by 5% for every 1% of gradient.

When landing downhill, add 10% to the landing roll for every 1% gradient.

Increase the landing distance by 4% for every 1000 ft of density altitude.

Soft or Rough Field

Performance

When using a soft field landing technique, multiply the short field distance for the same surface by 3 or 4.

Crosswind

Considerations

When crosswind limits are unpublished, use 20% of the stall speed.

Questions and Problems

NOTE: For questions that pertain to performance, assume that no manufacturers data is published beyond the information given.

1) What are the differences between a short and a soft landing?

2) How should the rudder and ailerons be used in a normal landing? Does this change in a crosswind landing? If so, how?

3) Explain how a normal flare is executed. Soft field flare?

4) An aircraft has a MGW of 2350 lbs. and a published short field approach speed of 52 knots. When flown at a weight of 1900 lbs., what should the short field approach speed be if the winds are at 15 knots gusting to 30 knots?

5) An aircraft is completing a soft field landing with a strong crosswind from the left. What directions will the rudder and ailerons be deflected in? Why?

6) If wet grass increases the takeoff roll due to increased rolling friction, why does it also increase the landing roll?

7) An aircraft is certified for a MGW of 1950 lbs. If this aircraft has a published landing distance of 430 ft. with an approach speed of 47 knots CAS at sea level, what will the landing distance be at a density altitude of 4000 ft. and a weight of 1820 lbs.? What will the approach speed be?

8) If the aircraft in Q7 was landing on a wet grass strip with a direct 8 knot tailwind at sea level and MGW, approximately what would the required landing distance be? the touchdown/stall speed is 39 knots.

Appendix I
Parts of the Airplane

The Airplane

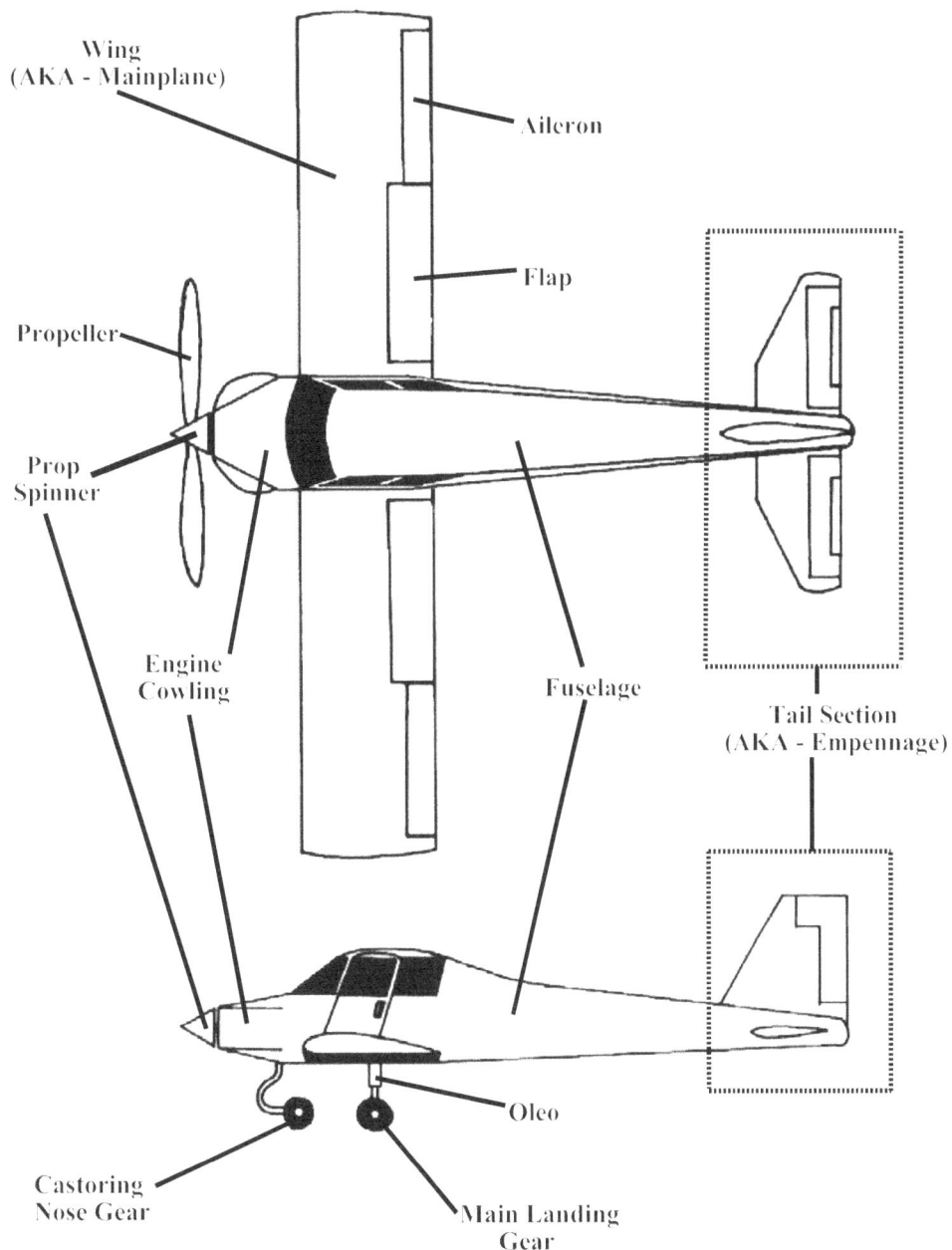

The Wing

Leading Edge

Wing Root

Wingtip

Left Wing

Right Wing

Chord

Trailing Edge

Aileron

Flaps

Wingspan

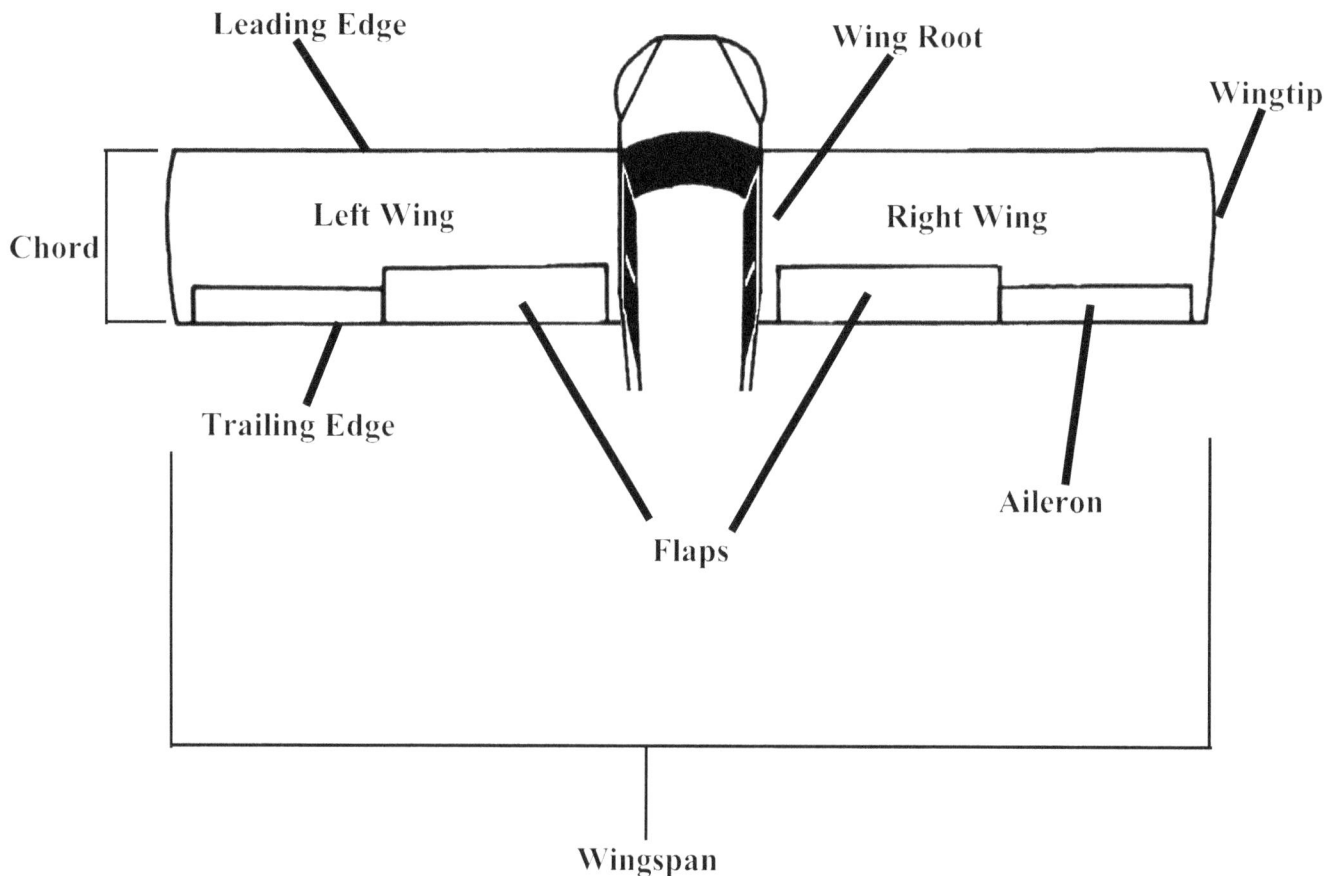

Airfoil - A two dimensional section of a wing (i.e. - a cross-section). For details on airfoil nomenclature and airfoil designation, see the next page.

Aspect Ratio - The ratio of the wingspan to the chord, also defined as the square of the span divided by the wing planform area.

Plan or Planform - The top view of the wing surface.

NOTE: A given wing can have various airfoils used across the wingspan.

The Airfoil

———————————————— Chord Line

- — — — — — — — Mean Line

. . . . Zero-Lift Line

Leading Edge   Trailing Edge

Absolute
Angle of Attack

Point of
Maximum Thickness

Relative
Airflow

Angle of Attack

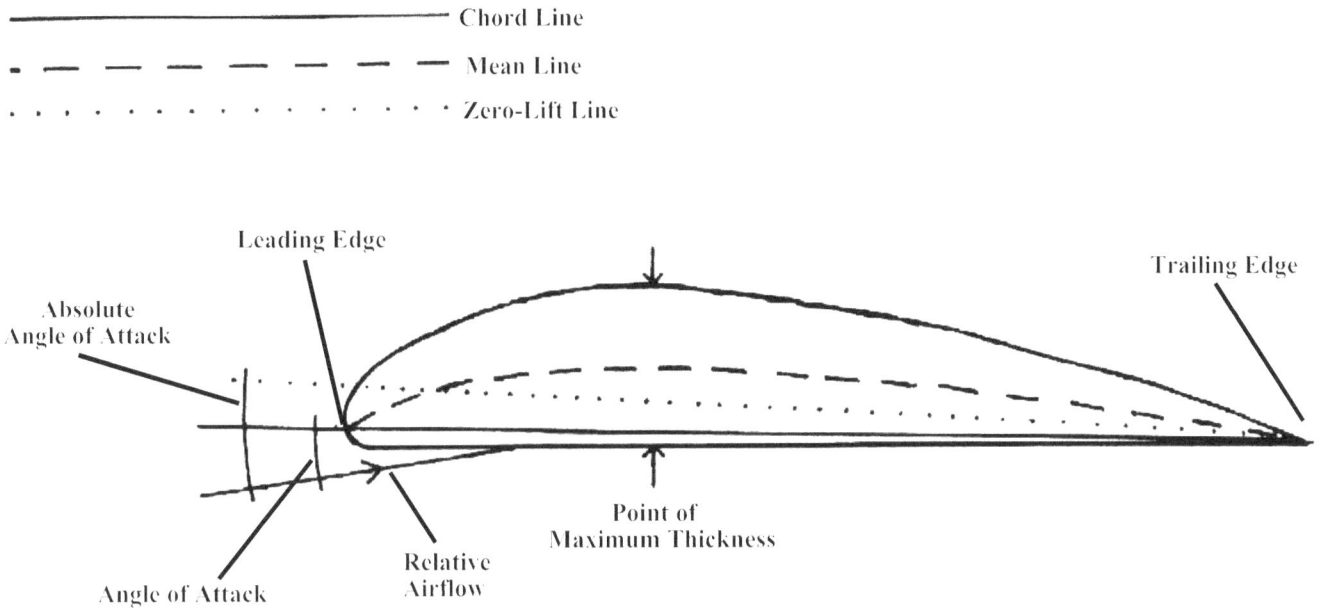

Camber - The difference between the mean line and the chord line.

NOTE: Airfoil shapes are designated by a set of numbers that define various values such as maximum thickness, maximum camber, locations of these maximum points, design lift coefficients, etc. Characteristic dimensions are given as fractions of the chord length so that the numbers can be applied to airfoils of various sizes. Details of the three more commonly used numbering systems are given below.

4-Digit Airfoils

Using the '2315' Airfoil
as an Example:

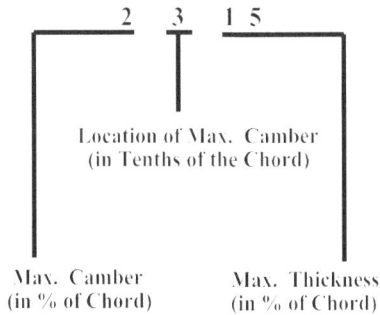

2   3   1 5

Location of Max. Camber
(in Tenths of the Chord)

Max. Camber
(in % of Chord)

Max. Thickness
(in % of Chord)

5-Digit Airfoils

Using the '23115' Airfoil
as an Example:

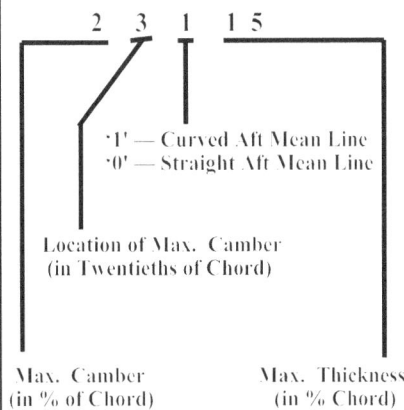

2   3   1   1 5

'1' — Curved Aft Mean Line
'0' — Straight Aft Mean Line

Location of Max. Camber
(in Twentieths of Chord)

Max. Camber
(in % of Chord)

Max. Thickness
(in % Chord)

Series Airfoils

Using the '64₁-512' Airfoil
as an Example:

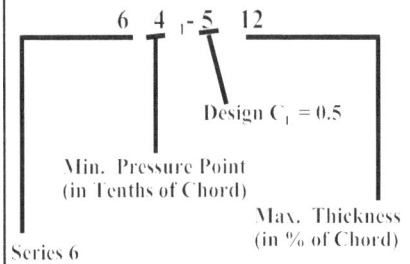

6   4 ₁-5   12

Design C_l = 0.5

Min. Pressure Point
(in Tenths of Chord)

Max. Thickness
(in % of Chord)

Series 6

NOTE: The subscript '1' indicates that the
low drag range is the design C_l
+/- 0.1.

The Empennage

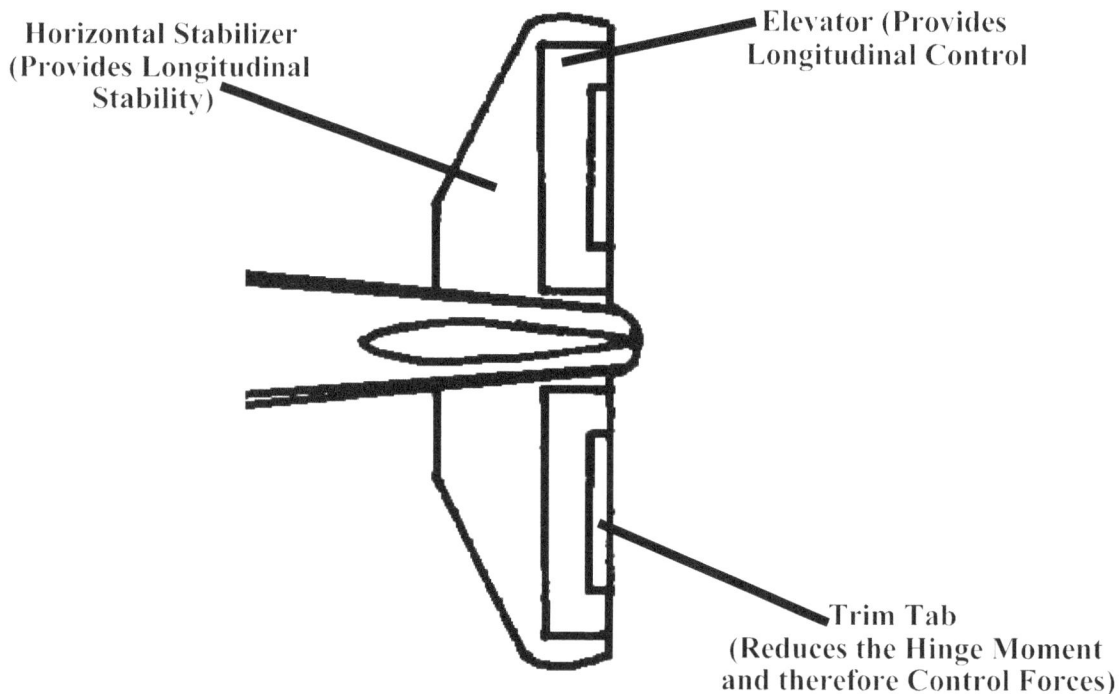

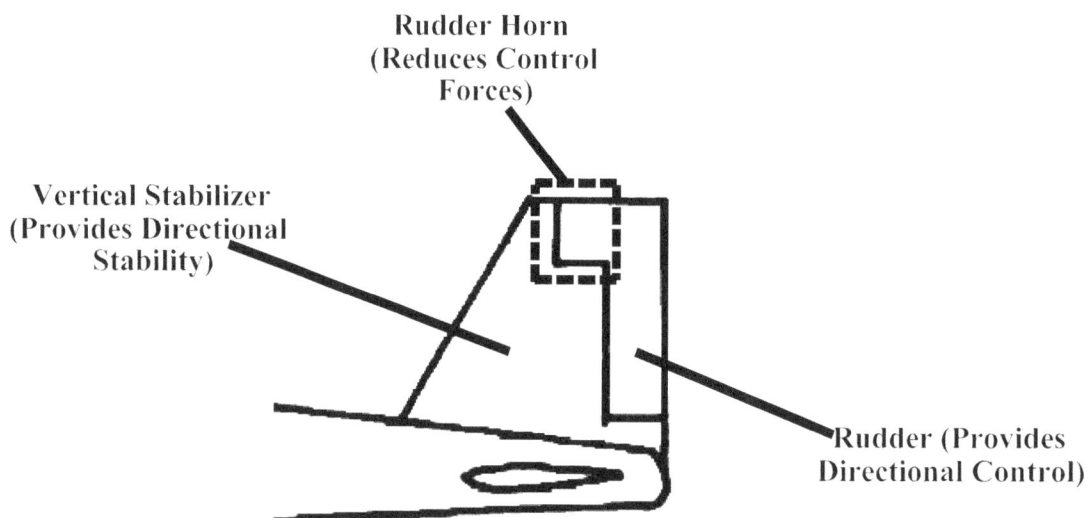

Horizontal Stabilizer
(Provides Longitudinal
Stability)

Elevator (Provides
Longitudinal Control

Trim Tab
(Reduces the Hinge Moment
and therefore Control Forces)

Rudder Horn
(Reduces Control
Forces)

Vertical Stabilizer
(Provides Directional
Stability)

Rudder (Provides
Directional Control)

NOTE: The horn method for reducing control forces
can also be used on other control surfaces.

Appendix II
The Standard Atmosphere

The Standard Atmosphere is an idealized representation of the real atmosphere. It is derived from average temperatures at various altitudes along with the application of various gas laws to calculate the pressure and density at these altitudes. The resulting numbers do not generally represent the actual conditions in the real atmosphere at a given point in time, but they reduce all of the variables to a common standard. This provides several advantages, but for pilots the main one is the ability to predict aircraft performance. To say, for example, that an aircraft will climb at 1250 FPM at an *altitude* of Sea Level (0 ft.), is meaningless unless it is referenced to the Standard Atmosphere. This means that the aircraft will climb at 1250 FPM at a *density altitude* of Sea Level (0 ft.). The aircraft doesn't actually care how *high* it is, but it will perform according to the *density* of the air.

In aviation, we deal with 5 different types of altitudes, depending on the context. We deal with:

Absolute Altitude – This is the height above ground level (AGL).

Indicated Altitude – This is the altitude read off of the altimeter when the barometric pressure is set properly. It is measured in feet above sea level (ASL).

True Altitude – This is the actual height above sea level. True altitude varies from Indicated Altitude due to temperature errors in the altimeter.

Pressure Altitude – This altitude is not an actual measure of height, but one of pressure. The Pressure Altitude is the height (ASL) that you would have to be at in the Standard Atmosphere in order to experience the same air pressure. When barometric pressure fluctuates due to meteorological phenomenon, the Pressure Altitude at a given height can change as well.

Density Altitude – Like Pressure Altitude, Density Altitude is not a measure of height, but one of air *density*. The density altitude is the height (ASL) that you would have to be at in the Standard Atmosphere in order to experience the same air *density*. As with the Pressure Altitude, Density Altitude can vary with meteorological phenomenon. Aircraft performance is a function of air density, and therefore density altitude.

Because the air density varies with temperature and pressure, it can change significantly from one day to the next. So an aircraft will perform differently on different days (even with the same weight and C of G). Peak performance is obtained in high density air—meaning high pressure and low temperature. This is the reason aircraft take-off and climb with noticeably better performance in winter than in summer.

The Standard Atmosphere is based on the following conditions:

1) Sea level pressure = 29.92 "Hg (1013.25 mb)(14.7psi),
2) Sea level temperature = 15°C (~288K),
3) Lapse Rate in the troposphere = 1.98°C/1000' (6.5°C/1000 m), and
4) Air is a perfectly dry gas.

The table on the next page gives values for temperature, pressure, and density at various altitudes in the Standard Atmosphere.

Altitude (feet)	Temperature (°C)	Pressure ("Hg / psi)	Density (Slugs/ft^3)
-1,000 ft	16.98 °C	31.02 "Hg / 15.23 psi	0.0024473 s/ft^3
-500 ft	15.99 °C	30.47 "Hg / 14.96 psi	0.0024119 s/ft^3
0 ft (Sea Level)	15.00 °C	29.92 "Hg / 14.69 psi	0.0023769 s/ft^3
500 ft	14.01 °C	29.39 "Hg / 14.43 psi	0.0023423 s/ft^3
1,000 ft	13.02 °C	28.86 "Hg / 14.17 psi	0.0023081 s/ft^3
1,500 ft	12.03 °C	28.35 "Hg / 13.92 psi	0.0022743 s/ft^3
2,000 ft	11.04 °C	27.82 "Hg / 13.66 psi	0.0022409 s/ft^3
2,500 ft	10.05 °C	27.33 "Hg / 13.42 psi	0.0022079 s/ft^3
3,000 ft	9.06 °C	26.82 "Hg / 13.17 psi	0.0021752 s/ft^3
3,500 ft	8.07 °C	26.34 "Hg / 12.93 psi	0.0021429 s/ft^3
4,000 ft	7.08 °C	25.85 "Hg / 12.69 psi	0.0021110 s/ft^3
4,500 ft	6.09 °C	25.37 "Hg / 12.46 psi	0.0020794 s/ft^3
5,000 ft	5.10 °C	24.91 "Hg / 12.23 psi	0.0020482 s/ft^3
5,500 ft	4.11 °C	24.44 "Hg / 12.00 psi	0.0020174 s/ft^3
6,000 ft	3.12 °C	23.99 "Hg / 11.78 psi	0.0019869 s/ft^3
6,500 ft	2.13 °C	23.54 "Hg / 11.56 psi	0.0019667 s/ft^3
7,000 ft	1.14 °C	23.10 "Hg / 11.34 psi	0.0019270 s/ft^3
7,500 ft	0.15 °C	22.67 "Hg / 11.13 psi	0.0018975 s/ft^3
8,000 ft	-0.84 °C	22.24 "Hg / 10.92 psi	0.0018685 s/ft^3
8,500 ft	-1.83 °C	21.81 "Hg / 10.71 psi	0.0018397 s/ft^3
9,000 ft	-2.82 °C	21.41 "Hg / 10.51 psi	0.0018113 s/ft^3
9,500 ft	-3.81 °C	21.00 "Hg / 10.31 psi	0.0017833 s/ft^3
10,000 ft	-4.8 °C	20.59 "Hg / 10.11 psi	0.0017556 s/ft^3
10,500 ft	-5.79 °C	20.18 "Hg / 9.91 psi	0.0017282 s/ft^3
11,000 ft	-6.78 °C	19.80 "Hg / 9.72 psi	0.0017011 s/ft^3
11,500 ft	-7.77 °C	19.41 "Hg / 9.53 psi	0.0016744 s/ft^3
12,000 ft	-8.76 °C	19.04 "Hg / 9.35 psi	0.0016480 s/ft^3
12,500 ft	-9.75 °C	18.68 "Hg / 9.17 psi	0.0016219 s/ft^3

Appendix III
Answers to Questions

Throughout this book, each chapter has ended with a series of questions and problems for the reader to work with. It is intended that you use these questions as an opportunity to learn the material presented in this book more thoroughly so that the information can be applied effectively in the aircraft. The purpose of this section is to provide the correct answers so you can evaluate your progress. In many cases, the answers can be taken directly from the text. Some questions, however, require some interpretation of the text or simple mathematical

workings. Answers for all of the questions are provided in this appendix. For the reader who chooses to complete the problems involving math, my working are completed to one or two decimal places. Feel free to work your answers more or less precisely if you wish, but if you do you can expect to see some rounding errors when you compare your answers to mine. Also, in the math workings, the final answers are underlined so that they will stand out from the remainder of the workings.

Chapter 1: Basic Physics

1) Inertia is the tendency of an object to remain in steady motion—or equilibrium. The amount of inertia possessed by an object can be quantified by the objects mass.

2) Using Newton's second law, F=ma, the acceleration of a given object is directly proportional to the force acting on it.

3) They will each accelerate in opposite directions—away from one another.

According to Newton's third law, forces occur in pairs that are equal and opposite to one another. In this case, one skater pushes on the other. The equal and opposite forces acts on the skater doing the pushing.

4) A vector has both quantity and direction and can be represented graphically by an arrow.

Vectors can be added graphically by lining them up head to tail and connecting the first tail to the last head. Mathematically, they can be added using some basic trigonometry.

5) When two forces are equal and opposite, but do not pass through the same point, they form a couple.

A couple creates a turning moment that will rotate the object affected.

6) Work is the transfer of energy. It is determined by the distance a force acts over.

Power is the rate of doing work, or the rate of transfer of energy. It is defined as the work done (energy transferred) per unit of time.

7) Friction is the result of molecular interactions at the surfaces of two objects on contact with one another.

The amount of friction is determined by the friction coefficient between two surfaces and by the force holding them together (the "normal" force).

8) The total energy in a closed system remains constant. Although the energy can change forms, the amount of energy will remain constant.

9) Potential energy results from an object's position or state. For example, gravitational potential energy is a function of height.

Kinetic energy results from an object motion.

10) The energy provided by a machine is the same amount of energy that is supplied to the machine—less frictional losses. The function of most machines is to convert the form that the energy takes and/or to transmit it to a different location.

11) Due to friction, the kinetic energy in this case becomes thermal energy—or heat energy.

12) The mechanical advantage is the increase in the force produced by a machine as compared to the force acting on the machine. It is determined by dividing the output force by the input force.

13) Centripetal force.

14) Thrust, Weight, Lift, and Drag.
Thrust is lower than drag, and thus creates a nose-up moment. Weight is forward of lift, and thus creates a nose-down moment.

15) First, we need to find out what net force is acting on the crate. The net force will be the 20 lb push minus the frictional force. To determine the weight,

$$w = mg$$

$$w = (2slugs)(32.2ft/s^2)$$

$$w = 64.4 \ lbs.$$

Now to determine the frictional force (on a level surface, the normal force is equal to the weight),

$$F_{fr} = \mu_k N$$

$$F_{fr} = (0.15)(64.4lbs)$$

$$F_{fr} = 9.66lbs.$$

So the total force is,

$$20lbs - 9.66lbs = 10.44lbs.$$

Now we can finally determine the acceleration,

$$F = ma$$

$$a = F/m$$

$$a = (10.44lbs)/(2slugs)$$

$$\underline{a = 5.22ft/s^2.}$$

16) The work done by you is,

$$W = Fd$$

$$W = (20lbs)(30ft)$$

$$\underline{W = 600 \ ft\text{-}lbs.}$$

In a perfect world, the crate would then have 600 ft-lbs of kinetic energy. However, some energy has been lost to friction (it becomes heat energy). To determine how much energy is lost to friction, we can consider the force of friction (found during the last question) and apply it over the distance traveled,

$$W = Fd$$

$$W = (-9.66lbs)(30ft)$$

$$W = -289.8ft\text{-}lbs.$$

Since this energy is lost to friction, the total kinetic energy of the crate is,

$$\underline{600ft\text{-}lbs - 289.8ft\text{-}lbs = 310.2ft\text{-}lbs.}$$

17) First. Let's determine the speed of the crate once it is released,

$$KE = \tfrac{1}{2}mv^2$$

$$v^2 = \frac{2(KE)}{m}$$

$$v^2 = \frac{2(310.2ft\text{-}lbs)}{(2slugs)}$$

$$v = 17.6fps.$$

Since we already know the frictional force acting on the crate (from the previous question), we can now determine the deceleration of the crate,

$$a = F/m$$

$$a = (-9.66lbs)/(2slugs)$$

$$a = -4.83ft/s^2.$$

Knowing the initial speed of the crate and the deceleration, the time to stop can be found,

$$a = (v_f - v_i)/t$$

$$t = \frac{(v_f - v_i)}{a}$$

$$t = \frac{0fps - 17.6fps}{-4.83ft/s^2}$$

$$\underline{t = 3.64s.}$$

Using the initial speed, the acceleration, and the time taken to stop, we can now find the distance covered by the crate from the release point.

$$d = v_i t + \tfrac{1}{2}at^2$$

$$d = (17.6\text{fps})(3.64\text{s}) + \tfrac{1}{2}(-4.83\text{ft/s}^2)(3.64\text{s})^2$$

$$\underline{d = 32.1\text{ft}}.$$

18) We know the force being applied and we know the final speed, so we can determine the power being produced at the release point as follows,

$$Pwr = Fv$$

$$Pwr = (20\text{lbs})(17.6\text{fps})$$

$$\underline{Pwr = 352\ \text{ft-lbs/s}}.$$

We could also convert this to horsepower,

$$HP = Pwr/550$$

$$HP = (352\text{ft-lbs/s})/550$$

$$\underline{HP = 0.64HP}.$$

19) First, let's convert 60 knots into feet per second. 1 nautical mile is equal to 6080 feet. So the aircraft is landing at (60)(6080feet)per hour = 364800ft/hr. 1 hour has 60 minutes, each of which has 60 seconds, so,

$$\frac{60\ \text{min}}{\text{hour}} \times \frac{60\ \text{sec}}{\text{min}} = \frac{3600\ \text{seconds}}{\text{hour}}.$$

So, the aircraft is traveling 364800 feet in 3600 seconds, or 101.33fps.

Now let's consider the decelerating forces acting on the aircraft. With maximum braking, the decelerating force is,

$$F_{fr} = \mu_s N$$

$$F_{fr} = (0.4)(2000\text{lbs})$$

$$F_{fr} = 800\text{lbs}.$$

For the locked brakes, the decelerating force is,

$$F_{fr} = \mu_k N$$

$$F_{fr} = (0.08)(2000\text{lbs})$$

$$F_{fr} = 160\text{lbs}.$$

In order to find the deceleration, we need the aircraft's mass. We can get this using,

$$w = mg$$

$$m = w/g$$

$$m = 2000\text{lbs}/32.2\text{ft/s}^2$$

$$m = 62.1\text{slugs}$$

Now to determine the deceleration,

$$a = F/m$$

$$a = -800\text{lbs}/62.1\text{slugs}$$

$$a = -12.9\text{ft/s}^2 \text{ for the aircraft with maximum braking.}$$

$$a = F/m$$

$$a = -160\text{lbs}/62.1\text{slugs}$$

$$a = -2.6\text{ft/s}^2 \text{ for the aircraft with the locked brakes.}$$

Note that the accelerations are negative. This is because friction always slows an object (i.e. – the acceleration is opposite the motion).

Knowing the accelerations and the initial velocity of the aircraft, we now need the time taken to come to a stop. From there, distances used can be determined.

$$a = (v_f - v_i)/t$$

$$t = (v_f - v_i)/a$$

$$t = (0 - 101.33\text{fps})/-12.9\text{ft/s}^2$$

$$t = 7.9\text{s} \text{ for the aircraft with maximum braking.}$$

$$a = (v_f - v_i)/t$$

$$t = (v_f - v_i)/a$$

$$t = (0 - 101.33\text{fps})/-2.6\text{ft/s}^2$$

$$t = 39\text{s} \text{ for the aircraft with locked brakes.}$$

Finally, we can find the distances.

$$d = v_i t + \tfrac{1}{2}at^2$$

$$d = (101.33\text{fps})(7.9\text{s}) + \tfrac{1}{2}(-12.9\text{ft/s}^2)(7.9\text{s})^2$$

$$\underline{d = 398\text{ft}} \text{ for the aircraft with max braking.}$$

$$d = v_i t + \tfrac{1}{2}at^2$$

$$d = (101.33\text{fps})(39\text{s}) + \tfrac{1}{2}(-2.6\text{ft/s}^2)(39\text{s})^2$$

<u>d = 1975ft</u> for the aircraft with locked brakes.

20) Let's first find the mass of the aircraft.

$$m = W/g$$

$$m = 3000\text{lbs}/32.2\text{ft/s}^2$$

$$m = 93.2\text{slugs}.$$

Now we need to convert 150kts into feet per second. Using the same method as in the previous question, we get 253.3fps.

And now we can determine the centripetal force.

$$F = mv^2/r$$

$$F = (93.2\text{slugs})(253.3\text{fps})^2/2000\text{ft}$$

<u>F = 2989.9lbs</u>.

Chapter 2: The Production of Lift

1) A streamline is a graphical representation of airflow. A line represents the path followed by a series of air particles in a flow. When the streamlines are close together, they indicate high flow speed and therefore low pressure. When the streamlines are spread wider, they indicate low flow speed and therefore higher pressure.

Two streamlines cannot cross because that would indicate that two particles of air are occupying the same space at the same time.

2) The continuity equation,

$$\rho_1 v_1 A_1 = \rho_2 v_2 A_2 \qquad (2.3),$$

dictates that the mass flow of air will remain constant across every cross section of a given flow. This means that as the cross-sectional area is changed, the flow velocity must change as well.

3) Bernoulli's Theorem,

$$P_s + \tfrac{1}{2}\rho v^2 = \text{Constant} \qquad (2.8),$$

dictates that the total pressure in a flow must remain constant. Total pressure consists of two elements, dynamic pressure due to motion of the air and static pressure due to the random motion of the air molecules. When one increases, the other must be decreased in order to maintain a constant total.

In a venturi, continuity leads to an increase in velocity at the restriction. This increase in velocity is also an increase in dynamic pressure. From Bernoulli's Theorem, the increased dynamic pressure results in a decrease in static pressure.

4) The curvature on top of the wing creates a "venturi effect". That is to say that a restriction to the airflow is formed. This restriction forces the air on top of the wing to accelerate.

5) Due to the change in flow speed at various points on the wing, the dynamic pressure also changes. From Bernoulli's Theorem, this leads to changes in static pressure. At places where the airflow is accelerated, the static pressure drops and at places where the airflow is slowed, the static pressure increases.

6) One of the main contributors to position error is the change in static pressure over the static port. This is caused by the fact that the static port, by the nature of an airplane, must have air flowing over it. From Bernoulli's Theorem, this airflow leads to variations in the static pressure acting on the port, and thus atmospheric pressure is almost never read accurately.

7) True airspeed (TAS) is the actual speed that an aircraft is traveling through the air. Equivalent airspeed (EAS) is a function of the dynamic pressure acting on the aircraft. EAS can be considered to be the speed that the aircraft would have to move in order to experience the same dynamic pressure at standard sea level. The difference between TAS and EAS is caused by variations in air density.

The EAS is the speed that determines how much lift a given wing can produce. This is because the dynamic pressure is a deciding factor in the production of lift.

8) The difference between EAS and calibrated airspeed (CAS) is due to the compression of air as it enters the pitot tube. At low airspeeds and low altitudes, the compressibility error is negligible. So on low and slow aircraft (e.g. – trainers) the error is ignored.

9) The angle of attack (AOA) is the angle between the chord line of the wing and the relative airflow. With an increase in AOA at a given EAS, the lift produced by a wing will be increased. Likewise, an increase in AOA as the EAS is reduced allows us to maintain a constant amount of lift at various airspeeds.

10) The lift equation,

$$P_s + \tfrac{1}{2}\rho v^2 = \text{Constant} \qquad (2.8),$$

tells us the variables that control lift and how they relate to one another. For pilots, it can be converted into the following rule of thumb,

$$L = AOA \times EAS \qquad (2.10).$$

11) The center of pressure is the "balancing point" for all of the distributed pressure acting on a surface. On a cambered wing, the center of pressure on the top of the wing is offset from the center of pressure on the bottom of the wing. This imbalance creates a pitching moment on the wing.

As the AOA is increased, the center of pressure on top of the wing moves forward.

12) The aerodynamic center (AC) is the point around which the wing pitching moment remains constant. On most subsonic wings, the AC is at approximately the 25% chord point.

Lift is often considered to act at the AC because it simplifies the picture of forces acting on the wing. There can now be a single lift force fixed in place with a pitching moment acting at the AC.

13) The pitching moment on a wing increases with an increase in airspeed and it decreases with a decrease in airspeed.

AOA will have no effect on the pitching moment at the AC.

14) Converting 135 knots to feet per second, we get 228fps. The planform area of the wing (span times the average chord) is 252 ft^2. Referring to Appendix II, the air density at 10,000 ft is 0.001756 slugs/ft^3.

Using the lift equation, we can determine the lift coefficient that is required to maintain sufficient lift.

$$L = C_L \tfrac{1}{2}\rho v^2 S$$

$$3000\text{lbs} = C_L \tfrac{1}{2}(0.001756\text{slugs/ft}^3)(228\text{fps})^2(252\text{ft}^2)$$

$$C_L = 0.26.$$

And now referring to Fig 2-21, the required AOA is 3°.

To determine the aircraft's EAS, we find the dynamic pressure and then find the speed required at sea level to encounter that same dynamic pressure.

$$P_d = \tfrac{1}{2}\rho v^2$$

$$P_d = \tfrac{1}{2}(0.001756\text{slugs/ft}^3)(228\text{fps})^2$$

$$P_d = 45.64 \text{ psf}.$$

$$P_d = \tfrac{1}{2}\rho v^2$$

$$45.64\text{psf} = \tfrac{1}{2}(0.002377\text{slugs/ft}^3)v^2$$

$$v = 196 \text{ fps}$$

In knots, 116 kts.

Chapter 3: Drag and Power Required

1) The two main types of drag for low speed aircraft are parasite drag and induced drag. Induced drag is caused by the production of lift while parasite drag is caused by air flowing over any surface.

Parasite drag increases as the square of the airspeed. Induced drag *decreases* as the square of the airspeed increases (i.e. – it is inversely proportional to the square of the airspeed).

2) Skin friction drag is caused by the interaction between the air flowing over the aircraft and the aircraft surface via the boundary layer. Form drag is caused by the pressure difference between the front and rear surfaces of the aircraft. This pressure difference is in turn caused by the airflow energy lost in the boundary layer.

Form drag and skin friction drag both contribute to the total parasite drag of the aircraft.

3) Interference drag can be reduced primarily by either blending parts of the airframe or by placing fairings at the joints between parts such as the wing and fuselage.

This method reduces the mixing (interference) of two boundary layers and thus reduces the interference drag.

4) If the boundary layer separates over a surface, the form drag will be increased due to the drop in pressure at the aft end of the object.

5) Downwash causes induced drag because lift always acts perpendicular to the relative airflow. As the downwash inclines the relative airflow downwards, the lift will be inclined aft. The aft component of the lift is induced drag.

The lift is reduced because of the incline in the total lift. With the lift acting to the rear, the upwards component is reduced—this component is the effective lift.

6) Total drag is a combination of parasite drag (increasing with speed) and induced drag (decreasing with speed). The total drag starts relatively high at low speeds, then it decreases as the speed increases, and finally—after reaching a minimum point—the drag increases with an increase in speed. This increasing trend continues indefinitely.

7) With a change in weight, the induced drag will change as well. Since power required is directly proportional to drag, the power curve will be affected. As weight is increased, induced drag—and therefore induced power required—will increase. This leads to an increase in total power required.

8) Power required is determined by the drag being produced and the speed of the aircraft,

$$Pwr = Dv/325 \qquad (3.9).$$

Due to this, the power curve is developed directly from the drag curve.

9) At high speeds, the total power required increases approximately with the *cube* of the airspeed. Because of this, large increases in power lead to only small increases in speed. As well, increasing the engine size will increase the weight of the aircraft as well as the weight of fuel that needs to be carried. The net result is a marginal improvement in cruise speed with a very large increase in engine size.

10) At the speed for best L/D, parasite drag is

equal to induced drag. This means that the total drag for the aircraft in this case consists of 120 lbs of each type. Using equations (3.1) and (3.7), it can be seen that the parasite drag will quadruple when the airspeed doubles, but the induced drag will quarter.

$$D_p = C_{dp}\tfrac{1}{2}\rho v^2 S \qquad (3.1)$$

$$D_i = \frac{L^2}{\pi\tfrac{1}{2}\rho v^2 b^2} \qquad (3.7)$$

So the 120 lbs of parasite drag becomes 480 lbs of parasite drag. As well, the 120 lbs of induced drag becomes 30 lbs of induced drag. So the total drag at 180 knots will be 510 lbs.
The horsepower required for these two speeds can be determined using equation (3.9),

$$Pwr = Dv/325 \qquad (3.9).$$

At 90 knots,

$$Pwr = (240lbs)(90kts)/325$$

$$Pwr = 66.5 \text{ HP}.$$

At 180 knots,

$$Pwr = (510lbs)(180kts)/325$$

$$Pwr = 282.5 \text{ HP}.$$

Chapter 4: Thrust and Power Available

1) Initially, an air particle is stationary. According to Newton's First Law, this air particle will remain stationary until acted upon by an outside force. The aircraft engine (or propeller) provides this outside force, accelerating the air particle backwards according to Newton's Second Law: F = Ma. According to Newton's Third Law, an equal and opposite force will act on the propeller. This force is thrust.

2) A piston engine is more efficient (i.e. – the specific fuel consumption is lower) than a jet at low speeds. Jets, on the other hand, are more efficient at higher speeds and are capable of producing more thrust.

3) Intake, compression, combustion (a.k.a. power), and exhaust.

4) Indicated Horsepower – The power actually produced in the cylinders of the engine.

Brake Horsepower – The power produced by the engine after accounting for frictional losses and power used up in the non power producing strokes.
Shaft horsepower – The power delivered to the prop shaft after losses at the reduction gear box.
Thrust Horsepower – The power actually available for the aircraft to use after aerodynamic losses at the prop.

5) As the airspeed changes, the relative airflow of the prop changes. This means that the AOA of the prop blade will change as well. At higher speeds, the AOA of the prop is reduced and thus the thrust is reduced. When considering the speed effect on the thrust horsepower (recall that power is a function of both thrust and speed), the overall trend is for the THP available to increase to a maximum and then decrease again.

6) A variable pitch prop will allow the prop blade to maintain a constant AOA through various airspeeds. This means that the prop can remain at a high efficiency over a wider range of airspeeds.

7) Slipstream – The rotation imparted on the airflow by the prop causes a sidewards force on the vertical stabilizer. The sidewards force leads to a left yaw.

Torque – The equal and opposite reaction (Newton's Third Law) caused by the engine turning the prop will tend to roll the aircraft to the left.

Asymmetric Thrust – When the aircraft is at a high AOA, the downgoing blade of the prop is at a higher AOA than the upgoing blade. This creates an imbalance of thrust with the excess of thrust being on the right side of the prop disc.

Gyroscopic Precession – Since the prop is a spinning mass, it will act like a gyroscope. When the aircraft is pitched, the result is for the prop to induce a yaw. Raising the tail (pitching down) will lead to a left yaw, while lowering the tail (pitching up) will lead to a right yaw.

8) The direction of the turning tendencies is determined by the direction of rotation of the prop. In North America, most aircraft have props that rotate to the right as seen from the rear. This leads to *left* turning tendencies. If the prop of an aircraft rotates to the left as seen from the rear, the turning tendencies would be to the right.

9) the left turning tendencies are the most pronounced at low airspeeds with high power settings.

This is because at high power settings, the torque of the engine is high and the effect of the prop on the airflow is increased—leading to more slipstream. At low airspeeds, the slipstream has more time to act on the aircraft and the AOA of the aircraft is higher. This higher AOA leads to an increase in asymmetric thrust.

10) Built in corrections for the left turning tendencies include (but are not restricted to) an offset vertical fin, an offset thrust line, a cambered vertical fin, and an offset wing.

11) The terms "climb prop" and "cruise prop" refer to fixed pitch propellers that are optimized for a particular phase of flight—either climb or cruise.

An aircraft with a climb prop will takeoff and climb very well, but cruise performance will be sacrificed. An aircraft with a cruise prop will have increased cruise performance, but the takeoff and climb performance will suffer. Most fixed pitch aircraft have a prop that is some happy medium between the two extremes.

Chapter 5: Flight Controls

1) The Longitudinal Axis runs nose to tail, the Lateral Axis runs wingtip to wingtip, and the Vertical/Normal Axis runs straight up and down.

Movement around the longitudinal axis is roll, movement around the lateral axis is pitch, and movement around the vertical/normal axis is yaw.

All three axis' pass through the center of gravity.

2) A primary control is a control that is intended to have a direct effect on the attitudes and/or the movements of the aircraft.

Secondary controls are intended to control various aspects of an aircraft's performance and handling, but they are not intended to have a direct effect on the aircraft's attitudes or movements (although they often have indirect effects).

3) Primary controls normally include the ailerons to control roll, the rudder to control yaw, and the elevator/stabilator to control pitch. Other controls can serve as primary controls as well. For example, differential spoilers are sometimes used to control roll.

4) When an aileron is deflected, it varies the camber of the wing. The downgoing aileron increases the camber and therefore increases lift. The upgoing aileron decreases the camber of the wing and therefore decreases the lift of the wing. Since the ailerons each pivot in opposite directions to one another, the lift differential induces a roll in the direction of the upgoing aileron.

5) An elevator is a movable control surface that forms the aft portion of a horizontal stabilizer. A stabilator is a horizontal stabilizer and an elevator that is all one piece. A canard is simply a stabilator or an elevator/stabilizer combination that is forward of the wing instead of aft of it.

6) When the ailerons are deflected, the wing with the downgoing aileron produces more lift than the wing with the upgoing aileron. Since lift is the cause of induced drag, the induced drag is increased on the downgoing aileron and decreased by the upgoing aileron. This imbalance of induced drag causes the aircraft to yaw towards the upgoing wing/downgoing aileron (i.e. – yaw is opposite the roll).

To counteract this problem, some aircraft have either differential or frise ailerons. As well, some aircraft have a mechanical linkage between the ailerons and rudder so that roll is automatically coordinated. In aircraft that are not equipped with some sort of compensation for adverse yaw, or if the built in corrections are insufficient under certain conditions, the rudder can be applied by the pilot to eliminate adverse yaw and coordinate roll.

7) In ground effect, the downwash from the wing is reduced. Because of this, the AOA of the horizontal stabilizer/stabilator will be reduced as well. The reduction of AOA leads to a corresponding reduction of tail-down force. So in order to apply the same pitch control in ground effect as out of ground effect, the elevator/stabilator must be deflected further.

With a canard, the control surface is in front of the wing and is therefore not affected by the wings downwash. For this reason, a canard arrangement will not be affected by ground effect.

8) To be in coordinated flight means to be flying with a sideslip angle of 0°. This means that the rate of yaw must correspond to the angle of bank and the airspeed. With the wings level, no yaw should be present, but with the wings banked, an appropriate rate of yaw should be maintained to maintain coordination.

9) Yaw will induce a roll because the two wings will be traveling at different speeds through the air. The wing to the outside of the yaw will be moving faster and will therefore be producing more lift than the wing on the inside of the yaw. This lift imbalance will cause a roll in the same direction as the yaw.

10) Unwanted or uncoordinated yaw has many origins. They include the left turning tendencies of the engine (torque, slipstream, asymmetric thrust, and gyroscopic precession), aileron drag, turbulence, and improper use of the rudder.

11) The trim tab serves the function of reducing the pilot's workload by reducing or eliminating sustained control forces.

The trim tab is basically a control surface that is mounted on another control surface. It is deflected opposite the corresponding control surface so that the air loads acting on the surface will hold it in place instead of tending to neutralize it. This means that the pilot doesn't need to forcibly hold the controls in place for long periods of time.

12) There are many different types of flaps. Some of the more common ones are plain flaps, split flaps, fowler flaps, slotted flaps, and leading edge flaps. Flaps can also be configured to be combinations of these types.

The "bottom line" function of flaps is to increase the lifting capacity and the drag of a wing. In practice, this ability serves several purposes including shortening the takeoff roll and landing roll, reducing the distance required to clear an obstacle on an approach, and increasing the margin between certain critical operating speeds and the stall.

13) Spoilers extend on the upper surface of a wing to disrupt the airflow and "spoil" lift.

Often, spoilers are used as "lift dumpers" after touchdown in order to shorten the landing roll. As well, they can be used to control the descent rate of an aircraft.

14) Speed brakes protrude into the airflow and increase the parasite drag (and therefore the total drag) of the aircraft.

In increasing the drag of the aircraft, speed brakes can be used to slow the aircraft down or to increase the descent rate.

Chapter 6: Stalls and Spins

1) A stall is a condition of flight in which an increases in AOA will lead to a decrease in lift being produced.

Stalls occur when excessive airflow separation occurs over the top of the wing due to the critical AOA being exceeded. As the AOA is increased, the adverse pressure gradient increases and leads to flow separation and reversal on top of the wing. The pressure distribution resulting from this separation results in a decrease in lift with increasing AOA after the critical AOA has been exceeded.

2) 17°. A given wing will always stall at the same AOA. The stall speed and the aircraft attitude can vary

significantly, but the AOA does not change.

3) Stall symptoms include:
– Mechanical stall warning – Visual or aural warning that is activated by a high AOA.
– Airframe Buffeting – Separated airflow off the wing causes the airframe to shake. Buffeting can often be heard as well as felt.
– Forward Pitch – As the aircraft enters the stall, the nose will usually drop—sometimes imperceptibly, other times sharply, depending on aircraft type and configuration. This nose drop is caused by the sudden change in the wing's pitching moment.
Altitude Loss – As the lift of the wing

suddenly decreases, the aircraft will be unable to maintain equilibrium in level flight. A loss of altitude is unavoidable in both the stall and the recovery from the stall.

4) Flaps will decrease the stall speed of the aircraft because of the increased camber and therefore increased lifting capacity of the wing. If the stall should become a spin, flaps can delay or prevent recovery by dampening the effectiveness of the rudder.

5) Most aircraft are either dihedral or effectively dihedral due to being high wing. These aircraft will have unequal angles of attack on each wing when established in a climbing or descending turn. Because of this, one wing will stall before the other. In a climbing turn, the outside wing will stall first and the aircraft will have a tendency to spin to the outside. In a descending turn, the opposite occurs—the inside wing will stall first and the aircraft will tend to spin into the turn.

6) Spin recovery is more favorable with a forward C of G because the rudder has a longer arm and is therefore more effective in producing the anti-spin yaw moment that is required.

7) In order for a wing to drop, the aircraft must roll. The downgoing wing will then meet the relative airflow at a higher angle (i.e. – it will be at a higher AOA) and the upgoing wing will experience a reduced AOA. In a stalled condition, this means that the downgoing wing will enter the stall more deeply and will tend to drop further. Meanwhile, the upgoing wing will partly or fully come out of the stall, leading to an increase in lift and a tendency to continue upwards. This is the beginning of autorotation and if left unchecked can lead to a spin.

8) The center of pressure moves forward prior to the stall as the AOA is increased. Once the stall is reached, the center of pressure rapidly moves aft. This shifting of the center of pressure is what leads to the nose drop in the stall (see A3).

9) As the weight decreases, the stall speed will decrease as well.

Stall Speed = $V_s \sqrt{(GW/MGW)}$

Stall Speed = $55kts \sqrt{(2000lbs/2500lbs)}$

Stall Speed = 49 knots.

10) Using the same equation, the stall speed will increase due to the increase in weight.

Stall Speed = $V_s \sqrt{(GW/MGW)}$

Stall Speed = $55kts \sqrt{(3750lbs/2500lbs)}$

Stall Speed = 67 knots

11) Working with the 2000 lb stall speed, we can adjust the speed for load factor.

Stall Speed = $V_s(\sqrt{n})$

Stall Speed = $49kts (\sqrt{3g})$

Stall Speed = 85 knots.

This same calculation can also be applied to the aircraft at MGW using the 55kt equilibrium stall speed.

Stall Speed = $V_s(\sqrt{n})$

Stall Speed = $55kts(\sqrt{3g})$

Stall Speed = 95 knots.

12) An incipient spin is the developing stage of a spin. The rotation has not yet been fully established and isn't yet repetitive.
Once the spin becomes fully developed, the rotation is generally repetitive/cyclic and the spin axis is vertical.

Chapter 7: Aircraft Stability

1) The trimmed position of an aircraft is the position in which no moments exist. Trim applies around all three axis'. Longitudinal trim (around the lateral axis) is determined as a trimmed AOA. Lateral (around the longitudinal axis) and directional trim are determined as a trimmed sideslip angle (usually 0°).
The stability of an aircraft relates to the trimmed

position because a stable aircraft will tend to remain in it's trimmed position while an unstable aircraft will not. Positive stability means that when an aircraft moves away from it's trimmed position, moments produced will tend to return it to the trimmed position. For example, if an aircraft has a particular trimmed AOA, increasing the AOA will create moments that tend to reduce the AOA while decreasing the AOA will create moments that tend to increase the AOA. This principle can be applied to a sideslip as well.

2) Static stability is the initial tendency of an aircraft to return to it's trimmed position.

Dynamic stability is the time history of the aircraft's response to being displaced. For example, if an aircraft encounters a gust that increases it's Angle of Attack, the presence of restoring moments means that the aircraft has positive static stability. However, these restoring moments may lead to oscillation to and from the trimmed AOA. If these oscillations eventually dampen out and the aircraft reestablishes equilibrium, the aircraft then has positive dynamic stability as well. However, if the oscillations get larger, the aircraft has negative dynamic stability—even though it has *positive* static stability.

3) Longitudinal stability is achieved primarily by locating the C of G such that the pitching moments are restoring when the aircraft's AOA is changed. This is done by keeping the C of G within certain tolerances that are determined largely by the wing and horizontal tail configuration. An increase in AOA should create nose down moments, while a decrease in AOA should create nose up moments.

The presence of positive longitudinal stability is important because it prevents sudden and uncontrollable fluctuations in pitch and AOA.

4) Lateral stability is achieved by configuring the aircraft so that it rolls away from a sideslip. Factors that will contribute to this stability include dihedral, wing location (high or low), wing sweep, and vertical stabilizer.

With negative lateral stability, the aircraft would

tend to roll into a sideslip. This rolling movement would aggravate the sideslip and lead to large slip angles.

5) Directional stability is achieved by configuring the aircraft so that it yaws into a sideslip. Factors that contribute to this yaw include C of G location, vertical stabilizer, wing sweep, and dihedral.

Directional stability reduces or eliminates undesirable sideslips. The presence of a sideslip causes problems with control and performance of the aircraft, so positive directional stability is a desirable characteristic.

6) With the C of G too far aft, longitudinal stability (pitch stability) would be reduced or eliminated. This would lead to difficulty in controlling the AOA of the aircraft.

In this condition, the forces acting on the wings and tail would create aggravating pitch moments. This means that an increase in AOA will lead to a pitching moment that tends to increase the AOA. A decrease in AOA will lead to pitching moments that tend to decrease the AOA further.

7) Lateral and directional stability are both dependant on a sideslip. As well, they are both affected by the same factors (i.e. – dihedral, wing sweep, vertical stabilizer).

Typical coupling effects between lateral and directional stability include spiral divergence and dutch roll. Another possibility that manufacturers will avoid is directional divergence.

8) Longitudinal stability is degraded when the elevator is floating freely (stick free). This is because after a change in AOA the air loads acting on the elevator will tend to displace it in the direction that contributes to the change instead of correcting it.

Chapter 8: Aircraft Performance

1) Weight – an increase in weight will increase the power required. This increase will be more pronounced at the low speed end of the aircraft's operating speed range.

Altitude – An increase in altitude will increase the power required at a given EAS. On a power curve drawn based on TAS, the curve will shift to the right and up.

Load Factor – Like weight, an increase in load factor will increase the power required. This increase will also be more pronounced at the lower end of the operating speed range.

C of G – As the C of G moves forward, the power required will increase. This increase will be more pronounced at the lower end of the operating speed range.

Configuration – High drag configurations (i.e. –

flaps extended or landing gear extended) will increase the power required. This increase will be more pronounced at higher speeds.

Wind – Wind will have no effect on the power curve.

2) Weight – Increasing weight will increase the induced drag of the aircraft. Since induced drag is predominant at lower speeds, the change in total drag—and therefore total power required—will be more significant at lower speeds.

Altitude – At higher altitudes, a given EAS requires a higher TAS due to the lower air density. Since power required is a function of drag (which is based on EAS) and speed (TAS), the power required will increase for a given EAS at higher altitudes.

Load Factor – At higher load factors, more lift is being produced. This increases the induced drag. Once again, since induced drag is predominant at lower airspeeds, the increase in power required will be more significant at lower speeds.

C of G – A forward C of G increases the required tail down force. In order for the aircraft to maintain equilibrium with the increased tail down force, lift must be increased. This of course increases the induced drag and will therefore increase the power required—particularly at lower airspeeds where induced drag is predominant.

Configuration – Any configuration that increases drag will produce a corresponding increase in the power required.

Wind – The power required is based on the speed (TAS) and the drag—which in turn dependant on the EAS. Wind does not affect either of these variables and will therefore not have any effect on the power required.

3) The term "rolling friction" refers to two phenomenon. The first is the friction between the wheel axle and the wheel itself. This friction tends to slow the rotation of the tire. In order for the tire to continue rotating, it must be kept moving by it's contact with the ground. The force applied by the ground contact has an equal and opposite force (Newton's Third Law) that tends to slow the aircraft.

The other phenomenon that falls under the term "rolling friction" is the drag created on soft or rough surfaces that the tires must roll over or push out of the way. These types of surfaces create a drag on the tires that will tend to slow the aircraft.

Rolling friction reduces the net force accelerating the aircraft. This means that more distance will be required in order to accelerate to the required lift-off speed.

4) A runway gradient means that a component of the aircraft's weight will be acting either with or against the takeoff. Taking off on a downgrade will mean that the weight will contribute to the acceleration and shorten the distance required. Taking off on an upgrade will mean that the weight is acting against the acceleration and increase the distance required.

5) First, we need to determine the power required.

$$Pwr = Dv/325$$

$$Pwr = (260lbs)(87kts)/325$$

$$Pwr = 69.6HP$$

Now we can determine the rate of climb based on the excess thrust horsepower.

$$ROC = \frac{33,000 \times ETHP}{W}$$

$$ROC = \frac{33,000 \times (P_a - P_r)}{W}$$

$$ROC = \frac{33,000 \times (120HP - 69.6HP)}{2400lbs}$$

$$\underline{ROC = 693 \; FPM}.$$

6) The best rate of climb occurs at the speed that provides the greatest distance between the power required and the power available—that is, the greatest *excess* power.

Recall from Chapter 1 that power is the rate of transfer of energy. In a climb, the aircraft is gaining potential energy (height). The faster that energy can be transferred to the aircraft, the higher the rate of climb will be. Excess power is used to either accelerate or climb (as opposed to the power required to maintain level flight, which is just used to replace energy lost to the atmosphere). In the case of a climb, the more excess power there is available, the more height will be gained per unit of time.

7) For some aircraft, the manufacturer will call for a climb speed that is *slower* than V_x because the distance required to accelerate to V_x outweighs the advantage gained by climbing at the steepest possible angle. With a lower climb speed, the climb can start earlier, therefore the shallower climb resulting from the lower speed can clear a standard obstacle better than a climb at V_x.

8) Aside from design characteristics of the aircraft, variables that will affect the climb performance include:

– Weight – Heavier aircraft will climb slower due to the higher weight and due to the increased induced drag present.

– C of G – With a forward C of G, the aircraft produces more induced drag due to the increased tail-down force. This increase in drag will cause a decrease in climb performance. With an aft C of G, the opposite occurs and climb performance is improved.

– Configuration – The lowest drag (and therefore power required) possible will provide for the best climb performance. This is because less power required will lead to a greater amount of *excess* power. To minimize drag, the flaps, landing gear, etc. should be retracted.

– Density Altitude – At higher density altitudes, the engine cannot produce as much power. As well, the increased TAS leads to an increase in power required. The overall effect is a reduction in excess power. Any reduction in excess power will of course lead to a corresponding reduction in climb rate and climb angle.

9) The maximum level cruise speed is the speed at which the power required and power available cross so that the P_a becomes lower than the P_r.

At speeds higher than the maximum level cruise speed, a descent must be established in order to maintain the speed. This is because a power deficiency is always present at higher speeds.

10) Maximum endurance (for propeller aircraft) occurs at the lowest point on the power required curve.

This speed is maximum endurance because level flight can be maintained at the lowest possible power setting and therefore the lowest possible fuel consumption rate. Obviously, the slower fuel is consumed the longer the aircraft can remain airborne.

11) Maximum range occurs at the speed indicated by the tangent point when a line is drawn from the origin to just touch the power curve. This speed corresponds to the best lift to drag ratio for the aircraft.

At the best range speed, the aircraft is fling at the highest speed to power ratio. Since fuel burn rate is proportional to power, this speed provides the highest speed for the lowest fuel consumption. This condition maximizes the range of the aircraft.

12) Endurance is simply the *time* that an aircraft can remain airborne. Since wind only affects the speed and not the amount of drag (or power required), the endurance will not be influenced by wind.

Range will be reduced by a headwind since the groundspeed is reduced and increased by a tailwind since the groundspeed is increased. A headwind can be *partially* compensated for by increasing the airspeed an appropriate amount. Likewise, a tailwind can be exploited further by decreasing the airspeed an appropriate amount.

13) Any drag producers will increase the descent angle because they will decrease the lift to drag ratio. This means that an aircraft in a glide must descend more

steeply in order for the forward component of weight to counteract the extra drag being produced.

14) At higher angles of bank, the total lift produced by the aircraft must be increased in order for the vertical component to remain equal to weight. The horizontal component, on the other hand, is unbalanced and creates an acceleration (centripetal acceleration). The greater lift force and the resulting acceleration is what causes the increase in load factor.

15) The rate of turn increases with an increase in AOB and decreases with an increase in airspeed.

The radius of turn decreases with an increase in AOB and increases with an increase in airspeed.

16) Depending on the approach and landing technique used, increased weight may have no effect on the distance required or it may have the indirect effect of increasing the distance required.

If the approach and touchdown speeds are held constant with varying weight, the weight will have no appreciable affect on the landing distance. On the other hand, if the operating speeds are adjusted for weight, an increase in weight will increase the distance required to land while a decrease in weight will decrease the distance required to land. The reason for this is the change in speed and therefore the reduction in distance required to decelerate to a stop. With a change in weight, the increased/decreased momentum of the aircraft is compensated for by the increased/decreased braking effectiveness during the landing roll.

17) Runway gradient affects the landing run in much the same way as the takeoff run, except the more favorable condition is reversed. That is to say that the landing distance will be decreased when landing uphill and the distance will be increased when landing downhill. The reason for this is the component of weight that acts along the landing path when the aircraft is operating on an upgrade or downgrade.

18) When landing on a soft surface, the braking effectiveness is decreased. This decrease in braking capacity will often overcome the increase in "natural" rolling friction.

Chapter 9: Aircraft Limitations

1) The limit load factor (LLF) is the maximum load factor that an aircraft is certified to fly at without any permanent structural deformations. The ultimate load factor (ULF) is the load factor that an aircraft is certified to fly at without structural failure. As a type certification requirement, the ULF must be at least 150% of the LLF.

2) As weight is decreased, the loads acting on the overall airframe will be decreased at a given load factor (i.e. – the LLF). This would imply that as the weight is decreased, the LLF would be increased. However, fixed weight components are normally only designed to the LLF. These components primarily include the engine(s), but can also include other items such as the landing gear, the pilot's seat (with the pilot in it of course), etc.

3) Many aircraft are prohibited from performing inverted maneuvers because of engine considerations. Even though the airframe may be capable of safely withstanding negative loads, the fuel supply may require positive g. Also, the oil supply system and any hydraulic systems may requires positive g in order to function properly.

4) Aircraft with higher wing loading are less susceptible to load factors due to turbulence.
On a given wing surface, a given gust will cause a fixed increase (or decrease) in the lift being produced. A heavier aircraft will experience a smaller acceleration if the lift force applied to it is the same as the lift applied to a lighter aircraft. Since wing loading is simply defined as the ratio of weight to wing area, the heavier aircraft is the one with the higher wing loading (still assuming that we are working with the same wing surface). So the aircraft with the higher wing loading will experience smaller accelerations—and therefore will be less affected by turbulence.

5) V_{le} is the maximum allowable speed with the landing gear in the extended (i.e. – "down and locked") position. V_{lo} is the maximum speed at which the landing gear can be extended or retracted. On some aircraft the two speeds are the same, but on others V_{lo} will be lower than V_{le}.
The reason for this is that the gear in transition cannot withstand the same air loads that the down and locked gear can. Sometimes this is because hydraulic or electrical actuators cannot take the resistance brought on by the aerodynamic loads. Other times, the reason is because doors must open to facilitate the extension and retraction of the gear. These doors may not be as strong as the gear itself.

6) V_a can often be determined as the stall speed at the aircraft's limit load factor,

$$V_a = V_s \sqrt{LLF}.$$

Unfortunately, this method will not always work. This method determines the minimum allowable speed for V_a. However, some aircraft will have a V_a that is higher than this speed.

7) V_{ne} is the maximum speed that the aircraft is permitted to fly at under any circumstances. V_{ne} is established by the manufacturer as 90% of the dive speed (V_d). The dive speed is the highest speed that can be flown without encountering flutter or control reversal.

8) Flutter is an unstable oscillation of a component. All objects have a natural resonance frequency that they vibrate at. Normally, these vibrations dampen out over a relatively short period of time. However, at high airspeeds, the airloads acting on a surface may amplify the vibrations. This is flutter. Flutter can destroy aircraft components (like the wing or tail) very quickly and as such, conditions leading to flutter should be avoided at all times.

9) The minimum radius turn speed (for a given configuration) is the aircraft's corner speed. This speed is determined by determining the stall speed at the LLF according to equation (9.3):

$$V_a = V_s \sqrt{LLF} \qquad (9.3).$$

This equation indicates that this speed is the maneuvering speed, but keep in mind that it is actually the *corner speed*, which is the minimum allowable maneuvering speed. This speed is a stall speed and can be adjusted for weight using equation (9.4):

$$\text{New } V_a = V_a \sqrt{(GW/MGW)} \qquad (9.4).$$

The new corner speed for the aircraft's actual weight is the one to be used for minimum radius turns. The AOB to be used is the one that corresponds to the aircraft's LLF.
Another factor to keep in mind is that this equation is for a particular configuration. Varying the configuration with high lift devices can decrease the turn radius in some cases.

10) At V_a, full control deflection can be used safely. At V_{ne}, $^1/_3$ deflection can be used safely. At speeds in between, the amount of safe deflection can be estimated by interpolating between these two limits.

11) Using equation (9.3),

$$V_a = V_s \sqrt{LLF} \qquad (9.3),$$

we can determine what the minimum allowable maneuvering speed is for this aircraft.

$$V_a = 45 \sqrt{4} = 90 \text{kts}.$$

The published maneuvering speed is 105 kts, so we know that even though full control deflection will not damage the control surfaces, elevator deflection can overstress the rest of the airframe. This same reasoning applies to 100 kts. It is below V_a, but it is above 90kts corner speed. So elevator deflection can possibly overstress the airframe.
So, full aileron deflection at 100kts would be safe. Full rudder deflection at 100kts would also be safe. However, full elevator deflection could damage the aircraft. For this reason, elevator deflection would have to be limited in order to remain within the aircrafts maneuvering envelope even though the airspeed is below

V_a.

Chapter 10: Weight and Balance

1) The weight of the aircraft will determine performance factors including stall speed, rate and angle of climb, cruise speed, takeoff distance, landing distance, etc.

The C of G will also affect these performance factors (a forward C of G will degrade performance). As well, the C of G will have an effect on the stability characteristics of the aircraft. As the C of G moves aft, the pitch and yaw stability will be reduced. The rudder and elevator also lose control authority as the C of G moves aft. With a forward C of G, control and stability are both improved up to a point. Beyond this point, elevator authority can be insufficient to pitch the aircraft up—particularly at low airspeeds in ground effect (i.e. – in the flare).

2) 0-Fuel:

      Weight   - 1907.7 lbs.
      Moment - 164668.24 in-lbs.
      C of G   - 86.32 in.

Landing:

      Weight   - 2027.7 lbs.
      Moment - 174736.24 in-lbs.
      C of G   - 86.17 in.

Takeoff:

      Weight   - 2147.7 lbs
      Moment - 184804.24 in-lbs.
      C of G   - 86.04 in.

3) The aircraft is within limits for all phases of flight.

4) Switching the front and rear passengers is equivalent to moving 70 lbs aft 33.6". Knowing this, we can determine C of G locations with the new loading.

0-Fuel:

$$\frac{W_{AC}}{w_o} = \frac{D_o}{d_{CG}}$$

$$\frac{1907.7 \text{ lbs}}{70 \text{ lbs}} = \frac{33.6 \text{ in}}{d_{CG}}$$

$$d_{CG} = 1.23 \text{ in. aft}$$

The original 0-fuel C of G location was 81.19in. So the new C of G is 86.32 in + 1.23 in = 87.55 in.

This C of G location remains within limits.

Landing:

$$\frac{W_{AC}}{w_o} = \frac{D_o}{d_{CG}}$$

$$\frac{2027.7 \text{ lbs}}{70 \text{ lbs}} = \frac{33.6 \text{ in}}{d_{CG}}$$

$$d_{CG} = 1.16 \text{ in. aft}$$

86.17 in + 1.16 in = 87.33 in.

This C of G location remains within limits.

Takeoff:

$$\frac{W_{AC}}{w_o} = \frac{D_o}{d_{CG}}$$

$$\frac{2147.7 \text{ lbs}}{70 \text{ lbs}} = \frac{33.6 \text{ in}}{d_{CG}}$$

$$d_{CG} = 1.09 \text{ in. aft}$$

86.04 in + 0.97in = 87.01 in.
This C of G location remains within limits.

Chapter 11: Taxiing

1) The wind is blowing from 50° to the aircraft's right. Because of this, the ailerons should be deflected fully to the right and the elevator should be held neutral. This input will reduce or even eliminate the rolling tendency due to the strong crosswind.

2) Having turned left to exit the runway, the wind will now be blowing from 140° to the right of the aircraft (i.e – from 40° to the right of the tail). A strong quartering tailwind requires a "dive away" control input. So the ailerons should be deflected fully to the left and the elevator should be deflected fully forward.

3) The increase in power at the beginning of a taxi is required to initiate the motion. This mean both overcoming the static friction of the stationary wheels (recall static friction from Chapter 1) and accelerating the aircraft from rest to an appropriate taxi speed. This requires more power than simply maintaining an already established motion. For this reason, the extra power needs to be removed once the aircraft is moving.

Chapter 12: Attitudes and Movements

1) From the cruise attitude, the control column is pulled back. This deflects the elevator upwards and therefore causes the aircraft to pitch upwards.

If no power change is made, pitching up will cause the aircraft to slow down and climb.

When looking outside, the horizon will move to a lower position in the windscreen as compared to the cruise attitude.

2) An attitude is the aircraft's position relative to the horizon. Attitude is defined by a pitch angle and a bank angle. A movement is a rotation of the aircraft around one of it's axis'. Movements are used to transition from one attitude to another.

3) The pitching movement is about the lateral axis and is controlled by the elevator. The rolling movement is about the longitudinal axis and is controlled by the ailerons. The yawing movement is about the vertical axis and is controlled by the rudder.

4) To return from a banked attitude to a cruise attitude, the ailerons are deflected in the direction opposite the bank. This initiates a rolling movement towards wings level. Once the wings are level with the horizon, the ailerons are returned to neutral and the roll is stopped.

5) Pitch attitude and AOA are related through the flight path. The AOA is the angle between the relative airflow and the chord of the wing. The relative airflow is determined by the flight path and the chord orientation is determined by the pitch attitude (ignoring wing angle of incidence). So the AOA is the sum of the flight path angle and the pitch angle.

6) The trim tab changes the trim of the aircraft so that various positions can be held without prolonged effort on the part of the pilot. The most common type of trim tab is the pitch trim. When a new AOA is established by the pilot, the trim can be adjusted so that the AOA can be held without control pressure. In practice, pitch trim is used to maintain a constant pitch attitude (not AOA) but the mechanics of the device actually affects the AOA.

7) When power is applied, most aircraft will respond with a tendency to pitch upwards. This occurs for a variety of reasons. First of all, the thrust/drag couple is increased. Second, the downwash on the tail is increased. And third, if the aircraft accelerates due to the increase in power, extra lift will be produced. This leads to an upward acceleration and a resulting reduction in AOA. The aircraft's longitudinal stability responds to this by tending to pitch up and increase the AOA to the trimmed value.

8) Configuration changes affect the airflow patterns over the airframe. In this way the trim of the aircraft can be changed. For example, if flaps are deflected, the nose down pitch moment of the wing is increased. This can cause the aircraft to pitch forward when the flaps are extended. On the other hand, if the airflow over the tail is directly affected by the wing and flaps (as in a high wing aircraft), the increased downwash on the tail can increase the overall pitch-up moment of the aircraft. In this case, extending the flaps will cause the aircraft to pitch upwards.

9) Undesirable—or uncoordinated—yaw is caused by influences such as the left turning tendencies of the engine, aileron drag, and turbulence. Also, corrections that are built into the aircraft by the manufacturer can become over-corrections under certain conditions.

Uncoordinated yaw can be corrected by the pilot by using the rudder.

Chapter 13: Cruise Flight

1) Straight flight is maintained through several control inputs. Directional control is maintained with the rudder and the wings are kept level with the ailerons. As long as the desirable direction is maintained, these control inputs will maintain straight flight. However, if the direction drifts due to distraction of the pilot or due to an misjudged control input, the correction should be made with the ailerons. Establishing a very shallow banked attitude will turn the aircraft back to the intended direction of flight.

Heading corrections are not made with the rudder because this leads to uncoordinated flight. When yaw is present with the wings level, the aircraft is flying in a sideslip.

2) Using the rule of thumb for power changes (100 RPM for 5 knots), the power can be reduced by about 600 RPM. As the aircraft slows down, the pitch attitude needs to be increased in order to maintain level flight. Once the speed has stabilized, smaller power adjustments can be made if necessary in order to get the aircraft to 100kts.

3) At a density altitude of 5000 ft, 55% power occurs at 2220RPM. The TAS resulting would be 109kts.

Chapter 14: Climbs and Descents

1) Best rate of climb gets the aircraft to altitude in the shortest period of *time*. Best angle gets the aircraft to altitude in the shortest *distance* possible. At best angle, the rate of climb is reduced, but the forward speed is also reduced. This leads to a steeper angle of climb.

2) During a climb recovery, the extra power that was used for the climb is used to accelerate to the cruise speed. If a power reduction is made early, the aircraft will be slow to achieve the desired cruise speed. Also, if the power reduction is made prior to pitching down to the cruise attitude, the aircraft will actually slow down.

3) The cruise climb's primary advantage over "performance" climbs is that ground is covered at a higher rate. This makes the cruise climb useful for cross-country flying (hence the reason that the cruise climb is often referred to as an en-route climb). As well, the cruise climb provides improved engine cooling and improved visibility along the flight path.

The cruise climb's disadvantage is in fact the same as it's main advantage. Since ground is covered faster, the angle of climb is reduced. This would clearly make the cruise climb undesirable during obstacle clearance maneuvers. As well, the rate of climb is reduced. This makes the cruise climb undesirable when trying to gain altitude quickly (i.e. – when trying to gain the advantage of a tailwind aloft or when trying to gain "contingency" altitude for safety).

4) During a glide descent, the power is set at idle (or perhaps the engine has failed or been shut down), and the aircraft is giving up altitude to maintain a constant airspeed by replacing energy lost to the atmosphere via drag. Glides are normally conducted at the best glide speed (best lift to drag ratio), but this is not always the case.

During an en-route descent, power is normally carried in order to allow a lower descent rate at the same time as a higher airspeed. Pitch is set to control the airspeed while power is varied as necessary to control the vertical speed.

During an emergency descent, the objective is to accomplish the highest possible descent rate that the aircraft can safely maintain. Generally, drag devices such as flaps and landing gear are used and a high airspeed is used with no power to maximize the vertical speed.

5) Using the level-off lead rule of thumb (10% of the vertical speed), the level off should be initiated 60 ft prior to the target altitude. This means beginning to level of at 1760 ft.

Chapter 15: Turning

1) Gentle and medium turns don't require as much back pressure as steep turns. As well, steep turns often require an increase in power in order to maintain altitude and airspeed.

Both of these differences occur due to the increase in lift required in a turn. This increase becomes more pronounced as the angle of bank increases. An increase in lift at a constant airspeed requires an increased AOA. This is the reason for the back pressure. The increased lift also results in an increase in induced drag. The increase in power required during a steep turn is due to this increase in induced drag.

2) Once an aircraft is established in a banked attitude, the lift produced by the wings is acting on an incline (perpendicular to the wingspan). The horizontal component of lift is unbalanced and causes an acceleration in the direction of the bank. Once the direction of flight changes, the relative airflow changes and the aircraft responds with directional stability to realign itself with the airflow. The combination of change in flight path and reorientation of the aircraft is a turn. Coordinated turns will always be in the direction of the low wing.

3) To deal with this question, we will have to

determine the minimum turn radius for both the clean configuration and the flaps extended configuration. We are assuming here that the aircraft has enough power available to maintain whatever airspeed is necessary once the turn is established.

Clean Configuration:

To find the corner speed,

Corner Speed = $V_s \sqrt{LLF}$

Corner Speed = 65kts $\sqrt{(4.4g)}$

Corner Speed = 137kts.

Then we need to find the AOB for the limit load factor,

n = 1/CosAOB

4.4g = 1/CosAOB

AOB = 76°

Knowing the speed to fly and the AOB to use, the radius of turn can be determined. We will assume here that the turn is being conducted at a density altitude of sea level. This means that the EAS stall speeds and corner speeds are equal to the TAS stall speeds and corner speeds. We also need to convert the corner speed to feet per second (137kts = 231.4fps).

$$r = \frac{v^2}{gTanAOB}$$

$$r = \frac{(231.4fps)^2}{(32.2ft/s^2)Tan75}$$

r = 446 ft.

Flaps Extended:

To find the corner speed,

Corner Speed = $V_s \sqrt{LLF}$

Corner Speed = 52kts $\sqrt{(3.5g)}$

Corner Speed = 98kts.

And to find the AOB for the limit load factor,

n = 1/CosAOB

3.5g = 1/CosAOB

AOB = 73°

Converting 98kts to feet per second, we get 98kts = 165.5fps. Knowing the speed to fly and the AOB to use, the radius of turn can be determined.

$$r = \frac{v^2}{gTanAOB}$$

$$r = \frac{(165.5fps)^2}{(32.2ft/s^2)Tan73}$$

r = 261 ft.

So this particular aircraft will be capable of a smaller turn radius with the flaps extended. For the minimum radius turn, the speed used should be 98 knots and the AOB should be 73°.

Some consideration must be made on the practical use of these numbers. 73° is a fairly precise AOB to maintain. Realistically, 70° is close to the desired value and it leaves some room for error. As well, many light aircraft will not be capable of producing enough power to maintain airspeed in the turn. Because of this, minimum radius turns are often entered at higher airspeeds with the airspeed decreasing during the turn. These facts need to be considered in actual flight operations. On paper, the aircraft can turn with a 261 ft turn radius. In reality, the turn radius will be larger than this due to the slightly decreased bank angle and increased airspeed. Other considerations include altitude (at higher altitude, the TAS is higher) and the winds (since ultimately, it is the turn radius *over the ground* that really counts).

These same considerations can be applied to the clean configuration. So all in all, the flaps extended is still the best option. A turn radius that would be more realistic (I won't do the calculation here) would be around 415 ft. This takes into account increased altitude (4000 ft in this example), increased airspeed, decreased bank angle and zero wind.

4) Using the rate 1 turn rule of thumb (10% of the airspeed plus 5 (in mph) or 7 (in knots)), the angle of bank to use is approximately 17°.

The load factor can be determined using equation (8.10),

n = 1/CosAOB   (8.10).

n = 1/Cos17°

n = 1.05g

Chapter 16: Aircraft Control Considerations

1) From an established cruise, a descent is normally established by reducing power and pitching down to either maintain an airspeed or descend at a reduced airspeed. However, if we wanted to establish a descent at an increased airspeed, simply pitching down would do it. Once established in the descent, pitch is used to control airspeed and power is used to control vertical speed.

2) "Attitude + Power = Performance" means that pitch and power combined control airspeed and vertical speed combined. In practice, we often separate the functions of pitch and power, but in reality they work hand in hand. In order to separate them, a "critical constant" must be selected—either airspeed or vertical speed/altitude.

3) The 4 types of energy manipulated when flying are kinetic energy (speed), potential energy (height), chemical energy (stored in the fuel), and heat energy (lost to the atmosphere via drag).

At the beginning of a flight, all of the aircraft's energy is stored in the fuel. As the aircraft accelerates down the runway, some of this energy is converted to speed (kinetic). Then, as the aircraft climbs, some more energy is converted into height (potential). Throughout the flight, energy is lost to drag and must be replaced by the fuel energy. As well, speed and height can be interchanged—diving can be used to build speed and slowing down can be used to gain altitude. In a descent, the energy provided by the engine is insufficient to replace energy lost to drag. So altitude is traded to compensate for drag.

4) When rudder is used to coordinate the aircraft, it's function is to control the rate of yaw so that it is appropriate for the angle of bank and airspeed. In straight flight, the wings should be level and no yaw should be present. However, in a banked attitude, the aircraft should be turning, so yaw is present. Rudder is used to increase, decrease, or hold steady the rate of yaw so that no sideslip is present in the turn.

When the ailerons are used to coordinate, they are used to very the angle of bank so that it matches the rate of yaw. If a slip is present, it means that the angle of bank is either too high or too low for the rate of yaw, and a correction can be made with the ailerons.

5) As we know, rudder is usually used to coordinate the aircraft. Once established in a turn, some aircraft will be coordinated with no inputs from the pilot. However, many aircraft will always need some

corrections. As well, under different conditions (i.e. – different airspeeds perhaps) aircraft will often need different amounts of correction.

The need for coordinating inputs comes from a variety of sources. These include turbulence, aileron drag when roll corrections are needed, the left turning tendencies of the engine, and the difference in airspeed of the two wings (and therefore the difference in drag).

6) When trim is used properly, the aircraft is established at the desired attitude, configuration, and power setting. Once established, the trim is adjusted to eliminate input forces on the control column. The attitude of the aircraft should not change with a trim adjustment. The control column is used to adjust the attitude and trim is used to adjust the control pressure.

Improper use of the trim would primarily include using the trim to control the attitudes/movements of the aircraft directly. If trim is used improperly, the result will be an oscillating flight path and constant readjustment of the trim. This is because when the trim is changed, the aircraft will no longer be flying in it's trimmed condition. The dynamic stability characteristics of most aircraft will lead to an oscillation back and forth beyond the desired attitude of the aircraft. Of course, once the target attitude has been overshot, the reaction is to retrim to return to the target attitude. When this is done, it is very likely that the pilot will get out of phase with the aircraft and the oscillations will not dampen out.

7) After an aileron failure, roll is controlled indirectly by the rudder. Yaw induced by the rudder will lead to a roll in the same direction. Entering a turn can be done by yawing in the desired direction. This yaw will lead to a roll, and the roll can be stopped by coordinating once the desired bank angle is established.

Once in the turn, pitch is controlled the same way as always—with the elevator. Roll and coordination are now both controlled with the rudder.

To get back to cruise, the rudder can be used again to roll. Once the wings are level, the yaw (and therefore the roll) can be stopped. Again, roll and coordination will be controlled by the rudder. Unfortunately, this means that only one (roll or coordination) can be controlled at a time. In order to roll, coordination must be sacrificed. And in order to remain coordinated, roll control is lost.

8) SLOWLY AND CAREFULLY!!! In this situation, the elevator will be floating freely. The oscillations resulting from pitch displacements will be prolonged as a result. The dynamic stability of the aircraft must be anticipated and accounted for.

Normally, a descent is entered using "PAT": Power, Attitude, Trim. This rule still works but must be applied a little differently.

Reducing the power will lead to the aircraft's trim AOA changing (due to the reduction in the thrust/drag moment and due to the reduction in downwash on the tail). The result will be a pitch down—this is the attitude adjustment, except as a general rule it will be too much. The power reduction should be very small in order to avoid large and sudden attitude changes and to avoid large and prolonged pitch oscillations. Once the power adjustment is made, the trim can be adjusted to prevent the overadjustment of the pitch (once again, inputs must be kept small).

If the descent that is established from these steps is not steep or fast enough, or if the airspeed is not the target airspeed, more control inputs can be made. Again, all inputs should be small and the aircraft's reaction must be anticipated. Plan the control inputs before they are made.

9) A rule of thumb is an approximation. In the case of aircraft, the laws of physics that dictate the behavior of the aircraft are simplified so that they can be applied during flight operations.

The main shortcoming of rules of thumb is that they are simplifications. This means that they are based on assumptions and approximations so that calculations can be made with relative ease. This inevitably leads to inaccuracies that could possibly be dangerous if rules of thumb were considered to be absolute.

Rules of thumb are still useful because the conditions that they are based on are generally known. This means that if actual conditions do not fit the assumptions made by the rule of thumb, the rule should not be used. As well, any time that a rule of thumb is used to determine critical information, the potential inaccuracies can be anticipated and accounted for.

Chapter 17: Range and Endurance

1) Altitude does not affect an aircraft's *maximum* range. However, range at a specific power setting will increase with an increase in altitude.

Maximum range doesn't change because as the TAS increases, the power required increases proportionally. The net effect is for the velocity to power ratio to remain constant when the aircraft is flown for maximum range.

The increase in range at a specific power setting is due to the higher TAS that results from the power setting. Also, one could look at this increase in range as being caused by the power curve shifting upwards—leading to a given power setting becoming closer to the optimum.

2) When flying for maximum range, a headwind can be compensated for by increasing the airspeed. This increase should be equal to approximately ½ of the headwind value.

This correction minimizes the losses to the headwind, but it does not completely restore the range of the aircraft to the zero-wind value.

3) Maximum endurance occurs at the lowest point on the power required curve.

The reason for this is in the definition of maximum endurance—the longest time aloft per unit of fuel. The fuel burn rate is proportional to the power setting, so the lower the power setting the slower fuel will be used. At the speed for maximum endurance, the power setting can be reduced to the lowest possible value that allows level flight. The minimized fuel flow rate will maximize endurance.

4) Wind has no effect on endurance because there is no need to go any distance. Wind will affect the groundspeed of the aircraft, but not the power required (since power required is based on airspeed). So the endurance will not be affected by the wind.

Altitude, on the other hand, will have a direct effect on maximum endurance. As altitude is increased, the power curve shifts upwards. Because of this, the minimum power point is higher—meaning that more power is required to maintain level flight. So maximum endurance is improved at lower altitudes.

5) When flying in turbulence, airspeed fluctuations will occur. At maximum endurance, a drop in airspeed will bring the aircraft into the slow flight speed range. This means that power must be increased in order to recover to the normal flight speed range or just to maintain altitude. As well, slow flight has other problems associated with it (detailed in Chapter 18). Either way, an increase in power will decrease endurance. In order to minimize the disadvantage of turbulence, a speed slightly higher than maximum endurance should be used.

6) All other things being equal, an aft C of G will improve both range and endurance, while a forward C of G will degrade both range and endurance.

The reason for this is the extra induced drag that is present with a forward C of G. In order to balance the increased nose down moment produced by the lift/weight couple, the tail down force must be increased. This in turn increases the lift that must be produced by the wings—which then increases the induced drag of the aircraft.

With an increase in induced drag, the minimum power required is increased—thus reducing endurance. As well, the velocity/power ratio is reduced—thus reducing range.

7) When reducing power in increments, maximum range is the power/speed combination that occurs just prior to the largest airspeed drop. In the example given, the largest drop is 12 knots and it occurs between 2000RPM/84KIAS and 1900RPM/72KIAS. So flight for maximum range will be achieved at 2000 RPM and 84 KIAS.

Maximum endurance occurs at the lowest power setting required to maintain level flight. In the example given, this is a power setting of 1900 RPM and an airspeed of 72 KIAS.

8) Maximum range with a headwind is found by increasing the airspeed by ½ of the headwind value. In this case, the speed for best range with no wind is 84 knots (we will assume that IAS and TAS are close enough for the difference to be insignificant in a rule of thumb). With a headwind of 20 knots, the new best range speed would be approximately 94 knots.

Maximum endurance is unaffected by the wind. So the speed for best endurance would still be 72 KIAS.

If the wind was a tailwind, the speed for best range would be approximately 74 KIAS. Note that when using this rule of thumb, the range speed should not be adjusted below the endurance speed.

The endurance once again is unaffected by the wind.

Chapter 18: Slow Flight

1) Slow flight is the range of airspeeds between the speed for maximum endurance and the stall.

Slow flight is considered to be an important phenomenon because the power required will actually begin to increase as the airspeed decreases. This leads to an unstable airspeed that can lead into a stall. As well, the symptoms of slow flight are considered to be warning signs of an approaching stall.

2) The intentional entry into slow flight is normally conducted from maximum endurance. From there, any method used to slow down will result in slow flight. Normally, simply pitching up slightly is the technique used. Once established in slow flight, an increase in power will be required in order to maintain altitude.

3) In slow flight, turns will often require an increase in power. As well, rudder will be required in order to coordinate the roll-in and roll-out of the turn. In the turn itself, more rudder than usual will often be required in order to coordinate the turn.

The increase in power required is due to the fact that induced drag is predominant in slow flight. So when a turn is established, the increase in lift has a significant impact on the total drag—and therefore the power required—of the aircraft.

The requirement for rudder during roll inputs is due to the increase in aileron drag and the decrease in the effectiveness of built-in corrections (i.e. – differential and frise ailerons). This increases the adverse yaw present and therefore increases the need for coordinating rudder.

The increased rudder in the turn is due to the difference in the speeds of the two wings. At lower airspeeds, this difference is more pronounced and requires a correction with the rudder.

4) Slow flight is recognized by using 5 main symptoms:

– Airspeed vs. Power/Vertical Speed – As the airspeed is reduced, the power required will increase. This will result in an increase in descent rate if no power adjustment is made, or an increase in power if the vertical speed/altitude is held. The airspeed will be below endurance.

– Speed Stability – Fluctuations in airspeed will not be self correcting. They will tend to get progressively worse unless the pilot actively intervenes.

– Decrease in Control Effectiveness – Due to the reduced airflow over the control surfaces, the controls will be less responsive in slow flight.

– Increase in Adverse Yaw – The predominance of induced drag will increase the aileron drag and therefore the adverse yaw. More rudder will be required to remain coordinated—especially during roll.

– Pitch Attitude – Not a reliable indication, pitch can be used to indicate the "odds" of being in slow flight. Slow flight is more likely in a nose high attitude—although it can still occur in a nose low attitude.

5) Using pitch to correct altitude deviations leads to airspeed fluctuations. In the normal flight speed range, this is acceptable since the airspeed is self correcting. However, in slow flight, airspeed changes tend to be self aggravating. Because of this, using pitch to correct the altitude in slow flight will lead to large airspeed changes and possibly a stall.

6) To recover from slow flight, full power is used

to accelerate the aircraft into the normal flight speed range. The pitch attitude is adjusted just enough to prevent a climb, but not enough to initiate a descent. If necessary, the aircraft should be returned to a clean configuration (flaps and gear retracted). Once a climb speed is reached, a climb can be initiated.

Minimum loss of altitude is important during the recovery because the critical areas for inadvertent entry into slow flight are all close to the ground. Practice at altitude shouldn't encourage diving for airspeed since slow flight practice is ultimately designed to prepare pilots for recovery at low altitude.

7) The critical areas for inadvertent entry into slow flight are all close to the ground. They include takeoff, landing, overshooting, and any other low level operations.

8) In some aircraft, certain conditions can lead to a power deficiency when recovering from slow flight—even when full power is being used. These conditions generally include flying in a low powered aircraft at a high density altitude (remember that this doesn't necessarily mean a high altitude) at high weight and in a "dirty" configuration (i.e. – flaps and/or landing gear extended). Without excess power, the engine cannot accelerate the aircraft, so a descent must be used to gain airspeed and recover from slow flight. However, this descent should be avoided unless it is absolutely necessary.

9) In the normal flight speed range, APT—Attitude, Power, Trim—is used for entry and recover from climbs. Adjusting the attitude first will lead to a change in airspeed. In the normal speed range, this is desirable since it brings us closer to the target climb speed. However, in slow flight, this change in airspeed can easily lead to a stall since pitching up for the entry to a climb will lead to a reduction in speed. Pitching down to recover from the climb will lead to an increase in speed and a subsequent recovery from slow flight. After an inadvertent entry into slow flight, this recovery is desirable. However, in the training environment, it is often the intent to remain in slow flight after the level-off. For these reasons, PAT—Power, Attitude, Trim—is used for climb entry and recovery in slow flight.

10) No. Slow flight is defined as the speed range between maximum endurance and the stall. Both limiting speeds (endurance and the stall) will vary under different conditions. With an increase in weight or a forward shift in C of G, both speeds will increase. With an increase in load factor, both speeds will increase. With a decrease in weight, a rearward shift in C of G, or a decrease in load factor, both speeds will decrease. Flaps will also change slow flight by decreasing the stall speed and the endurance speed. So under different conditions, the speed range for slow flight will change. Because of this, knowing "the numbers" for a given aircraft will only act as a guide. The symptoms of slow flight are still important indicators of entry into the slow flight range.

Chapter 19: Stalls and Spins

1) Stalls occur because of the separation that takes place on the top of the wing as the AOA is increased. Once the critical AOA is reached, the lift coefficient will decrease with an increase in AOA.

Stalls are avoided simply by keeping the AOA below the critical value. Since AOA and airspeed are so closely related, stalls are generally avoided in practice by maintaining a safe airspeed.

2) First, we will find the load factor of the turn,

$$n = 1/CosAOB$$

$$n = 1/Cos40°$$

$$n = 1.3g.$$

Now the stall speed can be found,

$$Stall\ Speed = V_s(\sqrt{n})$$

$$Stall\ Speed = 58kts(\sqrt{1.3g})$$

Stall Speed = ~66kts.

3) The stall speed for the aircraft at the reduced weight is,

$$New\ V_s = V_s\sqrt{(GW/MGW)}$$

$$New\ V_s = 58kts\sqrt{(2200lbs/2750lbs)}$$

New V_s = ~52kts.

This is the stall speed at the new weight and in straight flight. To determine the stall speed in the turn and at the new weight, we can either apply the weight change equation to the turning stall speed (A2), or we can apply the load factor equation to the lighter stall speed.

$$Stall\ Speed = V_s(\sqrt{n})$$

$$Stall\ Speed = 52kts(\sqrt{1.3g})$$

Stall Speed = ~60kts.

4) A stall can be recognized from the buffeting of the airframe, the nose drop, and the loss of altitude.

The approach to the stall can be recognized by recognizing passage through slow flight. Also, buffeting generally begins just prior to the stall.

5) In most aircraft, a power-on stall tends to give less warning and be more violent than a power-off stall. As well, since the stall occurs in a climbing turn, the top wing (outside the turn) should stall first.

So as this stall is entered, the nose will drop suddenly (violently in many cases) and the right wing will most likely drop. We will assume for our purposes here that recognition of the stall is immediate and therefore the recovery is initiated immediately. The power should already be set at climb power. If climb power is less than the full or emergency setting, power should be increased as the nose is lowered to reduce the AOA. At the same time, the ailerons should be returned to neutral and rudder should be applied against the roll (left rudder in this case). As the aircraft comes out of the stall and control is regained, a descent will be initiated. This descent must be ended by pitching up again—however, a secondary stall MUST be avoided. This means that the load factor must be kept low throughout the dive recovery.

6) Using ailerons during a stall can be counterproductive. If the stall is deep enough to affect the wingtips, the ailerons could induce roll in the opposite direction to that intended. Even if the wingtips are not stalled, the increased adverse yaw associated with the low airspeed and/or high load factor could cause control problems.

Undesirable roll during a stall should be corrected using the rudder. The secondary effect of yaw will assist in eliminating the roll and preventing a spin.

Without any correction, roll and/or yaw during a stall can easily lead into a spin. This will cause more recovery difficulties, as well as an increase in altitude lost—often critical during inadvertent stalls and spins.

7) Secondary stalls are stalls that occur during the recovery from other stalls.

Secondary stalls are avoided simply by keeping the AOA within limits after a stall recovery. This means keeping the load factor low during the pull out from post stall dive.

8) Critical areas for stalls and spins are takeoff, landing, the overshoot, and low level operations in general. These areas are more dangerous than others (e.g. – practice stalls which are completed at higher altitudes) because of the proximity of the ground. Since stalls and spins involve an uncontrolled loss of altitude, they can result in ground contact at a less than optimum time.

9) During a stall recovery, full power is added immediately. During a spin recovery, power is removed until control in regained and a climb attitude is established. At this point, full power is applied in order to climb.

The reason for the difference is in the dynamics of the two problems. In a stall, power helps to minimize the altitude lost in the recovery. However, in a spin, power can increase the spin rate and delay recovery. As well, with power on, the airspeed will build extremely rapidly. This means extra altitude will be lost in the dive recovery and there will be a risk of exceeding airspeed and load factor limitations in the pull out.

10) The reason for this is to ensure that the rudder arm is long enough to ensure recovery from the spin. The rudder is used to produce a yawing moment. Since the sidewards force produced by the rudder is approximately fixed, the moment can only be increased by increasing the arm. With the rudder arm based on the C of G, a forward C of G provides a longer arm and therefore a more effective rudder.

11) Certification requirements dictate that aircraft used for intentional spins must be able to complete 6 full rotations in the spin.

The number of turns required in the recovery must be no more than 2.

Chapter 20: Unusual Attitudes

1) Nose high and nose low unusual attitudes are distinguished primarily by the position of the horizon in the windscreen. However, in the event of disorientation, a cross check of the airspeed can be made. In a nose low attitude, horizon will be high and the airspeed will be increasing. In a nose high attitude, the horizon will be low and the airspeed will be decreasing.

2) The spin and the spiral can often be confused with one another. If there is an uncertainty about which is occurring, the airspeed is the key to differentiating between them. In a spiral, the airspeed will be high and increasing. In a spin, the airspeed will be low and fairly steady.

The recovery from the spiral is:

- power to idle
- level the wings
- ease out of the dive
- enter a climb

The spin recovery is:

- power to idle, flaps retracted, and ailerons neutralized
- full opposite rudder
- pitch forward enough to break the stall
- ease out of the dive
- enter a climb

3) Combining a high g pull out with a rolling movement can overstress the inside wing. This is because the aileron increases the lift and the twisting moments acting on the wing. Separating the pitch and roll movements allows the aircraft to be recovered from the spiral without the risk of overstress.

4) At V_a, full control deflection is allowed. At V_{ne}, $1/3$ control deflection is allowed. At intermediate speeds, the amount of control deflection can be interpolated. For example, if the V_a is 100kts and the V_{ne} is 200kts, a roll conducted at 150kts could be done with approximately $2/3$ deflection of the ailerons. Of course, this type of calculation would be very impractical in the middle of a spiral recovery. So being familiar with the aircraft and the appropriate speeds would be helpful.

Chapter 21: Slips

1) During a forward slip, the relative airflow is on an angle to the longitudinal axis as seen from above. The aircraft is pointed to one side of the flight path and the "into wind" wing is lower than the other. The sidewards airflow increases the drag on the airframe.

The forward slip is used to increase the descent rate and angle without increasing the airspeed. The extra drag created by the slip produces the need to increase the descent rate and angle in order to maintain airspeed.

2) During a crosswind landing, a sideslip provides a sideward component of aircraft velocity that counteracts the drift due to the crosswind. In a slip, the drift is eliminated while still keeping the aircraft oriented with the approach/landing path.

Eliminating this drift is important in order to avoid sideloads on the landing gear when the aircraft touches down. Uncorrected, these loads could cause a gear failure.

3) A slipping turn is simply a forward slip conducted simultaneously with a turn.

Typically, the slipping turn is used during the turn from base leg to final approach when there is extra altitude to be lost. However, a slipping turn can be used at any time that a slip would be used and a change in direction is needed.

4) Generally, forward slips should be conducted to the left. This provides improved visibility for the pilot as compared to a slip to the right.

One possible exception would be when a slip is needed to lose altitude *and* correct for a crosswind. A slip into the wind would be preferable in this case so that the slip doesn't need to be reversed just prior to the flare.

Chapter 22: Takeoff

1) Factors to be considered in selecting a runway include wind, runway surface condition, runway length and width, obstructions to be cleared on takeoff (or landing), density altitude, and anticipated aircraft performance.

As an aside, other factors which don't relate to the content of this text include traffic, ATC clearances, considerations for emergencies after takeoff (or during the approach), etc.

The runway selected is generally the one that has the best overall conditions prevailing. For example, ideal circumstances would be a long paved runway at a low density altitude (i.e. – low elevation, high pressure and low temperature) with a fairly strong and steady wind directly against the takeoff direction and no obstacles to clear during the climb out. However, these ideal conditions rarely occur. Often, one favorable condition will be traded off for another. If a long runway has a crosswind beyond the aircraft's limits, a shorter runway that is into the wind will often be selected instead.

2) Ground effect reduces the stall speed and the induced drag of an aircraft. During a soft or rough field takeoff, this phenomenon is used to lift-off at lower than normal airspeeds. However, during a normal takeoff, it is not actively used by the pilot. The induced drag reduction does contribute to the takeoff acceleration. When climbing out of ground effect, the increased downwash creates a tendency for the aircraft to pitch up—especially in high wing aircraft.

3) The normal rotation is generally conducted gradually as the aircraft accelerates down the runway. No specific target airspeed is used, but instead an airspeed range is used. Rotation begins near the low speed end of this range and lift-off occurs near the high speed end. Generally, some back pressure is applied and the aircraft flies off the ground as the airspeed increases.

The short field rotation is usually conducted at a lower airspeed and is more defined. That is to say that the rotation is conducted at a specific target airspeed and only covers the airspeed range that the aircraft accelerates through in the time it takes to pitch up to the takeoff attitude. The speed used for the short field rotation is usually lower than the low speed end of the normal rotation range.

4) This question has two correct answers. One that is based on keeping the rotation speed at the MGW value. The other that is based on changing the rotation speed to suit the new weight of the aircraft.

Keeping the MGW rotation speed, we get,

$$d = D(GW/MGW)$$

$$d = 1480ft(2800lbs/3275lbs)$$

$$\underline{d = {\sim}1266 \ ft.}$$

If we adjust the rotation speed for the new weight, we get,

$$d = D(GW/MGW)^2$$

$$d = 1480(2800/3275)^2$$

$$\underline{d = {\sim}1082 \ ft.}$$

As an aside to the actual distance required, if a new rotation speed is to be used it must be determined first. Since the rotation speed is a multiple of the stall speed, it can be adjusted for weight in the same manner,

$$\text{Rotation Speed} = RS\sqrt{(GW/MGW)}$$

$$\text{Rotation Speed} = 67kts\sqrt{(2800/3275)}$$

$$\text{Rotation Speed} = {\sim}62kts.$$

5) We can solve this problem by dividing it into two sections—the takeoff roll and the climbout.

For the 15 knot headwind,

$$d = D\frac{(v_{LO} - w)^2}{v_{LO}^2}$$

$$d = 1480ft\frac{(67kts - 15kts)^2}{(67kts)^2}$$

$$d = {\sim}892 \ ft.$$

892 ft. is the rolling portion of the takeoff.

The 0 wind distance to clear an obstacle is 2150ft-1480ft = 670ft. To account for the headwind,

$$d_o = D_o\frac{TAS-W}{TAS}$$

$$d_o = (670ft)\frac{(74kts-15kts)}{74kts}$$

$$d_o = {\sim}535 \ ft.$$

So the total distance to clear a 50 ft obstacle with a 15 kt headwind would simply be the sum of these two distances, $\underline{892ft + 535ft = 1427 \ ft.}$

With a 5 kt tailwind, similar calculations can be applied (but remember that a tailwind is negative, so the subtracted winds become added winds). The distance to clear a 50 ft obstacle with a 5 kt tailwind is <u>approximately 2426 ft</u>.

6) Flaps will reduce the aircraft's stalling speed and therefore allow lift-off at a lower speed. The reduced lift-off speed will shorten the ground run of the takeoff. Unfortunately, flaps will also create drag and therefore reduce the acceleration during the roll and the climb performance after lift-off. The overall effect of the flaps—whether they will increase or decrease takeoff performance—depends on the individual aircraft type.

7) For the short, dry grass, we increase the takeoff roll by 10% of the total distance to clear a 50 foot obstacle. So in this case, the ground run becomes 1695 ft. So the total distance required to takeoff over a 50 foot obstacle becomes 2365 ft. (the climb portion of the takeoff is unaffected by the grass and remains at 670 ft.).

Now the 10 knot headwind can be applied to the new distances. The equations used are the same as in Q5, only the numbers change. The total distance required becomes ~1227 ft. for the ground run and ~580 ft. for the climb portion. <u>This gives us a total takeoff distance of ~1807 ft. To clear a 50 foot obstacle</u>.

8) Takeoff distance increases approximately 15% per 1000 feet of density altitude, so the altitude correction gives us a takeoff roll of about 2146 feet.

Applying the weight correction to this distance (as was done in Q4—we will assume that the rotation speed was adjusted for the weight), <u>we get a new distance of ~1801 ft</u>.

The new rotation speed in this case would be 64 knots.

9) Referring to Fig 22-7, the crosswind and headwind components can be determined. The headwind component is 7.5 knots gusting to 12.5 knots. The crosswind component is ~13 knots gusting to ~22 knots. Performance calculations are done based on considerations for the worst case scenario, so the crosswind component should be considered to be a steady 22 knots (the stronger crosswind) and the headwind component should be considered to be a steady 7.5 knots (the lighter headwind).

The new rotation speed is based on the gust factor of the total wind. With the winds gusting 15 knots to 25 knots, the gust factor is 10 knots. Rotation speed should be increased by ½ of the gust factor, <u>so the new rotation speed will be 67 knots</u>.

To determine the takeoff distance, both the headwind and the increased rotation speed must be considered. In this case, the increase in rotation speed is 5 knots, so that will effectively cancel 5 knots worth of headwind. That gives us an effective headwind of approximately 2.5 knots. <u>Considering this, the new takeoff distance will be approximately 940 feet</u>.

10) According to the 20% rule, this aircraft has a crosswind limit of 10.8 knots. So this crosswind is well outside of the aircraft's limitations.

Luckily, the 20% rule is only a minimum. An aircraft may be capable of sustaining a stronger crosswind component in some cases. If the manufacturer tested and certified the aircraft to stronger crosswind limits, then these limits are the ones to apply. The 20% rule is useful when no manufacturers information on crosswind limits is available.

11) Operating on a grass strip after heavy rain would normally dictate the use of soft field procedures due to the "sponginess" of the ground.

The distance required can be determined by applying the density altitude rule of thumb and the wet grass rule of thumb. <u>The resulting distance is ~2659 ft</u>.

12) A decreased performance wind shear will cause a sudden drop in airspeed.

The hazard associated with decreases performance wind shear is the possibility of a stall.

13) The soft field takeoff technique is designed to get the aircraft off the ground at the lowest possible airspeed. This reduced stress on the undercarriage and reduces the chance of a loss of control due to the gear "digging in" at high speeds.

The only difference between a soft field and a rough field technique is the use of brakes at the beginning of the roll. During a takeoff on a soft field, the takeoff roll is started directly from the taxi roll. There is no intermittent stop as power is applied. This is done to prevent the wheels from sinking in during the stop. On a rough field, full power can be applied prior to brake release. This will shorten the ground roll and if the surface is rough but not soft, the wheels will not sink in.

During the soft/rough field technique, it is important to level off and accelerate in ground effect in order to achieve a safe climb speed. The lift-off speed during a soft/rough field takeoff is below the "out of ground effect" stall speed. Climbing in this condition will lead to a stall extremely close to the ground.

Chapter 23: Landing

1) During a short landing, the objective is to minimize the distance required to stop safely. For this reason, no power is used, maximum braking is used, and the flare maneuver is not overly prolonged. In comparison, the soft, rough field objective is to minimize stress on the landing gear and the probability of a loss of control due to "digging in". Because of this, some power is often used to control the descent rate upon touchdown, only minimum braking is used, and the flare is prolonged as necessary to achieve a safe and controlled touchdown.

2) During a normal landing, directional control is maintained with the rudder and lateral drift is corrected by rolling away from the drift (using the ailerons). Directional and lateral control are both maintained visually by aligning the longitudinal axis with the landing path (directional control) and eliminating any lateral drift.

During a crosswind landing, the actual methodology of the controls does not change at all. However, the magnitude of the inputs will become much more pronounced.

3) During the normal flare, the excess airspeed that is present in the approach is used to arrest the descent. The power remains at idle and the pitch attitude is adjusted to reduce the descent rate while simultaneously reducing the airspeed. As the airspeed bleeds off, the pitch attitude is increased gradually until the aircraft can no longer fly and the wheels touch down.

During the soft field flare, some power is usually used to improve the control of vertical speed. Pitch control works much the same way as the normal flare, airspeed is sacrificed in order to reduce the vertical speed upon touchdown.

4) First, we will correct the approach speed for the weight. Since the approach speed is a multiple of the stall speed, it can be adjusted in the same manner,

$$\text{App. Speed} = AS\sqrt{(GW/MGW)}$$

$$\text{App. Speed} = 52kts\sqrt{(1900lbs/2350lbs)}$$

$$\text{App. Speed} = \sim47 \text{ knots.}$$

Now we can adjust for the gusty winds. The approach speed should be increased by ½ of the gust factor (the gust factor is 15 knots in this case). So the new approach speed should be,

$$47kts + 7.5Kts = 54.5kts = \underline{\sim55kts}.$$

5) In order to land in a crosswind, the aircraft must be established in a slip into the wind. In this case, the crosswind is from the left, so the slip must be to the left. This means the ailerons will be deflected to the left and the rudder will be deflected to the right.

6) During landing, wet grass will reduce the effectiveness of the brakes. So even though the "brakeless" rolling friction will be increased, the overall rolling friction with the brakes applied will be decreased. Consider two of the same type of aircraft landing on two different surfaces—one on pavement and one on grass. If both aircraft roll to a stop without brakes, the one on grass will actually stop shorter. On the other hand, if they both used maximum braking, the one on the pavement would stop shorter.

7) First, we can calculate the effect of density altitude. Landing distance increases approximately 4% per 1000 feet. So the new distance will be ~499 ft.
Now the reduced weight can be considered,

$$d = D(GW/MGW)$$

$$d = 499ft(1820lbs/1950lbs)$$

$$\underline{d = \sim466 \text{ ft}}.$$

As in Q4, the approach speed can be determined in the same way as a new stall speed,

$$\text{App. Speed} = AS\sqrt{(GW/MGW)}$$

$$\text{App. Speed} = 47kts\sqrt{(1820lbs/1950lbs)}$$

$$\underline{\text{App. Speed} = \sim46 \text{ kts}}.$$

8) This surface will require a soft field technique, so when the final "short field" answer is determined, we will multiply it by a factor of 4.
First, we will account for the wet grass. This requires increasing the distance by 100%, so the new distance is 860 ft.
Next, we can determine the effects of the winds,

$$d = D\frac{(v_{td} - w)^2}{v_{td}^2}$$

$$d = 860ft\frac{(39kts - (-8kts))^2}{39kts^2}$$

$$\underline{d = \sim1249 \text{ ft}}.$$

Glossary of Key Words and Abbreviations

Abbreviations and Symbols

AC – Aerodynamic Center

A/C – Aircraft

AGL – Above Ground Level

AME – Aircraft Maintenance Engineer

AOA – Angle of Attack

AOB – Angle of Bank

ASL – Above Sea Level

ATC – Air Traffic Control

BAOC – Best Angle of Climb

BHP – Brake Horsepower

BROC – Best Rate of Climb

CAS – Calibrated Airspeed

C_d or C_D – Coefficient of Drag

CFS – Canada Flight Supplement

C_l or C_L – Coefficient of Lift

C of G – Center of Gravity

C of P – Center of Pressure

CRFI – Canadian Runway Friction Index

EAS – Equivalent Airspeed

FPM – Feet Per Minute

FSS – Flight Service Station

g – Gravity or Gravitational Acceleration (32.2 ft/s^2)

IAS – Indicated Airspeed

IFR – Instrument Flight Rules

IHP – Indicated Horsepower

KIAS – Knots Indicated Airspeed

Knots – Nautical miles per hour

L/D – Lift to Drag Ratio

LLF – Limit Load Factor

POH – Pilot's Operating Handbook

ROC – Rate of Climb

RPM – Revolutions Per Minute

SHP – Shaft Horsepower

TAS – True Airspeed

THP – Thrust Horsepower

VFR – Visual Flight Rules

V_a – Maneuvering speed

V_d – Dive speed

V_{fe} – Maximum flaps extended speed

V_{le} – Maximum landing gear extended speed

V_{lo} – Maximum landing gear operation speed

V_{ne} – Never Exceed speed

V_{no} – Normal Operations speed

V_s – Stall speed in cruise configuration

V_{so} – Stall speed in landing configuration

V_x – Best angle of climb speed

V_y – Best rate of climb speed

α – *Alpha* – Used to represent angles – Angle of Attack in this case

β – *Beta* – Used to represent angles – sideslip angle in this case

Δ – *Delta* – Used to represent a change in a quantity

θ – *Theta* – Used to represent general angles

ρ – Rho – Used to denote density

π – Pi – 3.14..., The ratio of a circle's circumference to it's diameter

Definitions

Absolute Angle of Attack – The angle between the zero lift line of the wing and the relative airflow.

Acceleration – A change in velocity. Acceleration is a vector quantity since it has both direction and magnitude. Note that since velocity is also a vector quantity, if a moving object changes direction while maintaining a constant speed, an acceleration has occured.

Ailerons – Control surfaces normally located at the trailing edges of the outboard portion of the wings. Ailerons pivot opposite one another to control roll.

Airfoil – A two dimentional section of a wing (i.e. – a cross-section of the wing).

Airspeed – The speed at which the aircraft is moving relative to the air. There are several types of airspeed, True, Calibrated, Equivalent, and Indicated.

Altitude – The height of the aircraft. Altitude can be measured in feet above ground level (AGL) or feet above sea level (ASL) depending on the context. As well, altitudes can be referenced to the Standard Atmosphere with density altitude, pressure altitude, etc.

Angle of Attack – The angle between the relative airflow and the chord line of the wing. Note : Be careful not to confuse angle of attack with pitch angle.

Aneroid Capsule – A flexible capsule that is designed to expand and contract with pressure changes so that the changes can be measured.

Anhedral – A condition where the wings are inclined along the span so that the wing tips are lower than the wing roots.

Anti-Servo Tabs – Secondary surfaces which are attached to a control surface to produce a moment which tends to return the surface to neutral.

Arm – The distance between an object and some reference point. Arms are normally used to determine moments.

Aspect Ratio – The ratio of the wingspan to the wing chord. For wings with irregular shapes, the aspect ratio can also be considered to be the ratio between the square of the wingspan and the wing planform area.

Attitude – The position of an aircraft in relation to the horizon. The attitude is defined in terms of pitch and bank angles.

Axis – A line about which an object rotates. In the case of an aircraft, there are three axis', the longitudinal, lateral, and vertical/normal axis'. All three axis pass through the C of G.

Balloon – A condition where the flare maneuver is overdone and the aircraft climbs by using the excess airspeed carried in the approach to land.

Boundary Layer – A thin region of air between the surface of an aircraft and the freestream airflow. Within the boundary layer, the air velocity increases as you move further away from the surface and towards the freestream.

Camber – The difference between the mean line and the chord line of the wing. The camber is a measure of the curvature of the airfoil.

Canard – A horizontal stabilizer/elevator combination or stabilator that is mounted ahead of the wing as opposed to the traditional aft position.

Ceiling – In the context of flight mechanics, the ceiling is the highest altitude to which the aircraft can climb. The absolute ceiling is the altitude which gives the aircraft a rate of climb of zero. The service ceiling is the altitude which allows a 100 FPM rate of climb.

Chord – The distance from the leading edge to the training edge of the wing.

Chord Line – An imaginary line running from the leading edge of the wing to the trailing edge of the wing.

Coefficient of Drag – A measure of the drag production of an object. The drag coefficient determines how much dynamic pressure is converted into drag. This coefficient can be further subdivided into the 'Induced Drag Coefficient' and the 'Parasite Drag Coefficient'.

Coefficient of Lift – A measure of the lift production of an object. The lift coefficient varies with angle of attack and determines the amount of dynamic pressure that is converted into lifting force.

Control Column – The yoke or wheel used by pilots to control the movement of the ailerons and the elevator. In some aircraft, the control column is replaced by a stick.

Coordinated Flight – Flight with a sideslip angle of zero.

Cruise Flight – Flight that is established at a cruising altitude and is maintained in straight and level.

Cowling – The portion of the airframe that covers the engine in order to reduce aerodynamic drag and optimize cooling efficiency.

Datum – A reference point used to take measurements. A standard datum for a given aircraft is published in the POH and weight and balance calculations and data are based on this datum.

Density Altitude – The altitude which you would have to be at in the Standard Atmosphere in order to have a given air density. Density altitude is the method which we use to reduce "real world" conditions to a common standard. During flight operations, density altitude is usually calculated by starting with the pressure altitude and correcting for a non-standard temperature if necessary.

Differential Ailerons – Ailerons that are designed to pivot upward farther than they pivot downward. This design is used to counteract adverse yaw.

Dihedral – A condition where the wings are inclined along the span so that the wing tips are higher than the wing roots.

Downwash – The downwards deflection of air behind a finite wing (in other words *all* real wings). Downwash is often misinterpreted as the cause of lift. In actual fact, it is the *result* of lift.

Drag – The force acting on an aircraft which opposes it's motion through the air. This force can be broken down into parasite drag and induced drag.

Drag Coefficient – See Coefficient of Drag.

Elevator – The control surface which is located on the trailing edge of the horizontal stabilizer. The elevator pivots up and down to control the AOA of the aircraft.

Empennage – The tail section of the aircraft.

Endurance – The amount of time an aircraft can remain airborne on a given amount of fuel.

Envelope – A defined set of limitations which an aircraft must be operated within. Two important envelopes which were introduced in this text are the *Weight and Balance* envelope and the *Velocity-Load Factor* envelope (V-g diagram).

Flaps – Secondary control surfaces normally located on the trailing edge of the inboard portion of the wings. Flaps are used to increase the lifting capacity of the wings as well as the drag produced by the wings.

Flaperons – Ailerons which can both be deflected downward simultaneously to replace or supplement flaps.

Freestream – the portion of an airflow that is unaffected by the passage of the aircraft.

Friction – A force which opposes the motion of an object resulting from the interactions of molecules at or near the surfaces of two objects in contact with one another.

Frise Ailerons – Ailerons that have an extension that protrudes into the airflow below the wing while the aileron is deflected upward. Frise ailerons are used to counteract adverse yaw.

Force – Defined by Newton's Second Law, force is equal to the mass of an object multiplied by the rate of acceleration of the object. Force can be considered to be the rate of change of momentum.

Fuselage – The main body of an aircraft.

Glide – A descent in which no power is used. All propulsive force comes from the forward component of the weight.

Ground Effect – In flight near the ground, wingtip vortices are disrupted by the presence of the ground and induced drag is reduced while the lifting capacity is increased.

Groundspeed – The speed of an aircraft in relation to the ground. The difference between the true airspeed and the groundspeed is caused by the wind.

Gust – A change in wind speed and/or direction that is of short duration. Gusts are often associated with turbulence and sometimes with wind shear.

Gust Factor – The difference between the steady wind speed and the peak wind speed. For example, if the winds are blowing at 20 knots gusting to 30 knots in the same direction, the gust factor is 10 knots.

Gyroscope – A spinning mass. Gyroscopes have a tendency to maintain a constant plane of rotation. They are used in aircraft instruments. As well, gyroscopic effects apply to the prop and must be anticipated under some conditions.

Horsepower – The transfer of energy at 33,000 ft-lbs per minute.

Inclinometer – A device that is used to indicate whether any sideward accelerations are occurring. The most common form of inclinometer is a ball inside a curved glass tube.

Induced Drag – Drag that is a byproduct of lift. Wingtip vortices cause downwash which in turn causes the lift of the wing to be inclined backwards. The aft component of the lift is induced drag.

Inertia – The resistance an object has to an acceleration. Inertia can be quantified in terms of mass. In the imperial system, mass is measured in slugs. In the metric system, mass is measured in kilograms.

Kinetic Energy – The energy possessed by an object resulting from it's motion.

Knot – A measurement of speed. 1 knot is 1 nautical mile per hour.

Lift – The force that normally supports the weight of the aircraft. Lift acts perpendicular to the relative airflow.

Lift Coefficient – See Coefficient of Lift.

Limit Load Factor – the maximum load factor that an aircraft can be operated at without the risk of sustaining permanent structural damage.

Load Factor – A factor which defines how much of a load the aircraft is experiencing due to an acceleration. Load factors are expressed in g's (multiples of gravity).

Manifold Pressure – The air pressure within the intake manifold of the engine. On aircraft with variable pitch propellers, the manifold pressure is controlled with the throttle (this is true for fixed pitch props as well, but on these aircraft, a change in manifold pressure leads directly to a change in engine RPM). On these aircraft, the prop RPM is controlled separately. The two controls combine to provide control over the power being produced.

Mass – A measurement of the amount of matter that an object contains. The mass of an object determines how much 'inertia' it has.

Mixture – The ratio (by mass) of fuel to air entering the engine. At a given power setting, engines can generally operate within a range of mixture settings. Manual mixture control is common on piston powered aircraft in order to allow more efficient operation.

Moment – A moment is some value multiplied by a distance. In this text, we've made some reference to inertial moments, but we've focused mainly on the moment of a force about some point.

Movement – In the context of flight mechanics, a movement is the rotation of the aircraft around one of it's three axis'.

Nautical Mile – Based on 1 minute of longitude at the equator, 1 nautical mile is equal to 6080 ft.

Overshoot – A transition from an approach descent to a departure climb without contacting the ground or leveling off. The term overshoot is also used to describe a condition where an approach is being flown too high to make the desired touchdown point.

Parasite Drag – Drag which is not caused by lift. Parasite drag includes skin friction, form drag (AKA pressure drag), and interference drag.

Payload – The load that an aircraft can carry after fuel, operating fluids such as oil and hydraulic fluid, and (by some definitions) the flight crew. The term 'payload' comes from the idea of how much weight in passengers or cargo a commercial operator can be paid to carry.

Pitch – The word pitch can refer to the pitch attitude of the aircraft, the movement of pitch, or the pitch of the propeller. The pitch attitude can be considered to be the angle between the longitudinal axis and the horizon as seen from the side. The pitching movement is a rotation around the lateral axis. Pitch is controlled by the elevator or stabilator. The pitch of the prop is determined by the angle of the blades. Prop pitch is measured in inches and is the distance that the prop will move forward in one revolution at a blade AOA of zero.

Planform – The planform of a wing is the shape as seen from above.

Porpoise – An unstable, pilot-induced oscillation that can occur on landing.

Potential Energy – The energy possessed by an object as a result of it's position. For example, gravitational potential energy is a result of height.

Power – The rate of transfer of energy. Power is a function of force and velocity.

Precession – The displacement of an applied force by 90° in the direction of rotation. Precession occurs with gyroscopes.

Pressure Altitude – The altitude you would have to be at in the Standard Atmosphere in order to experience a given atmospheric pressure.

Primary Control – A control surface that is intended to have a direct effect on the attitude and the movements of an aircraft. On most fixed-wing aircraft, the primary controls are the elevator (pitch), the ailerons (roll), and the rudder (yaw).

Range – The distance an aircraft can fly on a given amount of fuel.

Relative Airflow – The direction of the freestream airflow relative to the aircraft.

Roll – A rotation around the longitudinal axis. Roll is controlled by the ailerons and is used to transition the aircraft to and from different bank attitudes.

Rudder – The control surface on the trailing edge of the vertical stabilizer. The rudder varies the camber of the stabilizer and is used to control yaw.

Scalar – A measurement that has a quantity, but not a direction (i.e. – mass, energy, etc.).

Secondary Control – A control surface which is not intended to have a direct effect on the attitude or the movements of an aircraft, but that influences the flight characteristics of the aircraft.

Servo Tabs – Secondary control surfaces that are attached to the primary control surfaces. They deflect in a manner that encourages further deflection or lower control forces on the primary surface.

Sideslip – Flight with the relative airflow at an angle to the longitudinal axis as seen from above.

Slip – See sideslip.

Slipstream – The airflow that is shed from the propeller. This airflow is travelling backward over the airframe and rotating in a corkscrew fashion. Props in North America normally rotate clockwise as seen from the cockpit, so this is the direction that slipstream rotates in as well.

Speed Brakes – Control surfaces that deflect into the airflow for the purpose of producing drag.

Spin – An autorotation that can develop from an aggravated stall condition.

Spiral – A steep descending turn with an excessive nose-low pitch attitude. Spirals (AKA – spiral dives) can lead to excessive loads and/or speeds.

Spoiler – A control surface that disrupts the airflow over the top of a wing for the purpose of destroying lift.

Stabilator – A single surface that serves the purpose of both the horizontal stabilizer and the elevator.

Stability – The tendency for an object to maintain a certain condition. In the case of aircraft, stability normally refers to the aircraft's tendency to maintain a constant AOA (longitudinal stability) or to maintain a sideslip angle of zero (lateral/directional stability).

Stagnation Point – The point at the leading edge of an object where airflow stops or stagnates. This is also the point at which the airflow divides to flow over and under the object.

Stall – A condition of flight where an increase in AOA will decrease the lift coefficient.

Stall Strips – Strips placed at the leading edge of the wing to encourage airflow separation and therefore the stall. Stall strips are normally placed at the wing root to create a favorable stall pattern.

Sweepback – A wing design which places the wing tips further aft than the wing roots. On low speed aircraft, sweepback is used to influence the lateral and directional stability of the aircraft. As well, sweepback can be used to improve the high speed flight characteristics of the aircraft.

Taxiing – Maneuvering the aircraft on the ground.

Thrust – The force which propels the aircraft through the air. Thrust acts parallel to the flight path and in the direction of motion.

Torque – Technically, torque is a moment that will cause a rotational acceleration. In some contexts, however, it is also used to describe the action/reaction pair that accompanies all moments. For example, as the engine turns the prop, there is a tendency for the prop to turn the engine (and therefore the aircraft).

Trim – Trim describes the moments acting on an aircraft. If the aircraft is flying at it's trimmed position, no moments will be present to cause a rotation. Moving away from the trimmed position, moments will be produced. If these moments return the aircraft to it's trimmed position, the aircraft is stable. If the moments take the aircraft further away from it's trimmed position, the aircraft is unstable. The trimmed position is measured as an AOA and a sideslip angle (normally zero).

Trim Tab – A secondary control surface that is attached to a primary control surface to adjust the trimmed position of the aircraft. On light aircraft, trim tabs are normally only located on the elevator/stabilator. However, they can also be located on the rudder and ailerons.

Ultimate Load Factor – the maximum load factor that an aircraft can sustain without the risk of a major structural failure. Certification requirements dictate that the ultimate load factor must be at least 50% greater than the limit load factor.

Uncoordinated Flight – Flight with a sideslip angle other than zero. This will occur when the rate of yaw does not correspond to the angle of bank and the airspeed.

Undershoot – When an approach is flown too low to make the aiming point, it is said to be an undershoot.

Useful Load – the load that a given aircraft can carry after fuel, oil, and other operating fluids. Notice that unlike the payload, the useful load includes the flight crew.

Vector – A measurement that has both quantity and direction (i.e. – force, velocity, etc.).

Velocity – An objects speed *and* direction.

Wetted Area – The total surface area of the aircraft. The wetted area has a large influence on parasite drag characteristics.

Windshear – A sudden change in wind direction and/or velocity. Because of the inertia of an aircraft, wind shear causes sudden changes in airspeed.

Winglet – Vertical or nearly vertical extensions placed on the wingtips to reduce induced drag.

Wingspan – The distance from wingtip to wingtip.

Work – Work is the transfer of energy and is defined as force applied over a distance. Multiplying the force and the distance will give the amount of work done.

Yaw – Rotation around the vertical axis. Yaw is controlled by the rudder.

Bibliography

1) Anderson, J.D.Jr., *Fundamentals of Aerodynamics*, Mcgraw-Hill, New York, 1984.

2) Anderson, J.D.Jr., *Introduction to Flight*, Mcgraw-Hill, New York, 1978.

3) Dole, Charles E., *Flight Theory For Pilots*, IAP, Casper, 1989.

4) Halliday, D., Resnick, R., and Walker, J., *Fundamentals of Physics*, 4th Edition, Wiley, New York, 1993.

5) Hurt H. H. Jr. 1960, *Aerodynamics for Naval Aviators*, NAVWEPS 00-80T-80.

6) Kershner, W.K., *The Advanced Pilots Flight Manual*, 6th Edition, Iowa State University Press, Ames, 1994.

7) Langewiesche, W., *Stick and Rudder*, Mcgraw-Hill, New York, 1944.

8) Peery and Azaar, *Aircraft Structures,* Mcgraw-Hill, New York, 1982.

9) Raymer, Daniel P., *Aircraft Design : A Conceptual Approach*, 2nd Edition, AIAA, Washington, 1992.

10) Russel, J.B., *Performance and Stability of Aircraft*, John Wiley and Sons, New York, 1996.

11) Smith, Hubert C., *The Illustrated Guide to Aerodynamics*, 2nd Edition, Mcgraw-Hill, 1992.

12) Szurovy, G., and Goulian, M., *Basic Aerobatics*, Mcgraw-Hill, New York, 1994.

13) *Flight Training Manual*, 4th Edition, Gage Educational Publishing, Ottawa, 1994.

14) *From the Ground Up*, 26th Edition, Aviation Publishers, Ottawa, 1991.

15) A.I.P. Canada, Aeronautical Publications Services, Ottawa

16) FAR Part 23 - Airworthiness Standards: Normal, Utility, Acrobatic, and Commuter Category Airplanes

Index of Keywords

www.ingramcontent.com/pod-product-compliance
Lightning Source LLC
Chambersburg PA
CBHW081804200326

41597CB00023B/4135